Dahlem Workshop Reports
Life Sciences Research Report 39

The goal of this Dahlem Workshop is:
to formalize and clarify alternative hypotheses about sexual selection and to devise discriminating tests

Life Sciences Research Reports

Series Editor: Silke Bernhard

Held and published on behalf of the
Stifterverband für die Deutsche Wissenschaft

Sponsored by:

Senat der Stadt Berlin
Stifterverband für die Deutsche Wissenschaft

Sexual Selection: Testing the Alternatives

J.W. Bradbury and M.B. Andersson, Editors

Report of the Dahlem Workshop on
Sexual Selection: Testing the Alternatives
Berlin 1986, August 31–September 5

Rapporteurs: L. Heisler, T.D. Price,
M.J. West-Eberhard, G.S. Wilkinson

Program Advisory Committee:
J.W. Bradbury and M.B Andersson, Chairpersons
S.J. Arnold, G. Borgia, P. Hammerstein,
G.A. Parker, L. Partridge

A Wiley–Interscience Publication

John Wiley & Sons 1987
Chichester · New York · Brisbane · Toronto · Singapore

Copy Editors: J. Lupp, K. Geue
Photographs: E.P. Thonke

With 4 photographs, 4 figures, and 9 tables

Library of Congress Cataloging in Publication Data

Dahlem Workshop on Sexual Selection: Testing the
Alternatives (1986 : Berlin, Germany)
Sexual selection.

(Life sciences research report ; 39) (Dahlem
workshop reports)
Sponsored by Senat der Stadt Berlin, Stifterverband
für die Deutsche Wissenschaft.
Includes index.
1. Sexual selection in animals—Congresses.
I. Bradbury, J. W. II. Andersson, M. B. III. Heisler,
L. IV. Berlin (Germany : West). Senat.
V. Stifterverband für die Deutsche Wissenschaft.

ISBN 0 471 91624 2

British Library Cataloguing in Publication Data

Dahlem Workshop on Sexual Selection : Testing the
Alternatives (*1986 : Berlin*) Sexual Selection :
testing the alternatives : report of the Dahlem
Workshop on Sexual Selection : Testing the
Alternatives, Berlin 1986, August 31–September 5.
— (Dahlem Workshop reports. Life sciences research
report ; 39)
1. Sexual behavior in animals
I. Title II. Bradbury, J.W. III. Andersson,
M.B. IV. Heisler, L. V. Series
591.56′2 QL761

ISBN 0-471-91624 2

Typeset by Inforum Ltd, Portsmouth
Printed and bound in Great Britain by the Bath Press Ltd, Bath, Avon

Table of Contents

The Dahlem Konferenzen

Founders
Recognizing the need for more effective communication between scientists, the Stifterverband für die Deutsche Wissenschaft*, in cooperation with the Deutsche Forschungsgemeinschaft**, founded Dahlem Konferenzen in 1974. The project is financed by the founders and the Senate of the City of Berlin.

Name
Dahlem Konferenzen was named after the district of Berlin called *Dahlem*, which has a long-standing tradition and reputation in the sciences and arts.

Aim
The task of Dahlem Konferenzen is to promote international, interdisciplinary exchange of scientific information and ideas, to stimulate international cooperation in research, and to develop and test new models conducive to more effective communication between scientists.

The Concept
The increasing orientation towards interdisciplinary approaches in scientific research demands that specialists in one field understand the needs and problems of related fields. Therefore, Dahlem Konferenzen has organized workshops, mainly in the Life Sciences and the fields of Physical, Chemical, and Earth Sciences, of an interdisciplinary nature.

Dahlem Workshops provide a unique opportunity for posing the right questions to colleagues from different disciplines who are encouraged to state what they do not know rather than what they do know. The aim is not to solve problems nor to reach a consensus of opinion, the aim is to define and discuss priorities and to indicate directions for further research.

Topics
The topics are of contemporary international interest, timely, interdisciplinary in nature, and problem oriented. Dahlem Konferenzen approaches interna-

* *The Donors Association for the Promotion of Sciences and Humanities, a foundation created in 1921 in Berlin and supported by German trade and industry to fund basic research in the sciences*
** *German Science Foundation*

tionally recognized scientists to suggest topics fulfilling these criteria. Once a year, the topic suggestions are submitted to a scientific board for approval.

Program Advisory Committee

A special Program Advisory Committee is formed for each workshop. It is composed of 6–7 scientists representing the various scientific disciplines involved. They meet approximately one year before the workshop to decide on the scientific program and define the workshop goal, select topics for the discussion groups, formulate titles for background papers, select participants, and assign them their specific tasks. Participants are invited according to international scientific reputation alone. Exception is made for younger German scientists. Invitations are not transferable.

Dahlem Workshop Model

Since no type of scientific meeting proved effective enough, Dahlem Konferenzen had to create its own concept. This concept has been tested and varied over the years. It is internationally recognized as the *Dahlem Workshop Model*. Four workshops per year are organized according to this model. It provides the framework for the utmost possible interdisciplinary communication and cooperation between scientists in a period of 4½ days.

At Dahlem Workshops 48 participants work in four interdisciplinary discussion groups. Lectures are not given. Instead, selected participants write background papers providing a review of the field rather than a report on individual work. These papers, reviewed by selected participants, serve as the basis for discussion and are circulated to all participants before the meeting, with the request to formulate written questions and comments to them. During the workshop, each of the four groups prepares reports reflecting their insights gained through the discussion. They also provide suggestions for future research needs.

Publication

The group reports written during the workshop together with the revised background papers are published in book form as the Dahlem Workshop Reports. They are edited by the editor(s) and the Dahlem Konferenzen staff. The reports are multidisciplinary surveys by the most internationally distinguished scientists and are based on discussions of advanced new concepts, techniques, and models. Each report also reviews areas of priority interest and indicates directions for future research on a given topic.

The Dahlem Workshop Reports are published in two series:

1) Life Sciences Research Reports (LS), and
2) Physical, Chemical, and Earth Sciences Research Reports (PC).

Director
Silke Bernhard, M.D.

Address
Dahlem Konferenzen
Wallotstrasse 19
D-1000 Berlin (West) 33
Tel: (030) 891 5067

THE DAHLEM WORKSHOP MODEL

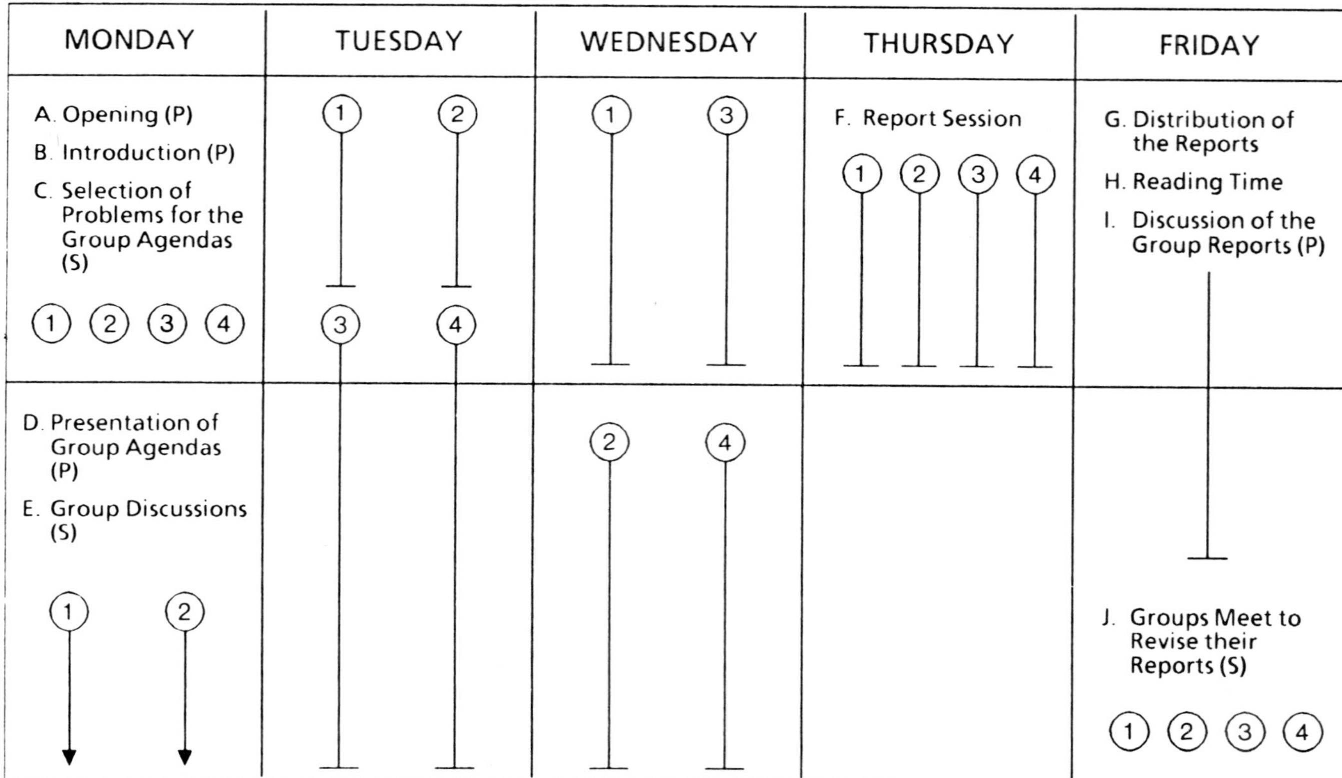

Key:
(P) = Plenary Session;
(S) = Simultaneous Sessions;
◯ = one discussion group

Explanation of the Dahlem Workshop Model

A. Opening
Background information is given about Dahlem Konferenzen and the Dahlem Workshop Model.

B. Introduction
The goal and the scientific aspects of the workshops are explained.

C. Selection of Problems for the Group Agenda
Each participant is requested to define priority problems of his choice to be discussed within the framework of the workshop goal and his discussion group topic. Each group discusses these suggestions and compiles an agenda of these problems for their discussions.

D. Presentation of the Group Agenda
The agenda for each group is presented by the moderator. A plenary discussion follows to finalize these agendas.

E. Group Discussions
Two groups start their discussions simultaneously. Participants not assigned to either of these two groups attend discussions on topics of their choice.
The groups then change roles as indicated on the chart.

F. Report Session
The rapporteurs discuss the contents of their reports with their group members and write their reports, which are then typed and duplicated.

G. Distribution of Group Reports
The four group reports are distributed to all participants.

H. Reading Time
Participants read these group reports and formulate written questions/ comments.

I. Discussion of Group Reports
Each rapporteur summarizes the highlights, controversies, and open problems of his group. A plenary discussion follows.

J. Groups Meet to Revise their Reports
The groups meet to decide which of the comments and issues raised during the plenary discussion should be included in the final report.

Sexual Selection: Testing the Alternatives
eds. J.W. Bradbury and M.B. Andersson, pp. 1–8
John Wiley & Sons Limited

Introduction

M.B. Andersson* and J.W. Bradbury**

*Department of Zoology, P.O. Box 25059
University of Göteborg
40031 Göteborg, Sweden
**Department of Biology, C—016
University of California, San Diego
La Jolla, CA 92093, USA

The concept of sexual selection was defined, and most of its implications spelled out, by Darwin over a century ago. After his masterly start, and given the steady maturation of evolutionary biology, one might wonder whether there are any burning issues on the subject still left to discuss. In fact, sexual selection is one of the most rapidly developing topics in modern biology and one that currently leads to the most divergent views and outspoken disputes. The problem in convening a conference on this subject was thus not insufficient substance but how to accommodate even a fraction of the differing viewpoints and the scientists representing them in a single setting.

The current turmoil certainly has historical precedents. The initial theory of sexual selection met strong resistance among Darwin's colleagues, particularly his suggestion that female choice may be responsible for much of the ornamentation in males (see O'Donald 1980). The disputes resurfaced in the 1930s as part of the general attempt to synthesize genetics and evolution: while Huxley (1938) reiterated the traditional doubts that females chose mates in nature, Fisher (1930) experienced one of his typically brilliant insights by recognizing that female choice and male display traits potentially constitute a powerful evolutionary feedback loop which can generate strong sexual dimorphism and even speciation. Fisher's argument, however, was not based on a formal model, and its predictions were sufficiently contrary to popular views that it was largely ignored for the next 30 years.

With the postwar boom in field studies of animal behavior, more and more examples of traits not easily explained by natural selection were described. Until the mid 1960s sexual selection was often invoked as an explanation for such traits, but in a post hoc manner without rigor. Thereafter, attitudes changed quickly with the debate over group selection, which led to new criteria for evaluating any proposed evolutionary process (Williams 1966). As before, the new critics accepted the idea of sexual selection via intersexual competition, but sexual selection by female choice remained suspect. In a timely and pioneering series of studies, O'Donald outlined a genetic basis for Fisher's proposed feedback loop between female preferences and male traits (see review in O'Donald 1980). Whether because of this theoretical progress or perhaps because of the liberal political and social trend of the late 1960s, increasing numbers of field studies looked for and found examples of female choice of mates. Gradually, after a century of debate, sexual selection became a respectable subject among evolutionary biologists. The acceptance of Darwin's ideas on sexual selection did not, however, end the dissent engendered by this topic but raised several new problematic issues. Four of these are treated in this volume.

MATING PREFERENCES AND ATTRACTIVE TRAITS

One of the most pervasive controversies stems from uncertainty about how "adaptive" we can expect the world to be. Identifying the adaptive benefits of traits is a grand charge handed down to evolutionary biologists by Darwin. Until the early 1960s, biologists usually assumed that what is good for the individual is also good for the group and even for the species. The debates over group vs. individual selection changed the scene by showing that what is adaptive at one level may be maladaptive at another. Thus populations might well suffer reduced fitness as a consequence of indivividual adaptations. More recently even the individual has been seen as a vehicle through which individual genes vie for replication.

Recently legitimized and with a historical legacy of debate, sexual selection has become a new battleground over the limits of adaptation. Williams (1966) and, especially, Zahavi (1975) suggested that female choice of mate ought to have beneficial genetic effects, as regards the general viability of her offspring, and proposed a mechanism by which this might occur (Zahavi's "handicap principle"). There followed a quick response of theoretical and genetic analyses arguing that the handicap principle could not or was unlikely to work (summarized by O'Donald 1980; Maynard Smith 1985). Subsequent models by Lande (1981), Kirkpatrick (1982), and Seger (1985) also showed that Fisher's feedback process can run perfectly well and suffices to produce the known diversity in male traits. But there is also evidence from genetic models that the evolution of mating preferences for viability-indicating traits is logically feasible (Andersson 1986). The debate continues: do females choose mates preferentially using cues that will lead to more adapted, viable offspring or not? Just because the Fisherian process *can* explain bizarre males, is that process really occurring? Some researchers who happily accepted the demoting of population fitness when it conflicted with individual dynamics balked at the idea of female choice leading a species off on random evolutionary trajectories. Sides were formed, models proliferated, and as advocates from the various sides became both more zealous and inventive, it became difficult for an agnostic to know which field studies or laboratory experiments would best help resolve the debates. The first section of this volume should help add some order and direction to such future tests.

This issue pervaded nearly all discussions at the conference, and it was brought into explicit focus as the first of our four major topics. To sort out the players and the teams, Maynard Smith begins the volume with a classification of models of sexual selection. As he points out, most of the models of adaptive female choice require that there be some additive genetic variance in total fitness. In the second chapter, Charlesworth examines the evidence for heritability of total fitness maintained by mutation pressure and other processes, and whether it would be sufficient to support adaptive mate choice.

Both adaptive and Fisherian models of sexual selection are based on a large number of assumptions. In the third chapter, Andersson lists the major assumptions for several of the better-known models, stressing that the reality of assumptions and predictions that differ between the models may be useful criteria in discriminating between the alternatives. Given the sharing of many common assumptions, it seems possible that handicap or indicator processes and Fisherian processes might co-occur within a population. If there are competing preferences, some adaptive and others arbitrary, Borgia argues that the benefits of the former ought to make adaptive preferences much more common in nature than arbitrary ones. When there is direct selection on females as a consequence of their choices, Kirkpatrick shows that Fisherian outcomes are more constrained than when choice is free of cost, but runaway dynamics still lurks in the system. He also shows that equilibria differ depending on whether the direct selection consists of benefits from mating (e.g., nuptial offerings or access to resource on territories) or costs of searching.

What about indirect costs of Fisherian mate choice? Females of species with extravagant male characters often show less developed versions of the same characters. Lande contrasts the rates at which such traits are elaborated in males with the rates at which their expression in females becomes sex-limited. He suggests that even at some cost to females, partial female expression might be an unavoidable consequence of mate choice where the additive genetic correlation between the sexes is high.

The group report by Heisler et al., which concludes this section, attempts to summarize where we stand in the basic debate. Several popular alternatives have not yet been formally modelled. The history of evolutionary biology is replete with verbal ideas that did not hold up when made explicit genetically, and to see whether these alternatives are sound, they must be developed formally. This ought to keep the theoreticians busy. Some issues are already sufficiently clear that subsequent progress will depend mainly on appropriate field and laboratory data. How do females choose? What traits do they use? What do the preference functions look like? To help granting agencies along, the group report lists research topics that urgently need attention.

CONFLICTS BETWEEN AND WITHIN THE SEXES

Darwin pointed out that females generally invest much resources directly in offspring, whereas males make smaller material contributions but larger expenditures on competition to inseminate females. Trivers (1972) generalized this notion by arguing that the sex with the largest parental investment becomes a limiting resource over which members of the other sex will compete. This has led to the widely held view that patterns of parental investment largely determine the nature and strength of sexual selection.

The second section of the volume challenges this idea. The chapter by Hammerstein and Parker notes that the parental investment game (or "who gets stuck with the baby") is only one of a set of games between the sexes. Others concern which sex will search for mates, which sex gives in first in a conflict over accepting each other as mates, as well as the relative roles of incest avoidance, risk of hybridization, uncertainty of paternity, and male quality in determining the outcome of mating encounters. The authors suggest that there is no necessary link between the outcomes of the parental investment game and these other games between the sexes. This implies that there may be several independent determinants of mating effort and variation in mating success within a sex. These notions are echoed in the chapter by Bradbury and Davies who survey a number of animal taxa but find no simple link between patterns of sexual selection and parental investment or mating system. Which sex performs the parental care does not predetermine which sex does the searching, and neither is well correlated with which sex gives in first when eager males meet coy females.

Can these ideas be applied to plants? Queller discusses the degree to which the options of each sex are constrained or expanded in plants as compared to animals. While the often biassed investment in male flowers or parts of flowers implies pollen scramble competition in plants, the options for mate choice seem much more limited than in animals. On the other hand, Queller points to much opportunity for postfertilization choice in plants, which might occur in relation to nonadditive genetic variance and heterozygotic advantages that cannot form the basis of sexual selection in animals.

The group report by West-Eberhard et al. reinforces the notion that different independent transactions between the sexes must be considered before the diversity of sexual selection patterns can be understood. This notion creates more degrees of freedom than expected if parental investment were the only determinant of sexual selection, and it may help us better understand the diversity seen in broad comparisons of taxa. The section ends with a plea for field data which more directly recognize the potential conflicts between the sexes. The shopping list of this report overlaps to an encouraging extent with those of the other group reports.

THE MEASUREMENT OF SEXUAL SELECTION

One of the traditional approaches to studying selection is the comparative method. In the present context, one might gain valuable insights about sexual selection by comparing measures of its strength with differences in taxon, ecological context, mating system, demography, and the like. Unfortunately, what constitutes an appropriate measure of the strength of sexual selection is a difficult and hotly contested point.

Intuitively, one expects species in which only a few males do all the matings to be under stronger sexual selection than those in which every male mates. Variance in mating success has thus been the traditional measure of sexual selection. Bateman (1948) used such a contrast in fruit flies to argue that sexual selection is stronger in males than in females. Payne (1984) reviewed variance in male mating success in birds in relation to mating system. Wade and Arnold (1980) formalized this approach and showed how such variance could be partitioned into components.

There are three basic complaints lodged against this measure. The first is that sexual selection means selection of a trait. Unless the variance in male mating success is correlated with particular values of some trait, nothing is being selected. This led to pleas for the use of selection differentials (the covariance of trait with relative fitness) and selection gradients (the partial correlations of traits with fitness, correcting for the correlations between traits) (Lande and Arnold 1983; Arnold and Wade 1984). These are much more difficult to measure than variance in male mating success, and not all field workers are happy that efforts to measure variance might have gone for naught. As Wade argues in this volume, the variance in male success still sets the upper limit on what selection can occur: it thus measures the opportunity for selection. That this opportunity is not always realized does not prevent its usefulness as a measure in certain cases.

The second critique is that random mate assignment (i.e., no sexual selection) in many cases can produce sex differences in the variance of mating success depending on how males and females are distributed, how much time it takes to prepare for reproduction, etc. Differences in variance between the sexes or between species must thus be tested in relation to these null expectations before any conclusions can be drawn about the strength of sexual selection. In his chapter, Sutherland argues that variances have been misused so often that they should be abandoned as measures of sexual selection. Instead, he proposes a measure based upon the relative rate with which a searching individual finds mates.

Grafen provides the third critique. Perhaps sophisticated quantification of sexual selection is not all that necessary anyway: Darwin did quite well without multivariate statistics. Whether there is a need for quantification depends very much on what questions are asked. For many issues, it may not be necessary to measure selection differentials and selection gradients. In a typically lucid and provocative chapter, Grafen reminds us that quantification for its own sake is no virtue. This is easy to forget in a field where new, exciting, and advanced methods may receive as much attention for their sophistication as for their ability to answer the original questions. There is also a trade-off in many cases between sophistication and practicality of usage.

The group report by Wilkinson et al. identifies some common ground in these debates. Whereas variance measures are now widely available, their limits are admitted and the need for field data on selection gradients and differentials is emphasized. Where examples are available, variances and gradients do not always lead to the same conclusions, which reinforces the need for care. Other measures such as Sutherland's encounter rate need to be explored, but they may be even more difficult to use in practice. The group noted that all the proposed measures are susceptible to inflation by secondary factors, such as copying by choosing females or cultural inheritance of preferences. In addition to requiring additional statistical controls, these processes may prove to be important in sexual selection.

CONSTRAINTS ON SEXUAL SELECTION

All models of sexual selection presume that the selection pressures favoring male secondary sex traits are eventually opposed by costs to males bearing them. The majority of models invoke increased mortality as the major cost limiting further elaboration of male traits. This mortality is usually ascribed to predation, and there are cases supporting this presumption (e.g. Burke 1982; Ryan 1985). Alternatively, male display might lead to mortality as a consequence of excessive energy expenditure. It is also conceivable that males are directly limited at the time of display by physiological constraints without there being any subsequent mortality. In his chapter, Halliday examines a variety of potential physiological constraits on sexual selection. He concludes that short-term energetic capacities, such as those measured by physiological scope, have not been successfully linked to male performance in available studies. On the other hand, long-term and cumulative costs appear to play a limiting role in some sexually selected systems without necessarily increasing mortality.

Partridge and Endler consider a variety of demographic factors, such as the operational sex ratio, which may set additional limits to sexual selection. Their review suggests several testable predictions, e.g., that semelparous species will show more extreme evidence of sexual selection than iteroparous forms, all else being equal. The group report by Price et al. compares the relative importance of these and several other possible limits, including correlated shifts in male traits governing feeding or mobility, biases in female sensory acuity, presence of other species with similar display traits, costs of expressing homologous display characters in females (see also Lande, this volume), and the degree to which there is heritable variation left in a population. Their general conclusion is that the cost components of sexual selection have been largely neglected and that it is important for many reasons to identify them in specific systems.

We feel that the conference and this volume were timely. Debates on sexual selection in the last decade have begun to show the same kind of righteous zeal that sometimes surfaced in the conflicts over the adaptive nature of allozymes or the role of null models in community ecology. Our primary aim in calling this meeting was to forestall any divisive chauvinism by bringing protagonists together at an early stage. We are extremly pleased at the constructive attitudes of the participants and by the serious attempts to bridge markedly different views. The flavor of cooperation and gentle reason which grew daily at the meeting is hopefully tangible in the group reports. It is our hope that one subtle benefit of this volume is a prevention of aggressive antagonism which might hinder intellectual progress. We know we have differences; resolving them should be a common goal.

The conference succeeded in clarifying where we differ and also where we do not. The aim of devising discriminating tests succeeded to varying degrees; hopefully, readers will find the various chapters useful in planning their own work. Our only regret about the conference and the volume is that we were able to invite only a small fraction of the many researchers who have worthwhile contributions to make to the topic. The Dahlem format is limited to under 50 participants, only a tiny fraction of the many qualified researchers working in this expanding field. We can only ask for indulgence from those who were not invited. Their work is part of the intellectual foundation for this meeting; hopefully, they too will gain something from this volume.

All of the participants owe an enormous debt of gratitude to the Stifterverband für die Deutsche Wissenschaft, the Deutsche Forschungsgemeinschaft, and the Senat of the City of Berlin for sponsoring this conference at a critical time in the development of the topic. It is also clear that without the Dahlem Workshop Model, so ably designed and cheerfully administered by Dr. Silke Bernhard and her staff, this "meeting of the minds" so necessary for constructive progress would not have been possible. Our sincere thanks to them all.

REFERENCES

Andersson, M. 1986. Evolution of condition-dependent sex ornaments and mating preferences: sexual selection based on viability differences. Evolution 40: 804:816.

Arnold, S.J., and Wade, M.J. 1984. On the measurement of natural and sexual selection: theory. Evolution 38: 709–719.

Bateman, A.J. 1948. Intra-sexual selection in *Drosophila*. Heredity 2: 349–368.

Burke, T. 1982. Evolutionary significance of predation on sexually signalling males. FLA Entomol. 65: 90–104.

Fisher, R.A. 1930. The Genetical Theory of Natural Selection. Oxford: Clarendon Press.

Huxley, J.S. 1938. The present standing of the theory of sexual selection. In: Evolution: Essays on Aspects of Evolutionary Biology, ed. G.R. de Beer, pp. 11–42. Oxford: Clarendon Press.

Kirkpatrick, M. 1982. Sexual selection and the evolution of female choice. Evolution 36: 1–12.

Lande, R. 1981. Models of speciation by sexual selection on polygenic traits. Proc. Natl. Acad. Sci. USA 78: 3721–3725.

Lande, R., and Arnold, S.J. 1983. The measurement of selection on correlated characters. Evolution 37: 1210–1226.

Payne, R.B. 1984. Sexual selection, lek and arena behavior, and sexual size dimorphism in birds. Ornith. Mono. 33: 1–52.

Maynard Smith, J. 1985. Sexual selection, handicaps and true fitness. J. Theor. Biol. 115: 1–8.

O'Donald, P. 1980. Genetic Models of Sexual Selection. Cambridge: Cambridge Univ. Press.

Ryan, M.J. 1985. The Tungara Frog. Chicago: Chicago Univ. Press.

Seger, J. 1985. Unifying genetic models for the evolution of female choice. Evolution 39: 1185–1193.

Trivers, R.L. 1972. Parental investment and sexual selection. In: Sexual Selection and the Descent of Man, ed. B. Campbell, pp. 136–179. Chicago: Aldine.

Wade, M.J., and Arnold, S.J. 1980. The intensity of sexual selection in relation to male sexual behaviour, female choice and sperm precedence. An. Behav. 28:446–461.

Williams, G.C. 1966. Adaptation and Natural Selection: A Critique of Some Current Evolutionary Thought. Princeton: Princeton Univ. Press.

Zahavi, A. 1975. Mate selection – a selection for a handicap. J. Theor. Biol. 53: 205–214.

Sexual Selection: Testing the Alternatives
eds. J.W. Bradbury and M.B. Andersson, pp. 9–20
John Wiley & Sons Limited

Sexual Selection — A Classification of Models

J. Maynard Smith

School of Biological Sciences
University of Sussex
Brighton BN1 9QG, England

Abstract. The main models of sexual selection are reviewed. The criteria that distinguish between the models are as follows:

1) Does only the competing sex (typically the males) evolve or do the preferences of the choosing sex (typically the females) also evolve?
2) Is female preference concerned with choice of a conspecific mate or with choice between conspecifics?
3) Does the female acquire resources? Is female choice directly selected: Is there a cost to choice?
4) Do females evolve a preference for some component of male viability or for a signal that does not itself increase viability?
5) Are there differences in male viability other than those arising directly from the presence of the trait chosen by females? If so, does a genetic correlation arise between genes for the chosen trait and other viability genes?

A BRIEF HISTORY OF IDEAS

The idea of sexual selection orginates with Darwin (1859, 1871). He was led to the idea because of the problem posed by such traits as the tails of peacocks or the antlers of male deer. It is not easy to account for the evolution of such characteristics by natural selection, because they do not contribute to survival (if they do, why are they not present in females?) nor to the survival of offspring (as do such organs as the placenta, which are present in one sex only). He suggested that they owe their presence to "the advantage which certain individuals have over other individuals of the same sex and species, in exclusive relation to reproduction."He thought that this advantage might arise in one of two ways: competition for mates between members of the same sex (typically, between males) and the active choice by one sex of particular individuals of the other (typically, the choice of males by females). There was peculiarly little interest in, or further development of, this idea for about a century, perhaps because Darwin was unable to offer a selective explanation for "choice," and because the main trend in behavioral studies was towards mechanistic explanations and away from anything that smacked of anthropomorphism. The next significant progress came with a brief but typically suggestive discussion by Fisher (1930), which made a number of points:

1) "The tastes of organism, like their organs and faculties, must be regarded as the product of evolutionary change, governed by the relative advantages which such tastes confer." I shall take it that this view is now widely, if not universally, accepted by biologists.
2) There will be selection favoring those members of one sex best able to discriminate "differences . . . which are in fact correlated with selective advantage." He thought that such preferences "may therefore be far more widespread than the occurrence of striking secondary sexual characters," and that they are a necessary initial condition for the evolution of sexual characters that are not so correlated.
3) Mate preference by one sex (typically females) can lead to the evolution of traits in the other to a degree that lowers fitness in contexts other than mating.
4) The joint evolution of preference, and of the preferred trait, can lead to a "runaway process," proceeding at a rate increasing exponentially with time, until checked by "severe counterselection in favor of less ornamented males."

Unfortunately, he gave no formal model, and it is not easy to see what led him to those conclusions and in particular to his runaway process.

These ideas had little immediate impact. The problem of sexual selection played little part in the "Modern Synthesis": there is no reference to it in the index of Huxley (1942) or of Dobzhansky (1951), and the only reference in Mayr (1942) says "many phenomena that have been recorded in the past as furthering intraspecific sexual selection are actually specific recognition marks." The main reason for this, I think, was that the central preoccupation of the founders of the Modern Synthesis lay in the nature and origin of species. Perhaps surprisingly, the growth of the ethology movement after the war did little to stimulate interest in the problem. In the extensive publications marking the centenary of the Origin of Species, the only explicit treatment of sexual selection was Maynard Smith (1958); although I did describe a possible mechanism of female choice in *Drosophila subobscura*, it is clear that I had not read or understood Fisher. Ten years later, the centenary of the publication of The Descent of Man did force the publication of a fuller treatment of sexual selection, in Campbell (1972), but the only real novelty in that volume was Trivers' (1972) treatment of the importance of parental investment.

The modern phase of the subject is characterized by the formulation of genetic models of the process, in particular by O'Donald (1962 and, subsequently, reviewed in 1980), Lande (1981), and Kirkpatrick (1982).

MALE-MALE COMPETITION AND FEMALE CHOICE

The distinction between these processes was fundamental for Darwin. I suggest that we speak of female choice when females have some behavior or structure causing them to be more likely to mate with some conspecific males than others. This is close to Halliday's (1983) definition: "any pattern of behavior, shown by members of one sex, that leads them to being more likely to mate with certain members of the opposite sex than with others."

When we come to classify models of sexual selection, however, we must distinguish between those in which female preference is genetically variable and is itself evolving, and those in which female characteristics are assumed to be constant. Some examples may help to explain the relevance of this distinction. Male red deer hold harems. If hinds remain with whichever male takes control of them, and do not try to move from one harem to another, there is no female choice. Consider a less obvious example (Arak 1983). Female natterjack toads are attracted by the croaks of males, or by loudspeakers playing those croaks. If a female is placed midway between two speakers of different loudness, she will go to the louder one, but if she is placed closer to the less loud speaker, she may approach it rather than the louder one. That is, she appears to move up a sound gradient. Such female behavior may well have evolved because it enables a virgin female to find a conspecific male. It is an *effect* of this behavior that females are most likely to mate with males giving louder calls, but there is no reason to think that the behavior evolved *because* it has that effect. Based on the definition given above, this is an example of female choice, because females have a behavior (moving up a sound gradient) which makes them more likely to mate with loud-calling males. But, if we wished to model this situation, we would treat it as a simple case of male-male competition.

Suppose, however, that Arak had found that a female placed between several speakers, approaches each of them in turn, and then moves to the loudest one. In that case, there would be grounds for thinking that the female's behavior had evolved because of the discrimination it made possible. To model this situation, we would have to assume that female behavior was genetically variable, and ask how it would evolve. I have earlier (Maynard Smith 1985) defined "female choice" so that it refers only to evolved preferences of this kind but have refrained from doing so here because of the almost universal opinion of those with whom I have discussed the matter that such a definition is confusing and unacceptable, essentially because it is difficult, if not impossible, to decide what the evolutionary cause of a behavior may be, and therefore foolish to include it in a definition.

There are, however, overwhelming reasons to expect genetic differences between females that influence which males they mate with, and there is some empirical evidence for such differences (Majerus et al. 1982). If so, this will affect the mating success, and hence the fitness, of males, and this in turn will cause changes in female preference. Hence, even if a preference is initially present for reasons that have nothing to do with choice between conspecific mates (as suggested by Kirkpatrick in this volume), its further evolution is likely to be influenced by its effects on mate choice.

A PROPOSED CLASSIFICATION OF SELECTIVE PROCESSES

In this section I give a classification of models, and later a commentary and explanation, but first a few introductory points:

1) The classification is of the selective processes that may occur, whether or not they have received formal mathematical treatment. I have shown where some of the more familiar models would belong in order to make the intended meaning of the classification clearer.
2) I have included "choice of a conspecific mate." although this would not come under the definition of female choice offered above, because it is an alternative selective force that should be borne in mind.
3) I have used "viability" to mean all components of fitness other than mating success. The word is far from ideal (successful parental care, for example, becomes part of viability, which is absurd), but I cannot think of one better.
4) The classification of "choice" assumes that it is the female that is choosing, although this need not be the case if male parental care is a limiting resource for females.

Mechanisms of sexual selection

1. Intrasexual selection (male-male competition)
2. Mate choice (female choice)
   - I. Choice of a conspecific mate
   - II. Choice between conspecifics
     - A. female gets resources
       (e.g., a territory, male parental care; typically monogamous species although courtship feeding can occur with polygyny)
     - B. female gets no resources
       - i) female chooses a male trait that itself contributes to viability. (e.g., Fisher's ". . . correlated with selective advantage": Maynard Smith's (1956) verbal explanation of *Drosophila* courtship dance)
       - ii) female chooses a signal that may itself reduce viability but which is not correlated with other components of viability. (e.g., O'Donald 1962, 1967; Lande 1981; Kirkpatrick 1982)
       - iii) female chooses a signal that is correlated with viability
         - a) the signal is a handicap (Zahavi 1975)
         - b) the signal reveals fitness (Hamilton and Zuk 1982)

c) the signal is conditional (West-Eberhard 1979).

In the next section I have a classification, orthogonal to the one above, of the choice process itself.

INDIVIDUAL AND POPULATION MODELS OF CHOICE

By an individual model of choice I mean one specifying the degree of preference a given female shows towards different males. A population model then converts such preferences into mating frequencies, given a frequency distribution of types of males. In polygynous models it has always been assumed that, for a given female, mating frequency depends only on the male frequency distribution and not on what other females are present. This assumes that there is no limit on the number of matings by a single male. This clearly cannot hold for monogamous models. Darwin argued that sexual selection would have similar effects in monogamous species, because the first females to be ready to breed would be in a position to choose the fittest males. O'Donald (1980, 1983b) has modelled Darwin's idea and concludes that the same equilibria are reached as in the polygynous case, but much more slowly.

Individual models

Lande (1981) defines $\psi\ (z/y)$ as being proportional to the chance that a male of phenotype z would be chosen by a female of preference y. He then defines three kinds of preference function:

1) Directional preference $\psi\ (z/y) = exp(yz)$. Thus all females prefer to mate with males with large z, but they differ in degree of discrimination.
2) Absolute preference; $\psi\ (z/y) = exp[-k(z-y)^2]$, where k = const.
Thus a y female prefers a male of value z and her willingness to mate falls off either side of this value.
3) Relative preference; $\psi(z/y) = exp\ \{-k[z-(\bar{z}+y)]^2\}$.
Thus a y female prefers a male of value y greater than the population mean.

Population models

How do these individual preferences convert into frequencies of matings in the population? There are various possibilities:

1) *Constant mating preference*. O'Donald (1962, 1973), and also Maynard Smith (1976) and Andersson (1982), assumed that a "choosy" female, C, mates with an A male with constant probability, P, (independent of the frequency of A), whereas with probability $1\text{-}P$ she mates randomly. This model is only suitable when the male trait is discontinuously distributed. Also, as O'Donald (1973) points out, it is unsatisfactory if A males are rare relative to C females

since it gives an unrealistically high mating success to A males.
2) *Fixed relative preference (open-ended search)*. A female encounters a random sequence of males until she mates. In a given encounter, her mating probablity is given by the value of $\psi(z/y)$, as defined above. If there are two types of male in the population, A_1 and A_2, for which the ψ values are P_1 and P_2, and if their frequencies are R_1 and R_2, it is easy to show that her chance of mating an A_1 male is $P_1R_1/(P_1R_1 + P_2R_2)$. This is the choice model adopted by Lande (1981) and Kirkpatrick (1982). The "encounter" model proposed by O'Donald (1980) is similar, but not identical.
3) *Best of N males (truncated search)*. Seger (1985) supposes that a female compares N males and mates with the one closest to her preferred type. He shows that this can lead to qualitatively different results, as discussed below.

COMMENTS AND EXPLANATIONS

Male-male competition

Maynard Smith and Brown (1986) analyzed a model based on the life history of male red deer, which grow without breeding until they reach size m. They then compete with other males for females; they may survive to breed for several years but do not grow any further. We simulated a model in which adult size, m, was polygenically determined. The fitness of an individual is given by $W(m) = s(m).\ B(m)$, where $s(m)$ is the probability that an individual survives to size m, and $B(m)$ its breeding success if it does so. We considered various models of $B(m)$, of which the simplest was $B(m) = a+bR$, where R is the fraction of the population smaller than m, and a and b are constants.

The important finding is that such model leads rather readily to a runaway process. In the simplest version, with m determined solely by genes and $s(m) = exp(-km)$, size increases until all high alleles are fixed, even though $\overline{W}$ is then less than 0.01. Of course, interspecific competition need not lead to a runaway. If there is environmental variance of size (as there certainly would be) *and* if $s(m)$ decreases more sharply with m than $exp(-km)$, then a stable size is reached. Thus a runaway can always be arrested by counteracting selection, but in this case it does so only by producing a population in which males are senescent before they breed. Parker (1983) has reached similar conclusions using an ESS approach.

Illogically, we simulated a model in which selection operated identically in the two sexes. However, I have recently repeated the simulations, with qualitatively identical results, on the assumption that the relevant genes are expressed only in males.

This model of male-male competition is relevant not only to cases in which males fight for females (e.g., red deer) but also to those in which they compete to attract females all of which are identical (e.g., Arak's toads). Why, then, do we find that runaway is a possible outcome, whereas Lande (1981) finds that runaway requires that males and females both evolve? Thus, in his model, if there is natural selection on the males for an intermediate optimum, runaway requires genetic variance for the female preference. I think that the reason for the difference is as follows. Lande's fitness functions are Gaussian; hence, as the trait value z increases, viability falls off as $exp(-kz^2)$, whereas in our simplest model it falls off as $exp(-kz)$. There are also differences in the way in which mating success, $B(z)$, is determined. I conclude that whether a model leads to a runaway, or to a fixed equilibrium, depends (among other things) on detailed assumptions about fitness and mating success functions. This conclusion is in line with that of Seger (1985), discussed below.

Choice of conspecific mate

Little need be said except to reiterate that the orthodoxy of the forties and fifties was that this is the major selective force influencing the evolution of mate choice. This is a view that badly needs testing by comparative methods.

Female gets resources

Again, little needs to be said. I do not doubt that selection will favor females able to select good territories or good baby-minders. There may be argument about how we should treat courtship feeding. Is the female trying to evaluate the male or is she just after a square meal?

Female chooses a male trait that contributes to viability

The major problem here lies in the nature of the genetic variance of fitness. Thus Maynard Smith (1956) argued that female *D. subobscura* engage in a sidestepping dance because this enables them to select as mates males of high "athletic ability." If the additive genetic variance of fitness is zero, however, would genes causing females to dance increase in frequency? It is worth listing various hypotheses (these are not mutually exclusive, but all may be true of some part of the genetic variance):

1) In Lande's models, genes affect viability only through their (additive) effects on phenotypic traits under stabilizing selection. Whether a particular allele substitution increases or decreases viability depends on its genetic background. In this model, females do not necessarily evolve to select as mates the most viable males.
2) Mutations may be unconditionally deleterious, independent of their background, and may be present only because of recurrent mutation. It is plausible that a gene causing a female to mate with a male with fewer-than-average

deleterious mutations (Manning 1984) would increase in frequency, although I know of no formal model. Such a choice could not depend on recognizing the presence of specific mutations but could depend on recognizing differences in health or vigor of a nonspecific kind. My 1956 argument could be justified along these lines.

3) There may be additive genetic variance of fitness because the "environment" is changing so that the species is under directional selection. It is easy to modify models, of the kind developed by O'Donald (1982) and Kirkpatrick (1982), to show that if a high viability allele B is spreading then a gene C causing females to mate with B males will increase in frequency, even when C is rare. The difficulty is to imagine how the same directly detectable trait could be variable for a protracted period. The strength of the host-parasite model of Hamilton and Zuk (1982) is to explain continuing additive genetic variance of viability. Direct detection by females of the disease-resistance genes is likely to be impossible, and would in any case be ineffective because a given genotype is sometimes selectively favored and sometimes not. Female choice would therefore have to be either of general health and vigor or of some "revealing signal," as discussed below.

4) The genetic variance of fitness may be nonadditive, e.g., due to heterosis. It seems plausible that selection will not favor female choice in this case, but I know of no formal model.

5) There may be additive genetic variance of particular components of viability but negative correlations between components (as observed by Rose and Charlesworth (1980) for longevity and fecundity in *Drosophila*). If so, presumably selection would not favor female choice of a particular component.

To conclude, I share Fisher's conviction that female choice of this type may be common. If so, it must depend on the presence of additive genetic variance of fitness arising from mutational (and perhaps migrational and recombinational) load or from continuing directional selection. Finally, if female preference evolves for some component of viability, such as athletic ability, then sexual selection can lead to the exaggeration of this component at the expense of others.

Females choose a signal that may reduce viability, but which is not correlated with other components of viability

Thanks to O'Donald (1962, 1967), Lande (1981), and Kirkpatrick (1982) this is the best understood type of sexual selection. In particular, it is encouraging that the basic finding of a line of equilibria should emerge both from a 2-locus and a polygenic model. A matter that remains to some degree controversial concerns the conditions for runaway selection. In Lande's model, this depends on the value of the genetic regression of female preference on male trait, B/G: if the slope is steeper than that of the line of equilibria, runaway ensues. The regression arises because of linkage disequilibrium between the two sets of genes as a result of nonrandom mating: its slope depends on the rates at which genetic variance in choice, and in the trait, are generated by mutation. O'Donald (1983a) has queried this conclusion, essentially because he does not think that the relevant variances and covariances will remain constant during selection. He argues that runaway depends on the nature of female choice, requiring that females choose the most extreme phenotype and not merely an optimum phenotype. I have argued above that such "directional preference," given an appropriate viability function, can lead to a runaway without female variance. However, this does not prove that Lande is wrong to claim that one can get runaway with "absolute preference" and female variance.

For the 2-locus model, the equivalent of runaway is the instability of the line of equilibrium. Seger (1985), in a modified version of Kirkpatrick's (1982) model, finds that the line does become unstable if a "best of N males" choice model is adopted. This has the effect of reducing the slope of the line of equilibria: thus Seger's result confirms Lande's argument that runaway depends on B/G being greater than the slope of the line of equilibrium.

Females choose a signal correlated with viability

Many biologists feel some discomfort with the conclusions of the last section and have argued that the signals chosen by females will, in some way, be connected with viability. One formulation of this view of Zahavi's (1975) "handicap principle," Maynard Smith (1976) attempted to formalize this in a 3-locus model, whose assumptions are shown in model (a) of Table 1. The essential point is that males with a "handicap," A, suffer a slight reduction in viability, (1-t), if they are of a high viability, B, but suffer a greater reduction,

TABLE 1 Viabilities of males and types of males selected by females, according to three models: (a) the handicap model (Zahavi 1975); (b) the revealing handicap model (Hamilton and Zuk 1982); (c) the conditional handicap model (West-Eberhard 1976)

A,a males with and without the handicap, respectively
B,b males with high and low viability, respectively
s,t,u positive constants (with $u<1$ and $t<1$) which can be chosen to give the required relative viabilities

Boxes indicate types of males preferred by choosy females

model (a)	A	a
B	$(1+s)(1-t)$	$(1+s)$
b	$(1-t)(1-u)$	1

model (b)	A	a
B	$(1+s)(1-t)$	(1+s)
b	$(1-t)$	1

model (c)	A	a
B	$(1+s)(1-t)$	$(1+s)$
b	1	1

(1-t) (1-u), if they are of low viability, b. Hence, by choosing an A male a female is more likely to be choosing a male of high viability.

There is an obvious difficulty in interpreting the results of such a model. We know, from the models discussed in the last section, that a handicap can spread even in the absence of viability differences, given female preference. How, then, are we to decide whether the correlation of the handicap with high viability genes is relevant? Maynard Smith (1976) argued that Zahavi's idea would be confirmed if it could be shown that the handicap would increase even if female preference was rare, because in that case the Fisher process would be ineffective. This is perhaps too severe a criterion. It would be sufficient to show that the handicap will increase in a wider range of circumstance (e.g., for lower frequencies of female preference) if viability differences are segregating than it they are not. In a recent review (Maynard Smith 1985) I concluded based on these criteria that the handicap model does not work, but recent papers (Andersson 1986a, b; Pomiankowski, unpublished manuscript) suggests that this conclusion may have been too negative.

It has never been quite clear whether model (a) of Table 1 properly represents what Zahavi (1975) had in mind. Two modifications, models (b) and (c), are shown. Although formally similar, they have difference biological motivations:

1) The "revealing handicap" model (model (b), Table 1) of Hamilton and Zuk (1982) supposes that the viability difference, B vs. b, can be directly detected in A males but not in a males.
2) The "conditional handicap" model (model (c) Table 1) was based on a verbal argument by West-Eberhard (1979); it has also been proposed by Williams (1966) and Zahavi (1997). Andersson (1986b) has investigated it with a formal model. It supposes that a male with the handicap gene A only develops the handicap if it also has the high viability gene, B.

An important difference between models (b) and (c) and that of model (a) is that a choice gene, C, does increase in frequency even when rare, provided that alleles at the A and B loci are segregating. But does the handicap increase? Kirkpatrick (1986) concludes that there will be some threshold value of the frequency of the choice gene, C, such that the handicap increases in frequency if C is above the threshold, and decreases otherwise. The level of the threshold is little affected by the segregation of the viability alleles B and b. Hence it is still not clear whether the correlation between the handicap and viability is all that relevant, although there is clearly room for further investigation.

There remains the problem of how female preference is supposed to originate. Thus what the models show, *given* an initial female preference, is that selection can further increase both the preference and the trait preferred. Some authors have argued that no explanation of the initial preference is called for: it could, for example, be an unselected correlate of a preference evolved in some other context, such as feeding. Fisher, as we have seen, thought that a preference would evolve at first because it was a preference for more viable males. It may be in this context that the models of this section are most relevant, because they can explain the increase of a choice gene when it is rare.

REFERENCES

Andersson, M. 1982. Sexual selection, natural selection and quality advertisment. Biol. J. Linn. S. 17: 375–393.

Andersson, M. 1986a. Sexual selection and the importance of viability differences. J. Theor. Biol. 120: 251–254.

Andersson, M. 1986b. Evolution of condition-dependent sex ornaments and mating preferences: sexual selection based on viability differences. Evolution 40: 804–816.

Arak, A. 1983. Sexual selection by male-male competition in natterjack toad choruses. Nature 306: 261–262.

Arnold, S.J. 1983. Sexual selection: the interface of theory and empiricism. In: Mate Choice, ed. P. Bateson. pp. 67–107. Cambridge. Cambridge Univ. Press.

Campbell, B. 1972. Sexual Selection and the Descent of Man. London: Heinemann.

Darwin, C. 1859. On the Origin of Species by Means of Natural Selection. London: John Murray.

Darwin, C. 1871. The Descent of Man and Selection in relation to Sex. London: John Murray.

Dobzhansky, T. 1951. Genetics and the Origin of Species, 3rd ed. New York. Columbia Univ. Press.

Fisher, R.A. 1930. The Genetical Theory of Natural Selection. Oxford: Clarendon Press.

Halliday, T.R. 1983. The study of mate choice. In: Mate Choice, ed. P. Bateson, pp. 3–32. Cambridge Univ. Press.

Hamilton, W.D., and Zuk, M. 1982. Heritable true fitness and bright birds: a role for parasites. Science 218: 384–387.

Huxley, J. 1942. Evolution, The Modern Synthesis. London: Allen & Unwin.

Kirkpatrick, M. 1982. Sexual selection and the evolution of female choice. Evolution 36: 1–12.

Kirkpatrick, M. 1986. The handicap mechanism of sexual selection does not function. Am. Natur. 127: 222–240.

Lande, M. 1981. Models of speciation by sexual selection on polygenic traits. Proc. Natl. Acad. Sci. 78: 3721–3725.

Majerus, M.E.N.; O'Donald, P.; and Weir, J. 1982. Nature 300: 521–523.

Manning, J.T. 1984. Males and the advantage of sex. J. Theor. Biol. 108: 215–220.

Maynard Smith, J. 1956. Fertility, mating behaviour and sexual selection in *Drosophila subobscura*. J. Genet. 54: 261–279.

Maynard Smith, J. 1958. Sexual Selection. In: A Century of Darwin, ed. S.A. Barnett, pp. 231–244. London: Heinemann.

Maynard Smith, J. 1976. Sexual selection and the handicap principle. J. Theor. Biol. 57: 239–242.

Maynard Smith J. 1985. Mini review: sexual selection, handicaps and true fitness, J. Theor. Biol. 115: 1–8.

Maynard Smith, J. and Brown, R.L.W. 1986. Competition and body size. Theor. Pop. Biol. 30: 166–180.

Mayr, E. 1942. Systematics and the Origin of Species. New York. Columbia Univ. Press.
O'Donald, P. 1962. The theory of sexual selection. Heredity 17: 541–552.
O'Donald, P. 1967. A general model of sexual and natural selection. Heredity 22: 499–518.
O'Donald, P. 1973. Models of sexual and natural selection in polygynous species. Heredity 31: 145–156.
O'Donald, P. 1980. Genetic Models of Sexual Selection. Cambridge: Cambridge Univ. Press.
O'Donald, P. 1983a. Sexual selection by female choice. In: Mate Choice, ed. P. Bateson, pp. 53–66. Cambridge: Cambridge Univ. Press.
O'Donald, P. 1983b. The Artic Skua. Cambridge: Cambridge Univ. Press.
Parker, G.A. 1983. Arms races in evolution: an ESS to the opponent-independent costs game. J. Theor. Biol. 101: 619–648.
Rose, M., and Charlesworth, B. 1980. A test of evolutionary theories of senescence. Nature 287: 141–142.
Seger, J. 1985. Unifying genetic models for the evolution of female choice. Evolution 39: 1185–1193.
Trivers, R.L. 1972. Parental investment and sexual selection. In: Sexual Selection and the Descent of Man, ed. B. Campbell. Chicago: Aldine Atherton.
West-Eberhard, M.J. 1979. Sexual selection, social competition and evolution. Proc. Amer. Phil. Soc. 123: 222–234.
Williams, G.C. 1966. Adaptation and Natural Selection. Princeton Univ. Press.
Zahavi, A. 1975. Mate selection – a selection for a handicap. J. Theor. Biol. 53: 205–214.
Zahavi, A. 1977. The cost of honesty (further remarks on the handicap principle). J. Theor. Biol. 67: 603–605.

Sexual Selection: Testing the Alternatives
eds. J.W. Bradbury and M.B. Andersson, pp. 21–40
John Wiley & Sons Limited

The Heritability of Fitness

B. Charlesworth

Department of Biology
University of Chicago
Chicago, IL 60637, USA

Abstract. Population genetics theory predicts that there will be no additive genetic variance in fitness in random-mating populations at equilibrium under natural selection, independent of the mode of selection, so that individuals of different genotype will all produce progeny with the same average fitness. This result holds to a good approximation even with linkage and epistatic interactions between loci, unless the latter are very strong. It may be violated if there are temporal fluctuations in relative genotypic fitnesses or if there are sufficiently strong forces in addition to selection acting on the population, such as migration between populations subject to differing selection pressures or mutation. Empirical evidence on the quantitative genetics of fitness-related traits is reviewed. It is shown that such evidence is hard to interpret, since it is rarely possible to analyse reproductive fitness itself. The frequent existence of significant additive genetic variance for components of fitness is compatible with the existence of little or no additive variance in fitness itself if there are negative correlations between components of fitness. Such correlations are often observed. The data do, however, suggest a negative correlation between the importance of a trait in determining net fitness and its heritability, in agreement with theoretical expectations. Comparisons of the genetics of fitness components in natural populations with the results of mutation studies in *Drosophila* suggest that there is little positive evidence for additive variance in fitness beyond that maintained by mutation pressure. It is concluded that this variance may in principle be sufficient to generate a pressure of selection in favor of mate choice, but some difficulties in accepting this as a realistic possibility are pointed out.

INTRODUCTION

"We may consequently state the fundamental theorem of natural selection in the form: *The rate of increase in fitness of any organism at any time is equal to its genetic variance in fitness at that time*" R.A. Fisher (1930).

It has frequently been suggested that choice of mates by females will be selectively advantageous if it enables them to select mates with high genetic quality, and hence produce offspring with high fitness (Borgia 1979; Hamilton

1982; Hamilton and Zuk 1982; Thornhill and Alcock 1983; Trivers 1972; Weatherhead 1984; Willson and Burley 1983) — the "good genes" hypothesis. This mechanism can produce a selective advantage only if there is additive genetic variance in fitness such that a correlation exists between offspring fitness and the male trait on which choice is exerted (Maynard Smith 1978, p. 170). Unfortunately for this hypothesis, it is a well-known result of population genetics theory that natural selection tends to exhaust the additive genetic variance in fitness (Fisher 1930, 1941). The intuitive reason for this is that a positive correlation in fitness between parent and offspring will lead to an increase in the mean fitness of the population from one generation to the next, so that a population in genetic equilibrium under the action of selection alone cannot exhibit such a correlation. This creates a serious difficulty for the good genes hypothesis (Maynard Smith 1978; Partridge 1983; Taylor and Williams 1982).

In this paper I shall review the general theory of quantitative and population genetics relating to the question of the existence of a parent-offspring correlation in fitness in natural populations, with particular reference to mechanisms that may help to maintain such a correlation. I shall also consider the relevant empirical evidence derived from quantitative genetics experiments on fitness components in a variety of species. I shall then briefly consider the relevance of this information to the evolution of mate choice.

THEORY

Definitions of heritability

The term "heritability" is somewhat vague, and we need to consider carefully how it is to be employed in the present context. At least three different usages are prevalent. Probably the most common is the definition of heritability as the ratio of additive genetic variance in a character to its total phenotypic variance (Falconer 1981, p. 148). The term "additive genetic variance" itself can be defined in two ways when multiple loci are considered. One definition is the sum over loci of the variance of the character removed by the linear regression of phenotype on allelic state at each individual locus controlling the character (Crow and Kimura 1970, p. 211). A second is the variance removed by the linear regression of phenotype on the multi-locus allelic state (Crow and Kimura 1970, p. 220; Ewens 1979, p. 218). In the absence of nonrandom associations between alleles at different loci (i.e., in the absence of linkage disequilibrium), these two quantities are the same in random-mating populations. In order to avoid confusion, the term "additive genetic variance" will be used here for the first of these quantities and the term "genic variance" for the second.

A second definition of heritability is the linear regression coefficient of offspring mean on the mean of the two parents. In the case usually considered in elementary treatments (random mating, no linkage, random combinations of alleles at different loci, no interactions between the effects of different loci, and no environmental causes of resemblance between parent and offspring), this is equivalent to the first definition (Falconer 1981, p. 151) using either measure of additive variance. If these assumptions are relaxed, however, this equivalence often breaks down (Bulmer 1980, chap. 8; Ewens 1979, chap. 7; Kempthorne 1957, pp. 308–399). This definition has the distinct utility that it applies to a parameter that can be estimated experimentally, whereas the previous one is purely theoretical. Empirical estimates of the additive genetic or genic variances are obtained from the covariances between relatives using these simplifying assumptions (Falconer 1981). They may, therefore, be somewhat biased. In order to avoid an unduly pedantic style, I shall refer to such empirical estimates as "additive genetic variances."

The third definition employs the concept of "breeding value" of a genotype (Falconer 1981, pp. 106–108). This is defined as twice the deviation from the mean value for the population of the mean of the progeny of individuals of the given genotype (the factor of two is due to the fact that individuals receive one allele at an autosomal locus from each parent.) Heritability can then be defined as the ratio of the population variance in breeding value to the total phenotypic variance. If the regression of offspring mean on parental mean is linear and homoscedastic, the heritability in this sense is equivalent to heritability in the second sense. This will be the case if the joint distribution of genotypic values among the parents and offspring is multivariate normal, which occurs if the character is controlled by a large number of additive, independent loci and if mating is random. Deviations from these conditions (such as dominance and linkage) will cause the two measures of heritability to differ (Bulmer 1980, pp. 131–143; Ewens 1979, pp. 225–231).

In the context of sexual selection and the evolution of mate choice, the third definition of heritability (or the more basic notion of the variance in breeding values) is in many ways the most relevant. If there is to be a direct selective advantage to females in choosing their mates, there must be a genotypic differences between males which lead to differences among the mean fitnesses of their progenies, i.e., there must be some variance in male breeding values with respect to fitness (see *DISCUSSION*). The magnitude of such an advantage is clearly related to the size of this variance (see *APPENDIX*). It is therefore important to clarify the conditions under which variance in breeding values in fitness is maintained in natural populations.

Effects of natural selection on components of variance in fitness

For simplicity, fitness will be discussed in terms of populations with discrete generations, where the relative fitness of a genotype can be given a precise definition in terms of the net contribution of new individuals to the next generation by members of that genotype (Charlesworth and Charlesworth 1978; Gregorius and Ross 1981; Lloyd 1977). Things are considerably more complex with overlapping generations unless selection is weak or the population is at equilibrium (Charlesworth 1980). In the latter case, lifetime reproductive success can be taken as a measure of fitness when the population is stationary in size, as is often approximately true.

Any theoretical discussion of the likelihood of the existence of significant heritability of fitness in natural populations must start with Fisher's Fundamental Theorem of Natural Selection (Fisher 1930, 1941), according to which the mean fitness of a population under natural selection increases at a rate equal to the additive genetic variance in fitness. It has since been recognized that the validity of this theorem depends on several assumptions that Fisher did not make explicit, most notably weak selection, constancy of genotypic fitness over time, and independence of genotype frequencies at different loci (Charlesworth 1980; Crow and Kimura 1970; Kimura 1958; Kojima and Kelleher 1961; Nagylaki 1979; Turner 1970). A more general result, which is highly relevant to the present discussion, is that a population at equilibrium under selection exhibits no genic variance in fitness. This is true even if there is nonrandom mating, frequency-dependent selection, epistasis, or linkage disequilibrium and when selection is arbitrarily strong, provided that there are no sex differences in fitness. The result follows from Kimura's generalization of Fisher's theorem (Kimura 1958) and can be seen to hold by equating the rate of change of population mean fitness (given by equation 5.6.19 of Crow and Kimura 1970) to zero. Consideration of the conditions for equilibrium at each locus separately yields the result that the additive variance contributed by each locus is zero, so that the additive genetic variance in fitness is also zero at equilibrium. Any genetic component of variance in fitness must therefore be due to dominance or epistasis.

Little attention seems to have been given to the analysis of variance in fitness when there are sex differences in genotypic fitnesses, a situation of obvious interest in the context of sexual selection. The work of Nagylaki (1979) implies that the effects of sex differences on the above conclusions must be of the order of the fourth power of the intensity of selection, as measured by the difference between the fitnesses of the best and worst genotypes, at least for the single locus case (see *APPENDIX*). This suggests that the above conclusions will hold to a good approximation with sex differences. An examination of the dynamics of selection at a single locus with arbitrarily strong selection reveals that gene frequency equilibrium implies that there is no additive genetic variance in fitness, if the net fitness of a genotype A_iA_j is defined as

$$w_{ij} = \tfrac{1}{2}\{(w_{ij}^f/\overline{w}^f) + (w_{ij}^m/\overline{w}^m)\} \quad (1)$$

where w_{ij}^f and w_{ij}^m are the female and male fitnesses of the given genotype, and

$\bar{w}^f$ and $\bar{w}^m$ are the mean fitnesses of females and males (see *APPENDIX*). This establishes that an equilibrium population exhibits no additive genetic variance in this measure of fitness. Similarly, it can be shown that sex differences in fitness at a single locus lead to a zero variance in breeding values at equilibrium, provided that breeding values are defined in terms of mates chosen prior to selection within a generation. If mates are chosen after the operation of selection, then there may be a small variance in breeding value at equilibrium, unless selection acts equally on the two sexes or acts only one one sex (see *APPENDIX*).

The nonexistence of additive genetic variance or genic variance in fitness at selective equilibrium does not necessarily imply that there is no offspring-parent regression or variance in breeding values in fitness. In a random-mating population with no sex differences, multi-locus selection theory tells us that this will be the case if there is no recombination between loci (when the dynamics of selection are equivalent to the single locus case), or when linkage is sufficiently loose and epistatic fitness effects are sufficiently small that the linkage disequilibrium is negligible (Crow and Kimura 1970, p. 221; Ewens 1979, chap. 7). Nonzero linkage disequilibrium implies, however, that there will be some variance in breeding values at equilibrium (Ewens 1979, chap. 7; Kojima and Kelleher 1961), since the opposing forces of selection and recombination result in the maintenance of gametes with unequal net fitnesss. The maintenance of strong linkage disequilibrium in random-mating populations seems to require close linkage and strong epistasis in fitness (Ewens 1979, chap. 7), so there is some reason to believe that the contribution of these factors to the variance in breeding values will usually be small. Experiments on the effect of recombination on fitness components of *Drosophila* have, however, suggested the existence of some variance in fitness due to this effect (Charlesworth and Charlesworth 1975; Mukai and Yamaguchi 1974; Wasserman 1972) and the selective maintenance of inversion polymorphisms is most easily explained in these terms (Charlesworth and Charlesworth 1973). Nonrandom mating due to inbreeding does not affect the conditions for equilibrium or the existence of breeding value variances (Kempthorne 1957, chap. 16), but nonrandom mating based on genotype or phenotype will often do so (Bulmer 1980, pp. 140–141).

The results discussed above have been derived on the assumption that the population exhibits no spatial and temporal heterogeneity in relative genotypic fitnesses. If there are spatial differences in fitness, associated with different "patches," but the population mates randomly across patches, as in Levene's (1953) model, the above conclusions are essentially unchanged. If there is restricted migration between subpopulations with different relative fitnesses, as in a clinal situation, then there will be some variance in fitness breeding value within each subpopulation. This variance is maintained as a result of immigration from populations with different equilibrium gene frequencies under selection (Bulmer 1980, pp. 180–184) and can lead to selection for premating isolation between subpopulations (Antonovics 1968; Fisher 1930; Lande 1982). Such variance could also create a selection for mate preference with respect to genotype at the loci responsible. It is difficult at present to assess its potential importance in relation to the topic of mate choice, in view of the lack of relevant quantitative data.

Temporal variation in relative fitnesses can also lead to the maintenance of genetic polymorphism (Gillespie 1973; Haldane and Jayakar 1963). This produces additive variance in fitness and variance in breeding values within each generation, since allele frequencies never reach equilibrium (Eshel and Hamilton 1984). Theoretical analysis shows that the long-term average fitness breeding value of the heterozygote at a locus with two alleles, under this kind of selection in a random-mating population, always exceeds that of the two homozygotes. This suggests that there will be an advantage to a gene causing an increase in the probability of females mating with heterozygotes at the selected locus, which is borne out by computer calculations (Charlesworth and Gimmelfarb, unpublished results). Although this conclusion is robust with respect to the nature of the temporally varying environment (periodic vs. stochastic, length of period, etc.), the strength of selection on the mate choice locus is usually small, unless choice is nearly absolute and selection on the primary locus is strong.

Probably the most important universal source of heritability of fitness is mutation, since all genes of functional significance must occasionally produce deleterious alleles by mutation. The theory of the variance maintained by mutation-selection balance at many loci has been developed by Mukai et al. (1974). If the fitness of the heterozygotes and homozygotes for a mutant allele at a locus are $1\text{-}hs$ and $1\text{-}s$ respectively, and the mutation rate from wild-type is μ, then (provided the mutations are not recessive, so that $h>0$, as is suggested by the *Drosophila melanogaster* data [Crow and Simmons 1983; Simmons and Crow 1977]) the additive and dominance variances in an equilibrium population are given approximately by

$$V_A = 2\mu hs \tag{2}$$

and

$$V_D = (1\text{-}2h)^2(\mu/h)^2 \tag{3}$$

Since μ is typically of the order of 10^{-5} or less, whereas average values of h and s for mildly deleterious mutations affecting *Drosophila* viability are approximately 0.2 and 0.03 respectively (Crow and Simmons 1983), the bulk of the variation contributed by mutation is additive rather than dominance variance. Since gene frequencies at each locus are low and effects of individual loci on fitness are mostly small, linkage, disequilibrium and epistatic effects can be neglected, so that the genic and additive genetic variances due to mutation can be equated if mating is random.

The net mutation rate for mildy deleterious genes affecting viability in *Drosophila* is estimated from the experiments of Mukai and co-workers to be about 0.25 per gamete per generation (Crow and Simmons 1983). Such minor mutations contribute the vast bulk of mutational variability. If egg-to-adult viability in the laboratory is a reliable indicator of net fitness (a highly problematic assumption, see below) then an equilibrium additive variance of $2 \times 0.25 \times 0.2 \times 0.03 = 0.003$ in fitness could be maintained at equilibrium, summing over all loci. The mean fitness $\bar{w}$ at equilibrium on this scale (where the fitness of the mutation free genotype is 1) is 0.75, so that the ratio of the additive variance to the square of the population mean is equal to 0.0053.

This ratio ($V_A/\bar{w}^2$) is dimension-free, so that predicted values can readily be compared with experimental measurements on different components of fitness (see below). It also has the advantage that it is independent of the contribution of environmental effects to the total variance in fitness, unlike heritability. The square root of $V_A/\bar{w}^2$ is the additive coefficient of variation in fitness and provides a useful measure of the spread of the distribution of breeding values in fitness (Charlesworth 1984); if the above prediction can be taken seriously, this has a value of 0.07.

An alternative way of obtaining information concerning the mutational contribution to additive genetic variance in fitness is through the models of mutational variation in additively inherited quantitative traits subject to stabilizing selection (Bulmer 1980; Fisher 1930; Lande 1975; Turelli 1984). Using the formulation of Bulmer (1980), it can be shown that the value of $V_A/\bar{w}^2$ contributed by thc loci controlling a single quantitative character under stabilizing selection in a random-mating population is the order of one-fifth of the ratio of the variance in the character generated by mutation in one generation to the phenotypic variance (see *APPENDIX*). Data on the mutational variance in *Drosophila* bristle characters suggest that this ratio is typically about 10^{-3} (Hill 1982; Lande 1975), so that mutations affecting a single quantitative character may contribute approximately 2×10^{-4} to $V_A/\bar{w}^2$. Since different characters may be both genetically correlated and interact in their effects on fitness (Cheverud 1982; Lande and Arnold 1984; Price et al. 1984), it is not possible to assess accurately the total amount of additive fitness variance arising from this source, but it seems likely to be at least of the order of 10^{-3} or so.

Heritability of components of fitness

As discussed above, natural selection exhausts variation in breeding values with respect to net fitness, so that such variation can only be maintained as a result of the interaction of selection with other forces such as linkage, nonrandom mating, migration, and mutation. It is technically very difficult to measure net genotypic or phenotypic fitnesses experimentally, even in the laboratory

(Christiansen 1984), and investigators usually have to content themselves with measuring components of fitness, such as egg-to-adult viability, female fecundity at a limited set of ages etc. (A fitness component is defined here as a character for which a small increase in its value, holding all other characters constant, leads to an increase in fitness [Robertson 1955]). As discussed below, there is often good evidence for heritability in components of fitness, but it would be unwise to jump to the conclusion that there is high heritability of fitness itself. If a suite of fitness components shows variances in breeding values in a population at equilibrium under selection, this may simply mean that the characters tend to be negatively genetically correlated. If fitness is controlled by a large number of loci of small effect, the additive variance in fitness is given by

$$V_A = \Sigma g_{ij} (\delta w/\delta z_i) (\delta w/\delta z_j) \tag{4}$$

where g_{ij} is the additive genetic covariance between fitness components z_i and z_j, and the sum is taken over all fitness components, and the derivatives are evaluated at the population means.

Since the g_{ii} are the additive genetic variances in the fitness components, and the derivatives $\delta w/\delta z_i$ are by definition positive for fitness components, negative genetic correlations between fitness components lead to the net additive variance in fitness being greatly reduced compared with the sum of the contributions of the variances in the individual characters to the additive variance in fitness (Dickerson 1955; Robertson 1955). Explicit models of antagonistic pleiotropic effects of genes that lead to the maintenance of high heritabilities of fitness components, in the complete absence of any heritability of fitness itself, have been worked out by Rose (1982). There is evidence in the literature for the existence of such negative genetic correlations between fitness components (Dickerson 1955; Mukai and Yamazaki 1971; Reznick 1985; Rose and Charlesworth 1981; Simmons et al. 1980).

DATA

The above considerations suggest that the variance in breeding value of fitness in populations of higher organisms will usually be low. It will not, however, be nonexistent, because of the universal occurrence of some mutational input of variation. As already mentioned, net fitness in sexually reproducing species is extremely hard to measure and requires the use of specialized laboratory techniques, effectively restricted to *Drosophila* (Haymer and Hartl 1982; Latter and Robertson 1962; Sved 1971; Yamazaki 1984), or painstaking demographic observations on natural populations (Christiansen 1984; Gustafsson 1986). It is probably fair to say that we have no precise estimate of the heritability of fitness for any organism, even under laboratory conditions. There is, however, a very large literature on the experimental quantitative genetics of fitness components. I do not propose to review this in detail but will try to highlight the most important conclusions. I shall concentrate on *Drosophila*, since this is where the largest data sets have been collected.

Drosophila results

By far the best body of evidence available comes from the analyses of variance in egg-to-adult viability due to loci located on the second chromosome of *D. melanogaster*, conducted over many years by Mukai and his colleagues (Mukai 1985). These studies utilize the technique of marked inversion-bearing chromosomes, which suppress crossing over when heterozygous, in order to isolate sets of independent second chromosomes from natural populations. By doing crosses of the type $Cy/i \times Cy/j$, where i and j represent two different extracted wild chromosomes and *Cy* represents an inversion chromosome bearing the recessive-lethal marker *Curly*, the viability of the combination i/j relative to that of heterozygotes for *Cy* can be estimated from the deviation of the proportion of wild-type progeny in a culture from the value of one-third expected with equal viabilities of *Cy* and wild-type. It is usually found that *Cy* has a mean viability equal to that of wild-type. By measuring the viability of a given chromosome i in combination with several other independent chromosomes, an estimate of the breeding value of i can be obtained, so that the total genetic variance in viability can be partitioned into variance in breeding value and residual variance. If there is no linkage disequilibrium and epistasis between the loci affecting viability, the two components of variance can be equated with additive and dominance components respectively. Table 1 summarizes the results of experiments conducted on several different Japanese and U.S. populations (Mukai 1985).

TABLE 1 Components of variance of egg-to-adult viability for the second chromosome of *D. melanogaster* (Mukai 1985)

Population	Latitude	Variance component	
		Breeding values	Residual
Aomori	40.6N	0.0027±0.0010	0.0001±0.0003
Akayu	38.1N	0.0023±0.0014	0.0009±0.0003
Raleigh, NC	35.8N	0.0096±0.0025	0.0012±0.0005
Orlando, FL	27.2N	0.0202±0.0038	0.0006±0.0057
Ogasawara	27.1N	0.0121±0.0040	0.0009±0.0005
Ishigaki-jima	24.3N	0.0183±0.0090	0.0011±0.0016

There are two points worthy of note. First, in every case the bulk of the variance in viability is due to variance in breeding values; indeed, there is no statistically significant evidence for any residual variance. This suggests that heterozygote advantage and other forms of balancing selection play a negligible role in maintaining variance in viability since such mechanisms would produce the opposite pattern for the ratio of variance components in equilibrium populations (see above).

Second, there appears to be a North-South gradient in the variance of breeding values. In the two most northerly populations, the variances are approximately the same as would be expected if the variation were maintained solely by mutation-selection balance. From the mutation data quoted above, the equilibrium value of additive genetic variance for the second chromosome (which constitutes 2/5 of the genome) would be approximately 0.002, assuming viability to be uncorrelated with other fitness components. This agrees well with the observed variances for these populations and the hypothesis that the variation is maintained solely by mutation cannot be rejected. The more southerly populations show a considerable excess in breeding value variance over the mutational prediction. There is no clear explanation for this pattern. Mukai and Nagano (1983) suggest that "diversifying selection" maintains more variation in the more southern populations;this would have to involve temporal fluctations in fitness and allele frequencies, or differentiation between partially isolated populations, since (as discussed above) niche differentiation within random-mating populations does not maintain variance in breeding values. Another possibility is that more linkage disequilibrium between interacting loci is maintained in the southern populations. Finally, negative genetic correlations between viability and other fitness components could result in the maintenance of additive variance in viability. These hypotheses do not, however, in themselves account for the absence of residual variance in these experiments. If the variation is mainly contributed by loci with small allelic effects, a high degree of additivity is understandable, however. This is more likely to apply to the heterogenous environment models than to the linkage disequilibrium model, since the maintenance of significant linkage disequilibrium usually requires strong selection (Ewens 1979, chap. 7). The greater variance in the more southern populations is suggestive of a role for biotic factors (such as host-pathogen interactions [Hamilton 1982]) in maintaining variation.

Data on the genetics of several other components of fitness in random-bred *Drosophila* populations have been reported in the literature (Roff and Mousseau 1986). A summary of the estimates of the additive genetic coefficients of variation ($\sqrt{V_A}/\overline{w}$) and heritabilities of some of these traits is given in Table 2. Female fecundity behaves very like egg-to-adult viability in that there is significant additive genetic variance but no evidence for significant dominance variance (Rose and Charlesworth 1981). Adult longevity, on the other hand, seems to have low additive variance but relatively high dominance variance (Maynard Smith 1959; Rose and Charlesworth 1981). Other *Drosophila* data sets show that fitness components generally exhibit lower heritabilities than

TABLE 2 Additive genetic coefficients of variation and heritabilities of *D. melanogaster* fitness components

Trait	Additive genetic coefficient of variation	Heritability	Source
Viability	0.165	—	(Mukai 1985; mean of 6 populations)
Adult female longevity	0.053	0.028	(Rose and Charlesworth 1981)
Adult female fecundity:			
Days 1–5	0.125	0.623	(Rose and Charlesworth 1981)
Days 6–10	0.115	0.296	(Rose and Charlesworth 1981)
Days 11–15	0.176	0.316	(Rose and Charlesworth 1981)
Days 16–20	0.259	0.434	(Rose and Charlesworth 1981)
Days 21–25	0.123	0.040	(Rose and Charlesworth 1981)
Development time	0.024	0.254	(Prout 1962)

morphological or physiological traits (Ross and Mosseau 1986), as is expected on the theoretical considerations discussed above (Robertson 1955).

Other species

Falconer (1981, Table 10.1) summarizes data from a number of domestic and laboratory animals which suggests that components of fitness tend to have lower, though nonnegligible, heritabilities than other characters. There are several examples of nonzero heritabilities for some fitness components in wild species (Berven and Gill 1984; Derr 1980; Istock 1978). Nonetheless, Roff and Mousseau (personal communication) have reviewed the literature on more than 1100 estimates of heritability for animals other than *Drosophila* and found a very similar general pattern to that for *Drosophila*, with lower heritabilities for fitness components than other characters. For plants, inheritance of yield in open-pollinated varieties of maize (Robinson and Comstock 1955) gives results remarkably similar to those for viability and female fecundity in *Drosophila* (Table 2): most of the genetic variance is additive rather than dominance and there is substantial heritability, such that artificial selection for increased yield is effective. Components of fitness in a wild population of *Impatiens capensis* also show significant heritabilities (Mitchell-Olds 1986). It should be borne in mind when considering the relevance of these data that estimations of variance components conducted under laboratory conditions may not represent accurately the natural situation. In particular, fitness differences between genotypes may be much greater in the field due to greater intensitites of competition, mortality from pathogens, etc.

Relatively few studies of the genetics of fitness components in natural populations have been reported; most of these are for bird populations. Van Noordwijk et al. (1980) found heritabilities of the order of 40% for clutch size in the Great Tit; this may, however, be largely under stabilizing selection (Lack 1954) and hence not a component of fitness in the sense used here. Gustafsson (1986) has analysed the quantitative genetics of several morphological and life-history traits in a Swedish population of the collared flycatcher *Ficedula albicollis*. Characters with high correlations with lifetime reproductive success, such as longevity and number of fledged young, showed near zero heritabilities, whereas characters with low correlations with reproductive success showed higher heritabilities, generating a strong negative association between the proportion of variance in lifetime reproductive success explained by a character and its heritability. This study provides strong empirical support for the notion that only a low level of additive genetic variance for net fitness is maintained in natural populations.

DISCUSSION

As is frequently the case in evolutionary biology, the theoretical basis for understanding the genetic basis of fitness is more developed than the empirical basis, largely because of the difficulty of estimating components of variance for net reproductive fitness, especially under natural conditions. The frequent existence of negative genetic correlations between different components of fitness (Reznick 1985) means that evidence for heritability of fitness components does not prove the existence of large quantities of heritable variation in net fitness. This is brought out graphically by the results of selection experiments in *D. melanogaster* for increased longevity (Luckinbill et al. 1984; Rose 1984); these have been highly successful in increasing longevity and the fecundity of older females, at the expense of reduced fecundity earlier in life, with the result that lifetime egg production is unchanged.

It is thus unclear at present how the data on variation fitness components relates to net fitness. In *Drosophila*, there is strong evidence that egg-to-adult viability only contributes a fraction of the concealed variability leading to inbreeding decline (Simmons and Crow 1977; Sved 1971), and it seems that male mating success is an important component of fitness (Anderson et al. 1979; Brittnacher 1981), but there is no information at present on the quantitative genetics of net fitness in random mating *Drosophila* populations. The large-scale breeding experiments necessary to obtain reliable estimates of variance components for traits with low heritabilities mean that information for most other species is even more unreliable. Nevertheless, there is certainly good empirical evidence that components of fitness tend to have smaller quantitites of additive genetic variance than other characters, both in *Drosophila* (Roff and Mousseau 1986) and other species (Falconer 1981; Gustafsson 1986). Direct experiments to test the idea that female choice may cause the production of offspring with higher fitness components have been performed on *Drosophila* (Schaeffer et al. 1984; Partridge 1980) and *Tribolium* (Boake 1986), but have yielded inconsistent results (two negative [Boake 1986; Schaeffer et al. 1984] and one positive [Partridge 1980]).

My suspicion is that there may be little more variance in breeding values with respect to fitness maintained in most populations than can be accounted for by the mutational input of deleterious alleles, but it is hard to be certain about this at present. As discussed earlier, there may be some additional contribution due to linkage disequilibrium between interacting loci. If the data on *Drosophila* per genome mutation rates for viability can be extended to net fitness and to other species, we might expect an additive genetic coefficient of variation in fitness of about 0.07 to be maintained by mutation. Individuals with a breeding value higher than the population mean by 14% or more would thus occur at a frequency of 2.5%, assuming a normal distribution. If one considers the fate of a mutant gene causing females to mate selectively with males of superior fitness, a significant pressure of selection in favor of mate choice can be shown to exist under these conditions (*APPENDIX*). This disagrees with the conclusion of Taylor and Williams (1982). They considered only the effect of a single locus subject to mutation to recessive deleterious alleles, whereas the present analysis takes into account the contribution of the entire genome and the fact that most mutations affecting fitness exert some heterozygous effects.

An alternative possibility, which has been vigorously championed by Hamilton (1982), is the host-parasite interactions may promote oscillations in allele frequencies at polymorphic host loci that determine resistance specificity. This is a special case of the maintenance of variability by temporal fluctations in fitnesses discussed earlier and could certainly lead to the maintenance of additive variance in fitness. As mentioned earlier, there are theoretical grounds for believing that this can lead to the evolution of female mating preferences with respect to genotypic differences at the selected loci. Furthermore, Hamilton and Zuk (1982) have proposed that male ornaments may evolve because they allow females to discriminate between males who differ heritably in their parasite loads and hence with respect to fitness when variation is maintained by cyclical changes in gene frequencies at resistance loci. There is no direct evidence as yet that known genetic polymorphisms are maintained by these processes.

The conclusion that sufficient additive genetic variance may exist to stimulate the evolution of female choice should be considerably qualified by the consideration that, especially in natural situations, environmental sources of variation in fitness may be much greater in magnitude than genetic influences (Mukai et al, 1972), greatly reducing the ability of females to select males with superior breeding values (see *APPENDIX*). Furthermore, it is questionable whether or not females could realistically monitor the net reproductive fitness of their potential mates, as opposed to some set of external characteristics related to components of fitness, e.g., the age of males. (Age has been suggested as an indicator of high survival ability [Halliday 1983; Weatherhead 1984], but the only empirical test has yielded negative results [Alatalo et al. 1986]). In view of the frequent existence of negative genetic correlations between fitness components, there may be little benefit to be gained from selective mating of this kind. Kirkpatrick (this volume) has pointed out other difficulties involved in models of the evolution of phenotypic mate choice. Nevertheless, it is conceivable that the prevalence of courtship rituals that apparently involve tests of males' athletic powers represents the outcome of selection for choice of mates with high genetic quality (Maynard Smith 1978, chap. 10).

Considerations of this kind apply with particular force to recent speculations concerning "mate choice" by plants, where it has been suggested that maternal selection of pollen grains or embryos with higher growth rates may be widespread (Willson and Burley 1983) and enable plants to produce offspring of higher genetic quality. The recent purported experimental support for this hypothesis relies on tests involving very small numbers of plants, which are inadequate to separate out additive genetic sources of variation in the male contribution to offspring performance from other factors such as environmental effects, genotypic differences with respect to self-incompatibility loci, and varying levels of inbreeding (Bertin 1985; Marshall and Ellstrand 1986). Many more carefully designed, large-scale genetic experiments are required if these issues are to be resolved.

APPENDIX

Sex differences in fitness in a random-mating population

Consider a single autosomal locus with two alleles, A_1 and A_2. The frequencies of the three genotypes A_1A_1, A_1A_2, and A_2A_2 before selection in a given generation are $P=p_fp_m$, $2Q=p_fq_m + q_fp_m$, $R=q_fq_m$, where p and q are the frequencies of A_1 and A_2, and the subscripts f and m denote female and male gametes respectively. Writing w_{ij}^f and w_{ij}^m for the fitnesses of A_iA_j females and males, we have the gene frequency change equations

$$\bar{w}_f\Delta p_f = Pw_{11}^f + Qw_{12}^f - \bar{w}_fp_f,\ \bar{w}_m\Delta p_m = Pw_{11}^m + Qw_{12}^m - \bar{w}_mp_m \qquad (A.1)$$

where $\bar{w}_f$ and w_m are the population mean fitness for females and males respectively. Noting that the net gene frequency p is the mean of p_f and p_m, and is equal to $P + Q$, we have

$$\Delta p = Pw_{11} + Qw_{12} - p \qquad (A.2)$$

where w_{11} and w_{12} are the net fitnesses defined by Eq. 1. From standard theory

(Kempthorne 1957, chap. 16), it is known that additive genetic variance in net fitness involves the *average excess* $(Pw_{11} + Qw_{12})/p-1$, noting that the population mean is unity on this scale. At equilibrium, Δp is zero so that the average excess and hence the additive variance in net fitness is zero (this is true even if mating is nonrandom). It can similarly be shown that the breeding values of all three genotypes are zero at equilibrium. For example, the expected fitness of the progeny of an A_1A_1 male mated with females drawn randomly from the population *prior to selection* is $pw_{11} + qw_{12}$, which from the above considerations is equal to the population mean of unity. The same result applies to the other two genotypes and to female breeding values. If, however, breeding values are defined in terms of mates drawn from the population *after selection*, then in general they will differ by an amount determined by the difference between p_f and p_m. This is zero only if selection acts equally on the two sexes or on one sex alone and, in general, is of the order of the square of the selection intensity (Nagylaki 1979). The variance in breeding values is thus of the order of the fourth power of the selection intensity. These results are easily generalized to multiple alleles at a locus.

Mutation-selection balance for polygenic characters under random mating

Consider a set of loci controlling an additively inherited quantitative character subject to stabilizing selection with an optimal value of zero, such that each locus segregates for two alleles with a difference $2a$ between the respective homozygotes (Bulmer 1980, pp. 179–180). If the mutation rate between the alleles is μ per locus per generation, then half the loci will have "plus" alleles present at low frequencies at equilibrium and half will have 'minus' alleles at low frequencies. In the absence of dominance, the equilibrium frequencies of the rare alleles are equal to $p=2\mu/(-Ba^2)$, where B is the ratio of the change in phenotypic variance within a generation to the square of the phenotypic variance and $p<<1$. The average excess of each rare allele with respect to fitness (on a scale with a population mean fitness of unity) is approximately equal to μ/p at equilibrium (Bulmer 1980), and so the additive variance is approximately equal to $2p\,(\mu/p)^2 = -\,\mu Ba^2$. If there are n loci controlling the character, the total additive genetic variance in fitness is thus $-\,n\mu Ba^2$. If selection is noroptimal, so that individuals with a phenotypic value z have fitness $\exp - \{z^2/2\omega^2\}$ (where ω determines the strength of selection), then we have $-\;B=1/(V_P+\omega^2)$ where V_P is the phenotype variance. It follows that the additive variance in fitness is given by

$$V_A/\overline{w}^2 = n\mu a^2/(V_P + \omega^2) \qquad (A.3)$$

The quantity $n\mu a^2$ is the total rate of generation of new variance in the quantitative character by mutation each generation, V_m, for which empirical estimates are available (Hill 1982; Lande 1975). The ratio ω^2/V_P determines the proportion of the population eliminated by stabilizing selection each generation; a typical value seems to be about 4, corresponding to elimination of 10% of the population (Lande 1975). $V_A/\bar{w}^2$ is therefore likely to be of the order of one-fifth V_m/V_P.

Selection for female choice with mutational variance in fitness

For simplicity, I assume that natural selection affects males and females in the same way. Consider a rare allele at a locus controlling female behavior, such that females heterozygous for the choice allele C mate with males whose breeding value with respect to fitness is a on a scale with a population mean fitness of unity. The choice locus is assumed to recombine freely with the loci controlling fitness. Let the fitness breeding value of C individuals be a^*_n in generation n, and the frequency of the allele C (before selection) be p_n. Assuming additive inheritance of the mutational contribution to fitness, as is justified by Eqs. 2 and 3, the fitness of C females is $1+a^*_n$, and so the frequency after selection becomes $p_n\ (1+a^*_n)$. The breeding value of the fitness of their progeny is given by one-half the breeding value contributed by the males (i.e., $\frac{1}{2}a$) and by one-half the breeding value they contribute themselves. Assuming free recombination, the latter is equal to $\frac{1}{2}a^*_n$, ignoring squared terms in a^*_n. Similarly, the breeding value of the progeny of C males is $\frac{1}{2}a^*_n$. Hence, $a^*_{n+1} = \frac{1}{4}a + \frac{1}{2}a^*_n$, and a^*_n rapidly approaches the asymptotic value of $\frac{1}{2}a$. It follows that p_n/p_{n-1} approaches the asymptotic value $1+\frac{1}{2}a$, so that the selective advantage of the choice allele C is $\frac{1}{2}a$.

The value of a is determined by the rules governing female choice and by the heritability of fitness (Kirkpatrick 1985, 1986). For example, if females only accept the top fraction x of the males they encounter and the heritability of fitness is h^2, we have $a=h^2\ z\sqrt{V_P}/x$, where z is the height of the ordinate of the standard normal deviate corresponding to the truncation point x and $\sqrt{V}$ is the standard deviation of fitness (Falconer 1981, p. 175). If females choose the top 50%, then for example, $a=0.8\ h^2\sqrt{V_P}$.

Acknowledgements

This work was supported in part by NSF grant BSR-8516629 and by a grant from the Louis Block fund of The University of Chicago. I thank W.D. Hamilton for reviewing the manuscript.

REFERENCES

Alatalo, R.V.; Gustafsson, L.; and Lundberg, A. 1986. Do females prefer older males in polygynous bird species? Am. Natural. 127: 241–245.

Anderson, W.W.; Levine, L.; Olivera, O.; Powell, J.R.; De La Rosa, M.E.; Salceda, V.M.; Gaso, M.I.; and Guzman, J. 1979. Evidence for selection by male mating success in natural populations of *Drosophila pseudoobscura*. Proc. Nat. Acad. Sci. USA 76: 1519–1523.

Antonovics, J. 1968. Evolution in closely adjacent plant populations. V. Evolution of self-fertility. Heredity 23: 219–238.

Bertin, R.I. 1985. Non-random fruit production in *Campsis radicans*: between-year consistency and effects of prior pollination. Am. Natural. 126: 750–759.

Berven, K.A., and Gill, D.E. 1984. Interpreting geographic variation in life-history traits. Am. Zoolog. 23: 85–97.

Boake, C.R. 1986. A method for testing adaptive hypotheses of mate choice. Am. Natural. 127: 654–666.

Borgia, G. 1979. Sexual selection and the evolution of mating systems. In: Sexual Selection and Reproductive Competition in Insects, eds. M.S. Blum and N.A. Blum, pp. 19–80. New York: Academic Press.

Brittnacher, J.G. 1981. Genetic variation and genetic load due to the male reproductive component of fitness in *Drosophila.* Genetics 97: 719–730.

Bulmer, M.G. 1980. The Mathematical Theory of Quantitative Genetics. Oxford. Clarendon Press.

Charlesworth, B. 1980. Evolution in Age-Structured Populations. Cambridge: Cambridge Univ. Press.

Charlesworth, B. 1984. The evolutionary genetics of life histories. In: Evolutionary Ecology, ed. B. Shorrocks, pp. 117–133. Oxford: Blackwell.

Charlesworth, B., and Charlesworth, D. 1973. Selection of new inversions in multilocus genetic systems. Genet. Res. 21: 167–183.

Charlesworth, B., and Charlesworth, D. 1975. An experiment on recombination load in *Drosophila melanogaster.* Genet. Res. 25: 267–274.

Charlesworth, B., and Charlesworth, D. 1978. A model for the evolution of dioecy and gynodioecy. Am. Natural. 112: 975:997;

Cheverud, J.M. 1982. Phenotypic, genetic and environmental morphological integration. Evolution 36: 499–516.

Christiansen, F.B. 1984. The definition and measurement of fitness. In: Evolutionary Ecology, ed. B. Shorrocks, pp. 65–79. Oxford: Blackwell.

Crow, J.F., and Kimura, M. 1970. An introduction to Population Genetics Theory. New York: Harper and Row.

Crow, J.F., and Simmons, M.J. 1983. The mutation load in *Drosophila.* In: The Genetics and Biology of *Drosophila*, vol. 3c, eds. M. Ashburner, H.L. Carson, and J.N. Thompson, pp. 2–35. New York: Academic Press.

Derr, J.A. 1980. The nature of variation in life history characteristics of *Dysdercus bimaculatus* (Heteroptera: Pyrrochoridae), a colonising species. Evolution 34: 548–557.

Dickerson, G.E. 1955. Genetic slippage in response to selection for multiple objectives. Cold Spring Harb. Quant. Biol. 20: 213–224.

Eshel, I., and Hamilton, W.D. 1984. Parent-offspring correlation in fitness under fluctuating selection. Proc. Roy. Soc. Lon. B. 222: 1–14.

Ewens, W.J. 1979. Mathematical Population Genetics. Berlin: Springer-Verlag.

Falconer, D.S. 1981. Introduction to Quantitative Genetics (2nd edn). London: Longman.

Fisher, R.A. 1930. The Genetical Theory of Natural Selection. Oxford: Clarendon Press.

Fisher, R.A. 1941. Average excess and average effect of a gene substiution. Ann. Eugen. 11: 53–63.

Gillespie, J.H. 1973. Polymorphism in random environments. Theor. Pop. Biol. 4: 193–195.

Gregorius, H–R., and Ross, M.D. 1981. Selection in populations of effectively infinite size. I. Realised genotypic fitnesses. Math. Biosci. 54: 291–307.

Gustafsson, L. 1987. Lifetime reproductive success and heritability: empirical support of Fisher's fundamental theorem. Am. Natural., in press.

Haldane, J.B.S., and Jayakar, S.D. 1963. Polymorphism due to selection of varying direction. J. Genet. 58: 237–242.

Halliday, T.R. 1983. The study of mate choice. In: Mate Choice, ed. P. Bateson, pp. 3–32. Cambridge: Cambridge Univ. Press.

Hamilton, W.D. 1982. Pathogens as a cause of diversity in their host populations. In: Population Biology of Infectious Diseases, eds. R.M. Anderson and R.M. May, pp. 269–296, Dahlem Konferenzen, Berlin: Springer-Verlag.

Hamilton, W.D., and Zuk, M. 1982. Heritable true fitnesses and bright birds: a role for parasites. Science 218: 386–387.

Haymer, R.S., and Hartl, D.L. 1982. The experimental assessment of fitness in *Drosophila*. I. Comparative measures of competitive reproductive success. Genetics 102: 455–466.

Hill, W.G. 1982. Predictions of response to artificial selection from new mutations. Genet. Res. 40: 255–278.

Istock, C.A. 1978. Fitness variation in a natural population. In: Evolution of Insect Migration and Diapause, ed. H.R. Dingle, pp. 171–190. Berlin: Springer–Verlag.

Kempthorne, O. 1957. An introduction to Genetic Statistics. New York. John Wiley.

Kimura, M. 1958. On the change of population fitness by natural selection. Heredity 12: 145–167.

Kirkpatrick, M. 1985. Evolution of female choice and male parental investment in polygynous species: the demise of the "sexy son." Am. Natural. 125: 788–810.

Kirkpatrick, M. 1986. The handicap mechanism of sexual selection does not work. Am. Natural. 127: 222–240.

Kojima, K.-I., and Kelleher, T.M. 1961. Changes in mean fitness in random mating populations when linkage and epistasis are present. Genetics 46: 527–540.

Lack, D. 1954. The Natural Regulation of Animal Numbers. Oxford: Clarendon. Press.

Lande, R. 1975. The maintenance of genetic variability by mutation in a polygenic character with linked loci. Genet. Res. 26: 221–235.

Lande, R. 1979. Quantitative genetic analysis of multivariate evolution applied to brain: body size allometry. Evolution 33: 402–416.

Lande, R. 1982. Rapid orgin of sexual isolation and character divergence in a cline. Evolution 36: 213–223.

Lande, R., and Arnold, S.J. 1984. The measurement of selection on correlated characters. Evolution 37: 1210–1226.

Latter, B.H.D., and Robertson. A. 1962. The effects of inbreeding and artificial selection on reproductive fitness. Genet. Res. 3: 110–138.

Levene, H. 1953. Genetic equilibrium when more than one ecological niche is available. Am. Natural. 87: 311–313.

Lloyd, D.G. 1977. Genetic and phenotype models of natural selection. J. Theor. Biol. 69: 543–560.

Luckinbill, L.S.: Arking, R.: Clare, M.J.: Cirocco, W.C.; and Buck, S.A. 1984. Selection for delayed senescence in *Drosophila melanogaster*. Evolution 38: 996–1003.

Marshall, D.L., and Ellstrand, N.C. 1986. Sexual selection in *Raphanus sativus*: experimental data on nonrandom fertilization, maternal choice, and consequences of multiple paternity. Am. Natural. 127: 446–461.

Maynard Smith, J. 1959. Sex-limited inheritance of longevity in *Drosophila subobscura*. J. Genet. 56: 1–9.

Maynard Smith, J. 1978. The Evolution of Sex. Cambridge: Cambridge Univ. Press.

Mitchell-Olds, T. 1986. Quantitative genetics of survival and growth in *Impatiens capensis*. Evolution 40: 107–116.

Mukai, T. 1985. Experimental verification of the neutral theory. In: Population Genetics and Molecular Evolution, eds. T. Ohta and K.–I. Aoki, pp. 125–145. Berlin: Springer-Verlag.

Mukai, T., and Nagano, S. 1983. The genetic structure of natural populations of *Drosophila melanogaster*. XVI. Excess additive genetic variance of viability. Gentics 105: 115–134.

Mukai, T., and Yamaguchi, O. 1974. The genetic structure of natural populations of *Drosophila melanogaster*. XI. Genetic variability in a local population. Genetics 76: 339–366.

Mukai, T., and Yamazaki, T. 1971. The genetic structure of natural populations of *Drosophila melanogaster*. X. Development time. Genetics 69: 385–398.

Mukai, T.; Schaffer, H.E.; and Cockerham, C.C. 1972. Genetic consequences of truncation selection at the phenotype level in *Drosophila melanogaster*. Genetics 72: 763–769.

Mukai, Y.; Cardellino, R.K.; Watanabe, T.K.; and Crow, J.F. 1974. The genetic variance for viability and its components in a population of *Drosophila melanogaster*. Genetics 78: 1195–1208.

Nagylaki, T. 1979. Selection in dioecious populations. Ann Hum. Gen. 43: 143–150.

Partridge, L. 1980. Mate choice increases a component of fitness in fruit flies. Nature 283: 290–291.

Partridge, L. 1983. Non-random mating and offspring fitness. In: Mate Choice, ed. P. Bateson, pp. 227–255. Cambridge: Cambridge Univ. Press.

Price, T.D.; Grant, P.R.; Gibbs, H.L.; and Boag, P.T. 1984. Recurrent patterns of natural selection in a population of Darwin's finches. Nature 306: 787–789.

Prout, T. 1962. The effects of stabilizing selection on the time of development in *Drosophila melanogaster*. Genet Res. 3: 364–382.

Reznick, D. 1985. Costs of reproduction: an evaluation of the empirical evidence. Oikos 44: 257–267.

Robertson, A. 1955. Selection in animals: synthesis. Cold Spring Harb. Quant. Biol. 20: 225–229.

Robinson, H.F., and Comstock, R.E. 1955. Analysis of genetic variability in corn with reference to probable effects of selection. Cold Spring Harb. Quant. Biol. 20: 127–136.

Roff, D.A., and Mousseau, T.A. 1987. Quantitative genetics and fitness: lessons from *Drosophila*. Heredity, in press.

Rose, M.R. 1982. Antagonistic pleiotropy, dominance and genetic variation. Heredity 48: 63–78.

Rose, M.R. 1984. Laboratory evolution of postponed senescence in *Drosophila melanogaster*. Evolution 38: 1004–1010.

Rose, M.R., and Charlesworth, B. 1981. Genetics of life-history in *Drosophila melanogaster*. I. Sib analysis of adult females. Genetics 97: 173–186.

Schaeffer, S.W.; Brown, C.J.; and Anderson, W.W. 1984. Does mate choice affect fitness? Genetics 107: 94.

Simmons, M.J., and Crow, J.F. 1977. Mutations affecting fitness in *Drosophila* populations. Ann. Rev. Genet. 11: 49–78.

Simmons, M.J.; Preston, C.R.; and W.R. Engels. 1980. Pleiotropic effects on fitness of mutations affecting viability in *Drosophila melanogaster*. Genetics 94: 467–475.

Sved. J.A. 1971. An estimate of heterosis in *Drosophila melanogaster*. Genet. Res. 18: 97–105.

Taylor, P.D., and Wiliams, G.C. 1982. The lek paradox is not resolved. Theor. Pop. Biol. 22: 392–409.
Thornhill, R., and Alcock, J. 1983. The Evolution of Insect Mating Systems. Cambridge, MA: Harvard Univ. Press.
Trivers, R.L. 1972. Parental investment and sexual selection. In: Sexual Selection and the Descent of Man 1871–1971, ed. B.Campbell, pp. 136–179. Chicago: Aldine-Atherton.
Turelli, M. 1984. Heritable variation via mutation-selection balance: Lerch's zeta meets the abdominal bristle. Theor. Pop. Biol. 25: 138–193.
Turner, J.R.G. 1970. Changes in mean fitness under natural selection. In: Mathematical Topics in Population Genetics, ed. K.-I. Kojima, pp. 32–78. Berlin: Springer-Verlag.
Van Noordwijk, A.J.: Van Balen, J.H.; and Scharloo, W. 1980. Heritability of ecologically important traits in the great tit (*Parus major* L.). Ardea 68: 193–203.
Wasserman, M. 1972. Factors influencing fitness in chromosomal strains in *Drosophila subobscura*. Genetics 72: 691–708.
Weatherhead, P.J. 1984. Mate choice in avian polygyny: why do females prefer older males: Am. Natural. 123: 873–875.
Willson, M.F., and Burley, N. 1983. Mate Choice in Plants. Princeton: Princeton Univ. Press.
Yamazaki, T. 1984. Measurment of fitness and its components in six laboratory strains of *Drosophila melanogaster*. Genetics 108: 201–211.

Sexual Selection: Testing the Alternatives
eds. J.W. Bradbury and M.B. Andersson, pp 41–53
John Wiley & Sons Limited

Genetic Models of Sexual Selection: Some Aims, Assumptions, and Tests

M.B. Andersson

Department of Zoology, P.O. Box 25059
University of Göteborg
40031 Göteborg, Sweden

Abstract. Two classes of models are distinguished: 1) simple, general models that help show consequences of the assumptions and derive new hypotheses; 2) more specific models that permit quantitative predictions and tests. Most genetic models of sexual selection have been of the first kind. Proving the plausibility of the Darwin-Fisher process, they created an open attitude to sexual selection by mate choice and stimulate work on crucial problems, but many of their predictions and basic assumptions remain untested. A review of four models shows that they are all based on so many assumptions that comprehensive testing of entire models may not be feasible. Critical tests are needed that will discriminate between mechanisms and estimate their relative roles. Some examples are discussed where models have helped promote illuminating empirical work (on mate choice and its genetic basis), and where they have led to differing conclusions (on 'handicapping' viability indicators).

INTRODUCTION

The increasing development of genetic models of sexual selection has led to many new insights. However, testing of the models is often difficult, and empirical support for several assumptions and predictions is therefore weak. This paper discusses aims and assumptions of the models, the kind of knowledge they provide, and what might be done to bridge the widening gulf between theory and data (or at least to maintain ferry traffic between the two shores). Bridging requires testing of the assumptions and predictions of the models, and building into them factors found important in empirical studies. Theory and data seem equally essential; it is of interest to know both a) what mechanisms are logically capable, alone or in combination, of leading to the evolution of sex traits and mating preferences, and b) what mechanisms actually do so in nature and what are their relative roles within and among species? These questions call for a combination of modelling and empirical work.

This review focusses on the testing of models of ongoing, present-day selection; an equally important problem concerns the mechanisms behind the origin and early evolutionary history of traits and preferences. Such historical questions call for partly different approaches such as comparative methods (e.g., Ridley 1983; Thornhill and Alcock 1983). Comparisons also provide a powerful means of testing for present function (Clutton-Brock and Harvey 1984).

Mathematical models in evolutionary biology can help to improve understanding in several ways (e.g., Levins 1966; Maynard Smith 1968, 1982; Lewontin 1974).

1) Compared to a verbal model, a mathematical model forces more explicit and precise statements of assumptions which makes for greater clarity, coherence, and avoidance of logical errors.
2) Formulating the model and exploring its consequences helps point out what must be known about the system for its behavior to be understood and predicted, and hence what needs to be measured.
3) Simple postulates often have complex outcomes, so a mathematical model may be necessary to make sure which consequences follow from the assumptions.
4) Mathematical models can yield nonobvious or counterintuitive results that give rise to new, interesting hypotheses and suggest how they might be tested.
5) Mathematical formulation helps show similarities between systems that on the surface have little in common. This can lead to an exchange of ideas by analogy, and to broader insights.
6) Comparison between predictions and reality can show whether some crucial aspect has been overlooked or misunderstood and how the mistake can be corrected and understanding improved through modification of the model followed by renewed testing. Lack of fit between predictions and data can therefore lead to rejection of faulty ideas, and to new discoveries.

Not all important ideas can be suitably represented by mathematical models; for example, their complexity may be overwhelming. There is then a temptation to simplify the idea until it is mathematically tractable, with a risk that the end product becomes little more than a mathematical exercise, void of biological insight. But this does not diminish the value of models; any tool can be misused. Formal genetic models have clearly helped increase the understanding of sexual selection (see below).

This review is confined to genetic models. Game theory models have gained increasing use in evolutionary biology, but they seem less suitable than genetic models for exploring sexual selection by mating preferences (Maynard Smith 1982). One reason is that the fitness of offspring may depend on the magnitude of genetic covariances (linkage disequilibrium) between the preferred trait and the preference; such a situation requires genetic modelling. Applications of game theory to other aspects of sexual selection are discussed by Parker (1979) and Hammerstein and Parker (this volume).

GENERAL VERSUS SPECIFIC MODELS

At least two classes of mathematical models can be distinguished in evolutionary biology (Maynard Smith 1982): simple, general models aiming to clarify the qualitative behavior of a system, and more complex, specific models applied to a particular example, aiming for quantitative predictions and tests. In practice, many models fall somewhere in between. Usually no single model contributes in all six respects mentioned above, and the two classes of models differ in regard to testability.

Simple, general models

Levins (1966) suggested that models in population biology preferably should have high *generality*, *precision* and *realism*. To this might be added *simplicity* in order to permit an intuitive grasp of the behavior of the model. These goals are not all compatible. Models of this first kind mainly aim for generality and simplicity; they are logical tools which help show the consequences of the assumptions and derive new insights and hypotheses. Most genetic models of sexual selection are of this kind (e.g., O'Donald 1962, 1967, 1980; Maynard Smith 1976, 1985; Bell 1978; Lande 1980, 1981, 1982; Lande and Arnold 1985; Andersson 1982, 1986a,b; Kirkpatrick 1982, 1985, 1986; Heisler 1984, 1985; Engen and Saether 1985; Seger 1985). They are not realistic and testable if that is taken to mean an ability to give accurate quantitative predictions for particular examples. Probably none of these models can be so tested without further elaboration. Some of their qualitative assumptions and predictions can be tested, however, and several of the models have greatly improved understanding of sexual selection by showing what is and what is not possible under different conditions and by predicting things that were not all foreseen.

One way of checking whether the predictions from a general model are robust is to examine the same basic process in several models with different simplifying assumptions. Each model leaves out much existing complexity and in this sense is false. If, however, the models all give the same qualitative conclusion, the risk is low that it is an artifact caused by the simplifications rather than an outcome of the basic process. In Levin's (1966) words, 'Our truth is the intersection of independent lies.'

One example from sexual selection is the qualitatively similar results from few-locus models and polygenic models of the Fisherian process (O'Donald 1967, 1980; Lande 1981; Lande and Arnold 1985; Kirkpatrick 1982; Seger 1985; Engen and Saether 1985; Heisler 1985). Both kinds of models confirm that the process can propagate mating preferences for traits that reduce survival, and that for any size or frequency of a male trait a strong enough female preference can balance the process. The possible equilibria fall along a line that under certain conditions represents stability. Under other conditions, the equilibria are unstable, and a slight displacement makes the system evolve away from the line at an increasing pace: the "runaway" stage envisaged by Fisher (Lande 1981; Seger 1985). The similarity in conclusions of few-locus and polygenic models suggests that these results are not artifacts but arise from fundamental common properties of the systems (see Seger 1985). Another case where different models support each other is Kirkpatrick's (1982) use of a haploid model for analytical derivation of rates of change and stability conditions for the trait and the preference; these results were then verified by simulations based on a diploid model.

Specific models

At the cost of giving up simplicity and generality, a model can be made sufficiently realistic and precise to represent a particular case in enough detail to yield quantitative predictions. These can be tested against the behavior of the real system through observations or experiments to show whether the main parts and mechanisms have been assigned correct roles and weights in the model. If the tests give predicted outcomes within the tolerance limits, the model is provisionally taken as valid for the system. Important gaps between predictions and results suggest further work that may clarify what was not understood. Few attempts have been made to test models of sexual selection in this comprehensive fashion (see below).

Empirical testing

Among the different, complementary ways of testing models, the comparison of *predictions* with the behavior of real systems often receives most attention. This is a natural predilection as one of the most fascinating aspects of models is their often nonobvious predictions. It is, however, equally important to test the basic *assumptions* as directly as possible. (In some cases the distinction between assumption and prediction is obscure, but usually this is no problem). The two kinds of test should not compete, but rather complement each other: both direct tests of assumptions and comparisons between predictions and the behavior of the real system can help make sure that its major parts and mechanisms have been identified an understood. Further, *comparative methods* are likely to give additional insights, especially concerning the evolutionary background of traits and preferences (Thornhill and Alcock 1983; Heisler et al. and Wilkinson et al., this volume).

One major task is to find out whether the assumptions are true or false. For instance, do animals have heritable preferences for mates with certain traits?

Do extravagant secondary sex traits lower survival, as Darwin (1871) and Fisher (1930) thought? These assumptions, which were questioned by many of Darwin's collegues, both form part of Fisher's (1930) verbal model of the evolution of female choice. In order to judge their validity it may not be necessary to test predictions from a mathematical model of the entire process; it seems more apt to test the assumptions as directly as possible. The more direct the test, the more powerful it should be for separating alternatives. Indirect tests via predictions may be less decisive: low precision of quantitative estimates in evolutionary genetics (Lewontin 1974) often prevents discrimination between hypotheses. Even if an assumption is wrong the outcome might still be close to that predicted by the model simply because of random variation or because a similar predicition follows also from other assumptions. Conversely, even if the result is not as predicted the assumption of interest may be correct, but some unknown factors affects the outcome. Each assumption therefore needs to be tested directly, with as few intervening steps as possible; this is rarely stressed by the theoretical literature with its focus on predictions (Taylor and Williams 1981 is an exception). There may, however, be little reason to expect assumptions concerning the early evolutionary history of a trait to still hold today, even if they once did hold.

When an assumption cannot be directly examined, comparison between predictions and the real system may be the only feasible test. For example, Fisher's idea of heritable mating preferences in females can probably not be fully tested by observing the behavior of females: that they prefer certain males is necessary but not sufficient for the idea to be correct. Whether the preference also show heritable variation might best be tested indirectly, for example, by examining the prediction that artificial selection can alter the proportion of females expressing the preference (Majerus et al. 1982; O'Donald and Majerus 1985). Generation of predictions for testing assumptions is therefore an important function of models. Their main value lies, however, perhaps not so much in indirect tests of explicit assumptions (since they can often be better tested directly) as in their ability to show if some fundamental aspect has been overlooked or misunderstood. Some example are given below.

ASSUMPTIONS AND TESTABILITY OF SOME MODELS OF SEXUAL SELECTION

Haploid and diploid few-locus models, as well as quantative genetic (polygenic) models, have been used to study both Fisherian sexual selection and viability-indicator ('handicap') mechanisms (for a classification see Maynard Smith, this volume). To assess the complexity and testability of the different kinds of models, the number of assumptions and quantities involved

TABLE 1 Assumptions of some genetic models of sexual selection†

Author, type of model	A. Structural and relational assumptions	B. Variables and parameters that must be assigned numerical values for quantitative predictions	C. No. of genotypes, variables and parameters that must be identified or measured for testing the model empirically	D. Total number of assumptions to test ($A+C$)
(1) O'Donald (1967) 2-locus, diploid	14	11	21	35
(2) Kirkpatrick (1982) 2-locus, haploid	14	9	9	23
(3) Andersson (1986a) 3-locus, haploid	18	13	15	33
(4) Lande (1981) polygenic, diploid	22 26*	10	10**	32**

† The estimates are minimal numbers of assumptions required for the models to work; the original papers consider additional alternatives. For example, O'Donald's (1967) models are represented here by the version assuming sex-limited loci and large population size; the mating rules in Lande (1981, eq. 8a–c) are represented by only one of the three alternatives, etc. Only positive assumptions have been included, not statements about effects, relations or differences that are assumed to be negligible; unless a factor is explicitly given a role in a model it is taken here to be assumed negligible.

* The higher figure includes the additional assumptions required for the genetic covariance: variance ratio (B/G) to be constant.

** This does not include detailed testing of the mechanism assumed to make B/G constant; such testing requires much additional measurement, but this assumption may not be necessary for Lande's (1981) main conclusions to apply (see e.g., Seger 1985).

in four models are listed in Tables 1 and 2. Four classes of assumptions and quantitites are distinguished.

A) Structural and relational assumptions (Table 2), e.g., one locus determines the male trait, a second locus determines the female preference, and viability selection takes place before mating.

B) Factors that must be assigned numerical values for the model to yield quantitative predictions, such as recombination fraction, selective coefficients, sex ratio, and allele starting frequencies.

TABLE 2 Structural and relational assumptions of the models in Table 1

Model no. (see Table 1)	(1)	(2)	(3)	(4)
Sexual reproduction	x	x	x	x
Generations do not overlap	x	x	x	x
Population size very large	x	x	x	x
Recombination between loci	x	x	x	x
Dominance relationships	x			
Initial linkage equilibrium	x	x	x	
Viability differences between genotypes	x	x	x	x
Viability selection before mating	x	x	x	x
Sex-linked loci	x			
Autosomal loci		x	x	x
Haploid inheritance		x	x	
Diploid inheritance	x			x
One locus determines male trait	x	x	x	
Another locus determines female preference	x	x	x	
A third locus affects viability			x	
Many loci determine male trait				x
Many loci determine female preference				x
Unlimited polygyny	x	x		x
Monogamy			x	
Mating preference absolute	x		x	
Mating preference relative		x		x
Additive genetic variance in vaibility is maintained			x	
Interaction between loci			x	
Choosy females mate before others			x	
There is mutation–selection balance				x
Phenotypic distribution of male trait is approximately normal				x
Phenotypic distribution of female preference is approximately normal				x
Shape of natural selection function is approximately normal				x
Variance of male trait is independent of its mean				x
Variance of female preference is independent of its mean				x
Male trait has permanent additive genetic variance				x
Female preference has permanent additive genetic variance				x
Additional assumptions about the form of relationships		1*	1**	2†
Total no. of structural and relational assumptions	14	14	18	22

* Kirkpatrick (1982), assumption on mate choice, p.2
** Andersson (1986a), assumption on survival of females with preference, p.810
† Lande (1981), forms of equations 3a and 8

C) The number of genotypes, variables, and parameters that must be identified or measured for testing the model.
D) The total number of assumptions and quantities (A+C) that must be checked for a complete empirical test of the model.

The least precise category is the structural and relational assumptions (Table 1, A; Table 2). It is not entirely clear how to define and count such assumptions. In principle it might be done via an axiomatic system from which the entire model can be deducted, but this formalistic approach would be clumsy and tedious. Like most other models, none of those examined here was derived in such strict fashion: 'It is probably true that no model ever states all of its assumptions explicitly' (Maynard Smith 1978). It is therefore difficult to determine how many structural assumptions are hidden beneath the surface of a model, and the figures in the tables are only approximate, probably erring on the low side. They should, however, give some idea of differences in complexity between the models.

The tables reveal some points that were perhaps not obvious beforehand. First is that all the models, even the simplest two-locus one (Table 1, no. 2), rest on surprisingly many assumptions that must be checked for a full test of the model. A few assumptions obviously are correct but the remainder will be difficult to test. For example, the wide confidence limits that mar most estimates of genotypic viability, mating success, and other aspects of fitness (Lewontin 1974) cast doubt on the feasibility of testing any of the models in a comprehensive, quantitative fashion with a natural system. For example, the models that assume haploidy (Table 1, nos. 2 and 3) cannot possibly be applied in detail to the organisms in which sexual selection is likely to occur. These models, however, were mainly made for logical deduction, so the untestability should come as no shock.

A second notable point is the low number of assumptions of the haploid two-locus model compared to the diploid one (cf. Table 1, nos. 1 and 2). The simplicity of the haploid model enabled Kirkpatrick (1982) to derive analytic solutions and greatly clarify the dynamics and stability conditions of this type of two-locus system; Seger (1985) made further important generalizations. This underlines the value of simplicity in models that aim to show the logical consequences of the assumptions.

Third, and perhaps most surprising, the step from few-locus models to quantitative genetics (Table 1, no. 4) does not greatly raise complexity as regards the number of assumptions or requirements for testing. Moreover, the number of necessary assumptions may be lower than shown in the table: Engen and Saether (1985) recovered some of Lande's (1981) conclusions with a simpler model that did not require mutation-selection balance (but see Lande, this volume). This suggests that the main results also apply under more general conditions. As probably most behavioral and morphological traits are polygenic, quantitative genetics should often be more realistic than a few-locus assumption.

Given the difficulties of testing, it is important to focus on critical assumptions or predictions that differ between the models. For example, Fisherian models do not require that male trait size increases with male phenotypic condition, whereas this assumption is crucial for the indicator mechanism modelled by Andersson (1986a). It also assumes that, other things being equal, females should favor the males with the most well-developed trait. If these assumptions are not met, the latter model can be refuted. On the other hand, quantitative-genetic models of the Fisherian process predict a positive genetic correlation between the magnitudes of female mating preference and male traits (Lande 1981). If this prediction does not stand, these models can be refuted (for further tests, see Arnold 1985 and Heisler et al., this volume).

DISCUSSION

Theoretical results

Genetic models have greatly clarified possible mechanisms of sexual selection, in particular the Fisherian process based on mating preferences. Darwin's (1871) idea that such preferences often lie behind the evolution of conspicuous secondary sex traits was met with scepticism. Fisher's (1930) explanation of how the system might evolve first gained wide attention after O'Donald (1962, 1967) modelled the idea mathematically, showing that the Fisherian mating advantage can indeed create powerful, self-reinforcing selection. More recent models have further clarified the dynamics and equilibria of such systems, revealing aspects that might not have been understood without mathematical models (e.g., Lande 1980, 1981, 1982; Lande and Arnold 1985; Kirkpatrick 1982, 1985, 1986; Heisler 1984, 1985; Engen and Saether 1985; Seger 1985; Seger and Trivers 1986; reviewed by Arnold 1985).

Formal models have also helped clarify other aspects of sexual selection. Using quantitative genetics Kirkpatrick (1985) showed that male contribution of material resources other than genes to females can profoundly affect the evolution of mating preferences and male traits. For example, the line of equilibria between trait and preference in models of the Fisherian type then shrinks to a point such that female reproductive success is maximized (Lande 1981; Kirkpatrick 1982; Seger 1985). Kirkpatrick also concluded that the 'sexy son' hypothesis (Weatherhead and Robertson 1979) does not work if formulated in a mechanistic genetic model.

The hypothesis suggests that selection can favor mating with attractive ('sexy') males even if they give females inferior resources and lower immediate fecundity than do less attractive males: females mating with attractive males are compensated for low fecundity by having sons with high mating success. Such a mechanism does not, however, seem to be evolutionarily stable: the preference evolves to a state where immediate female fecundity is maximized (Kirkpatrick 1985).

Genetic models have also been used to examine indicator (handicap) mechanisms in which a trait that lowers survival may become the target of female preferences, because a male that survives with the trait is likely to carry other favorable genes (e.g., Williams 1966; Zahavi 1975). Indicator mechanisms can be regarded as sexual selection because there is indirect male competition over certain female genotypes. There are also arguments for a contrary view (see Heisler et al., this volume), but the important issue concerns the effects of the mechanism, not what we call it. The first versions of this idea did not function as proposed (Davis and O'Donald 1976; Maynard Smith 1976, 1985). Later models suggested the mechanism might play a role if coupled with a Fisherian mating advantage (Bell 1978; Andersson 1982), but this conclusion was also criticized (Maynard Smith 1985; Kirkpatrick 1986). Renewed analysis, however, verifies that the indicator mechanism is logically capable of enhancing sexual selection together with the Fisherian mechanism (Andersson 1986b). Moreover, if there are genetic interactions such that the trait develops only in males of otherwise favorable genetic constitution, the mechanism even in the absence of any Fisherian mating advantage can lead to a high frequency of both male traits that lower survival and female preferences for such traits (Andersson 1986a).

Maynard Smith (1985) and Kirkpatrick (1986) used models of unlimited polygyny, whereas Andersson (1986a) used a monogamy model (without any Fisherian mating advantage) to see whether an indicator mechanism can suffice for the evolution of trait and preference. In addition, Kirkpatrick (1986) did not consider the kind of interaction dealt with by Andersson (1982, 1986a), who assumed that the handicapping trait is conditional and reflects variation throughout the genome: those males with the most favorable total genotype have the highest viability and develop the largest ornaments. In one version of his model, Kirkpatrick (1986) assumed that the additive genetic variance for viability becomes depleted. That this does not happen is, however, necessary for the indicator mechanism to be effective; these and other differences in assumptions, and an emphasis on other questions, may explain why Kirkpatrick (1986) found no effect of the mechanism. Hamilton and Zuk (1982) suggested that heritable viability differences might be maintained through coevolution between parasites and hosts: there are also several other possibilities (e.g., Charlesworth, this volume). The results of Andersson (1986a,b) suggest that indicator mechanisms might sometimes play a role in the evolution of female preferences and male traits that lower survival. The quantitative-genetic model of Heisler (1985) also suggests that indicator mechanisms can be important for the early evolution of mating preferences. In many cases, however, addition of indicator selection to a Fisherian model does not enhance the evolution of trait and preference. Indicator mechanisms therefore are still controversial and they, as well as Fisherian mechanisms, remain largely untested except for a few necessary but insufficient assumptions.

Empirical tests

Perhaps the main impact of genetic models of sexual selection so far has come from their proof that the Darwin-Fisher theory is logically plausible. This has helped to create a more open attitude to Darwin's theory of female choice than that of, for example, Huxley (1938) and Lack (1968), and to convince field workers that the theory is sound enough for testing to be worthwhile. This has led to important empirical discoveries, and there is now evidence from several species that females to some degree base their choice of mate on his external morphology or behavior (reviewed by Bradbury and Davies, this volume). Majerus et al. (1982) and O'Donald and Majerus (1985) have also shown that mating preferences show heritable variation and can be amplified by selection. These time-consuming experiments would hardly have been done had not mathematical models shown that heritable mating preferences are indeed capable of leading to sexual selection. As an added bonus the models have helped clarify, among other things, the conditions for stability vs. runaway selection (e.g., Lande 1981; Kirkpatrick 1982; Engen and Seather 1985; Heisler 1985; Seger 1985), but these deductions have not yet been tested.

The two classes of models described above are not equally testable. General, simple genetic models do not even have to be testable in order to be useful (see above), but in many cases it has been possible to check basic assumptions, such as the existence of mating preferences and mortality risks associated with secondary sex traits.

Comparison between quantitative predictions and real systems requires more realistic, specific models. So far no study has come close to testing all assumptions and predictions of a sexual selection model adapted to a particular case. O'Donald (1983, 1986) has, however, tested several aspects of a two-locus model in the polymorphic arctic skua (*Stercorarius parasiticus*). The study has provided valuable information and supportive evidence, but some crucial assumptions remain to be examined by direct behavioral observations, e.g., the existence of female preferences for particular male phenotypes. Nevertheless, O'Donald's (1983) study points the way towards further testing of genetic models of sexual selection in natural populations.

Acknowledgements.

I am grateful to A. Arak, S. Arnold, B.O. Bengtsson, G. Borgia, M. Kirkpatrick, R. Lande, and especially R. Thornhill for helpful comments and discussions, and to the Swedish Natural Sciences Research Council for support.

REFERENCES

Andersson, M. 1982. Sexual selection, natural selection and quality advertisement. Biol. J. Linn. 17: 375–393.

Andersson, M. 1986a. Evolution of condition-dependent sex ornaments and mating preferences: sexual selection based on viability differences. Evolution 40: 804–816.
Andersson, M. 1986b. Sexual selection and the importance of viability differences: a reply. J. Theor. Biol. 120: 251–254.
Arnold, S.J. 1985. Quantitative genetic models of sexual selection. Experientia 41: 1296–1310.
Borgia, G. 1979. Sexual selection and the evolution of mating systems. In: Sexual Selection and Reproductive Competition, eds. M. and A. Blum, pp. 19–80. New York: Academic Press.
Bell, G. 1978. The handicap principle in sexual selection. Evolution 32: 872–885.
Clutton-Brock, T.H., and Harvey, P.H. 1984. Comparative approaches to investigating adaptation. In: Behavioural Ecology: and Evolutionary Approach (2nd ed.) eds. J.R. Krebs and N.B. Davies, pp. 7–29. Oxford: Blackwell.
Darwin, C. 1871. The Descent of Man and Selection in Relation to Sex. London: John Murray.
Davis, J.W.F., and O'Donald, P. 1976. Sexual selection for a handicap: a critical analysis of Zahavi's model. J. Theor. Biol. 57: 345–354.
Engen, S., and Saether, B.E. 1985. The evolutionary significance of sexual selection. J. Theor. Biol. 117: 277–289.
Fisher, R.A. 1930. The Genetical Theory of Natural Selection. Oxford: Clarendon Press.
Hamilton, W.D., and Zuk, M. 1982. Heritable true fitness and bright birds: a role for parasites: Science 218: 384–387.
Heisler, I.L. 1984. A quantitative genetic model for the origin of mating preferences. Evolution 38: 1283–1295.
Heisler, I.L. 1985. Quantitative genetic models of female choice based on 'arbitrary' male characters. Heredity 55: 187–198.
Huxley, J.S. 1938. The present standing of the theory of sexual selection. In: Evolution: Essays on Aspects of Evolutionary Biology, ed. G.R. de Beer, pp. 11–42. Oxford: Clarendon Press.
Kirkpatrick, M. 1982. Sexual selection and the evolution of female choice. Evolution 36: 1–12.
Kirkpatrick, M. 1985. Evolution of female choice and male parental investment in polygynous species: the demise of the "sexy son". Am. Natural. 125: 788–810.
Kirkpatrick, M. 1986. The handicap mechanism of sexual selection does not work. Am. Natural. 127: 222–240.
Lack, D. 1968. Ecological Adaptations for Breeding in Birds. London: Methuen.
Lande, R. 1980. Sexual dimorphism, sexual selection, and adaptation in polygenic characters. Evolution 34: 292–305.
Lande, R. 1981. Models of speciation by sexual selection on polygenic traits. Proc. Natl. Acad. Sci. USA 78: 3721–3725.
Lande, R. 1982. Rapid origin of sexual isolation and character divergence in a cline. Evolution 36: 213–223.
Lande, R., and Arnold, S.J. 1985. Evolution of mating preference and sexual dimorphism. J. Theor. Biol. 117: 651–664.
Levins, R. 1966. The strategy of model building in population biology. Am. Scient. 54: 421–431.
Lewontin, R.C. 1974. The Genetic Basis of Evolutionary Change. New York: Columbia Univ. Press.
Majerus, M.E.N.; O'Donald, P.; and Weir, J. 1982. Female mating preference is genetic. Nature 300: 521–523.

Maynard Smith, J. 1968. Mathematical Ideas in Biology. Cambridge: Cambridge Univ. Press.

Maynard Smith, J. 1976. Sexual selection and the handicap principle. J. Theor. Biol. 57: 239–242.

Maynard Smith, J. 1978. Optimization theory in evolution. Ann. Rev. Ecol. Syst. 9: 31–56.

Maynard Smith, J. 1982. Evolution and the Theory of Games. Cambridge: Cambridge Univ. Press.

Maynard Smith, J. 1985. Sexual selection, handicaps and true fitness. J. Theor. Biol. 115: 1–5.

O'Donald, P. 1962. The theory of sexual selection. Heredity 17: 541–552.

O'Donald, P. 1967. A general model of sexual and natural selection. Heredity 22: 499–518.

O'Donald, P. 1980. Genetic Models of Sexual Selection. Cambridge: Cambridge Univ. Press.

O'Donald, P. 1983. The Arctic Skua. A Study of the Ecology and Evolution of a Seabird. Cambridge: Cambridge Univ. Press.

O'Donald, P. 1987. Polymorphism and sexual selection in the arctic skua. In: Avian Genetics, eds. P. Buckley and F. Cooke, in press.

O'Donald, P., and Majerus, M.E.N. 1985. Sexual selection and the evolution of preferential mating in ladybirds. I. Selection for high and low lines of female preference. Heredity 55: 401–412.

Parker, G.A. 1979. Sexual selection and sexual conflict. In: Sexual Selection and Reproductive Competition in Insects, eds. M. and A. Blum, pp. 123–166. New York: Academic Press.

Ridley, M. 1983. The Explanation of Organic Diversity. Oxford: Clarendon.

Seger, J. 1985. Unifying genetic models for the evolution of female choice. Evolution 39: 1185–1193.

Seger, J., and Trivers, R. 1986. Asymmetry in the evolution of female mating preferences. Nature 319: 771–773.

Taylor, P.D., and Williams, G.C. 1981. On the modelling of sexual selection. Q. Rev. Biol. 56: 305–313.

Thornhill, R., and Alcock, J. 1983. The Evolution of Insect Mating Systems. Cambridge, Massachussetts: Harvard Univ. Press.

Weatherhead, P.J., and Robertson, R.J. 1979. Offspring quality and the polygyny threshold: 'the sexy son hypothesis'. Am. Natural. 113: 201–208.

Williams, G.C. 1966. Adaptation and Natural Selection: A Critique of Some Current Evolutionary Thought. Princeton Univ. Press.

Zahavi, A. 1975. Mate selection – a selection for a handicap. J. Theor. Biol. 53: 205–214.

Sexual Selection: Testing the Alternatives
eds. J.W. Bradbury and M.B. Andersson, pp. 55–66
John Wiley & Sons Limited

A Critical Review of Sexual Selection Models

G. Borgia

*Department of Zoology
University of Maryland
College Park, MD 20742, USA

Abstract. The evolution of extreme sex-limited male display traits is a poorly understood process. Here I suggest that the claims for acceptance of some runaway models of sexual selection are not supported by the available data. Moreover, there exist other plausible models for the evolution of male display characters that have not been considered in detail. Many of the models discussed as alternatives are not necessarily in conflict and may help explain the same phenomena. I discuss constraints on the evolution of display characters, especially the effect of competition among coevolving male trait and female choice genotypes, and how such effects can influence the predictions of models. It is suggested that an understanding of the evolution of male display characters requires detailed empirical studies. Methods for evaluating models for the evolution of male display characters in natural populations are considered.

INTRODUCTION

A strong divergence of opinion has recently developed over how female choice can shape patterns of male sexual display. Most attention has focused on species in which males have no role in parental care and cannot directly control female mating decisions. It is in these species, such as peafowl and many birds of paradise, that some of the most extreme sex-limited male displays have evoked. A variety of models have been proposed that attempt to explain the evolution of these male display traits, but two have received the most attention: the 'good genes' models and runaway models.

Fisher (1980) initiated the modern discussion of the evolution of extreme male display characters when he proposed the 'runaway process.' According to his model, a consistent mating preference by even a small proportion of females in a population in which males initially show no strong sexual dimorphism can lead to (a) a spread in the choice pattern among females, and (b) continued enhancement of male characters chosen by females to the point of extreme sex-limited development. It predicts no enhancement of offspring fitness outside the context of sexual selection. Recent modelers (Lande 1981; Kirkpatrick 1982, 1986) suggest that characters with negative effects on the fitness of offspring may evolve by processes similar to that first described by Fisher.

Good genes models of female choice propose that female discrimination among males can lead to heritable viability gains for offspring. Extreme male display traits are commonly envisioned to have evolved as indicators that enhance the females' ability to select high quality sires. Currently popular models propose that male traits are expressed conditionally that males with the fullest expression are the most likely to produce vigorous offspring. Different models focus on different qualities that male characters are supposed to indicate. Some of these are the ability to survive to old age (e.g, Howard 1979; Halliday 1978), disease resistance (Hamilton and Zuk 1982); overall vigor (e.g., Trivers 1972; Andersson 1982, 1986), and male dominance in aggressive encounters (Borgia 1979; Borgia et al. 1985).

Apart from these two types of models there exist additional alternative explanations for the evolution of extreme male display characters. Parker (1983) proposed that extreme displays evolve because they enhance male advertisement and allow displaying males to be more readily discovered. LeCroy et al. (1980) suggest that male plumage displays evolve in the context of inter-male competition and that female choice is unimportant in their evolution. Finally, there may be instances where male displays provide some immediate proximate benefit to choosing females. For example, female bowerbirds may prefer males with bowers because it gives them a safe place from which they can observe a courting male (Borgia et al. 1985).

Advocates of the runaway model have been particularly outspoken in asserting the correctness of their views. For example, Davis and O'Donald (1976) argued that there is no other suitable model for the evolution of these characters, and there are no important criticisms of the runaway model; concluding with the view that '. . . the evolution of extreme sex-linked characters is fully explained by Fisher's model.' In a critique of what he calls the 'Panglossian view,' Arnold (1983) claims that good genes models cannot work because '. . . there is no direct force of selection on mating preferences.' This conclusion is based on the assumption that '. . . the number of progeny produced by the female is unaffected by mate choice.' More recently Kirkpatrick (1986) subsumed all good genes models under the term 'handicap' and, when the specific model failed, asserted that none of these models work.

The extreme certitude of these statements belies our ignorance of how exaggerated male displays evolve. Recent models have established the plausibility of the runaway process (e.g., Lande 1981) and good genes models (Andersson 1986) as mechanisms for the evolution of extreme male display characters. Even so, evidence that either of these processes is the exclusive or even a contributing cause for the evolution of exaggerated display is lacking.

Moreover, current versions of these models are relatively simplistic and are based on numerous unproven assumptions (e.g., O'Donald 1983; Seger 1985; Andersson and Charlesworth, both this volume). It is possible that none of the currently popular models is correct and that other mechanisms are operating. There are numerous other plausible models for the evolution of extreme characters that have not yet been explored in detail. This is coupled with an almost complete ignorance of the basic life history of species with extreme male sexual displays and no measurement of quantitative genetic and other parameters that have figured prominently in the discussion of recent models. So, while recent models have fueled an interesting and often intense debate over the plausibility of various explanations for the evolution of exaggerated male display traits, given their limitations and our lack of knowledge about what happens in nature, it is clear that this problem is far from resolved.

My objective in this paper is to discuss some distinguishing features of current models of sexual selection, present some plausible alternatives, and consider promising methodologies for testing these models. Our sparse knowledge of how sexual selection operates in species with extreme male display makes it difficult to determine how these male characters evolved. Without more information, we are left to determine if different models are plausible and to devise tests that might distinguish among these models in future studies. Solving the difficult problem of the evolution of extreme display will require an open-minded search that carefully evaluates the full range of theoretical possibilities and has a solid basis in the biology of species with extreme male displays.

GOOD GENES MODELS

There are several types of results supporting the possibility that selection for good genes occurs. Maynard Smith (1956) has shown that in *Drosophila*, females discriminate against inbred males based on the vigor of male display. He interpreted this as female choice for avoiding poor quality sires and, in an extreme version, this may be viewed as selection for good genes. If selection has favored females who choose males with vigorous displays, it is also possible that males have morphological characters with similar function.

There are currently no solid demonstrations of good genes effects. Mayr (1963) and others have suggested that male display characters often arise as species identification characters. It is costly for females to produce hybrids so there is selection on females to choose males with characters distinctive of their own species. Although this can be viewed as an extreme version of selection for good genes, it is not the same as demonstrating that finer within-species discriminations are also beneficial. Partridge (1980) has shown that females able to choose among males have higher fitness than females who are assigned mates. It is not clear if the observed effect was due to male interactions or to female choice, and Schaffer et al. (1984) were unable to replicate the original effect.

A theoretical difficulty seen for good genes models is the possibility that heritable variation in the benefits for choosing males is exhausted (Williams 1975; Borgia 1979; Charlesworth, this volume). Zahavi (1975) pointed out that for these models to work, they should be based on assessment of genome-wide characteristics of males. Given this, it is possible that shifting (Borgia 1979; Andersson 1986) or cycling (Hamilton and Zuk 1982) conditions can maintain sufficient variability in males so as to produce heritable gains for discriminating females in each generation (see also Charlesworth, this volume).

Selection for good genes in models of polygynous populations has usually been overshadowed by Fisherian runaway selection for the male trait (Maynard Smith 1985). Andersson (1986) circumvented this problem by modeling a monogamous population where the male advantage from polygynous matings could be eliminated. However, the most interesting cases of elaboration of male characters occur in polygnous populations, so it is reasonable to ask if the type of selection Andersson modeled has any important effect on the male display characters that evolve in polygynous species? Fisher (1930) suggested that what I call 'good genes selection' may be important in initially increasing the number of preferred traits. Good genes selection may partly determine the types of traits that evolve, but it may have little effect on the extent to which they are elaborated.

RUNAWAY MODELS

Much attention has been directed at demonstrating Fisher's premise that extreme male displays evolve as a result of a coevolution of male display and female choice. These models differ from Fisher's in not assuming an initial advantage for these traits. Both quantitative genetic (e.g., Lande 1981) and two-locus models (e.g., Kirkpatrick 1982) have been used to determine if the runaway process can operate. Most attention has focused on Lande's quantitative genetic model in which he was able to show that the self-reinforcing pattern suggested by Fisher could develop between female preference and male display characters. In the model, female choice in neutral and propels changes in male display characters until they are checked by the costs of extreme displays to males. The model produces a line of equilibrium that represents a balance between the effects of female choice and the cost of male characters.

These models differ in two important respects. The quantitative genetic models require that female choice and male display characters be highly polygenic; under this condition the models can explain an elaboration of male traits and female preferences as suggested by Fisher. The two-locus models describe the increase in frequency of genes coding for female choice and male display traits, but they do not produce the coevolutionary escalation in display and choice that was described by Fisher (1930). Hence, these models are not really equivalent as has been claimed (Arnold 1985). Moreover, the differences raise questions about the usefulness of these particular models as explanations for the evolution of extreme male displays.

O'Donald and Majerus (1985) have shown in studies of ladybird beetles that only a few loci are responsible for controlling female choice and male traits selected by females. This simple genetic system would prevent the ongoing type of selection envisioned in quantitative genetic models and because few-locus models do not explain extreme elaboration of male characters, it is not clear that the evolution of extreme male displays can be explained by available models if display and choice traits are not polygenic.

EXAGGERATION WITHOUT A GENETIC CORRELATION

Recent quantitative genetic models have assumed that a genetic correlation between the male trait and female preference is required to produce extreme male displays. This is not necessarily so. Consider the case in which as open-ended (supernormal) female preference is established in a population where the choice trait is controlled by only a small number of loci that become fixed. At the time the preference is established, no males show extreme displays, but with time sufficient mutations occur among males to allow this persistent preference to select for males with exaggerated displays. This result will be an elaborated male character without any coevolution of the female preference with the male trait. Unlike the claims of quantitative genetic models (e.g., Arnold 1983), the nature of female preferences can have a dramatic effect on evolutionary outcomes. The female preference that selects for extreme male character development could evolve in a variety of contexts. For example, it may be a result of selection for a character that allows females to obtain immediate benefits or from the evolution of a female preference as described by Kirkpatrick (1982). The key point is that the ultimate elaboration of the male character in this model is dependent on the type of female preference that is already established and not on the existence of a genetic correlation between the female choice and male display traits.

INITIATING THE SELECTION PROCESS

Fisher envisioned a preference initially linked to real benefits where a genetic correlation developed and led to a runaway process that produced extreme male characters. Female preferences that initially provide some real benefits may have a slight advantage over other less advantageous preferences. This may occur because a rare beneficial preference would be less subject to loss by drift and because benefits associated with these choice patterns may be sufficient to overcome the costs of choosing that might otherwise halt the evolution of the female preference. In recent gentic models (e.g., Kirkpatrick 1982), the evolution of preferences is typically frequency-dependent. New choice traits must reach a threshold frequency in the population before they can replace an established alternative. So while it is possible that these traits reach threshold frequencies by drift, it is more likely that differences in the initial benefits produced by traits affect their success in becoming established as suggested by Fisher (1958; see also Heisler 1984). This implies a bias in which the display traits most likely to evolve are those that have, at least at some time in their evolution, proven advantageous to choosing females.

ARE MALE DISPLAY CHARACTERS ADAPTIVE?

Much of the interest in the evolution of extreme male display traits results from the suggestion that runaway selection can lead to nonadaptive female choice (e.g., Lande 1981; Kirkpatrick 1982). Such treatments emphasize that a female preference for elaborated male traits can produce characters that lower the survivorship of sons and decrease the vigor of daughters. While this is not a necessary result, it has been used to distinguish these models from adaptive (good genes) models. This prediction of nonadaptive outcomes contrasts with empirical observations of well studied resource-based mating systems where females consistently evolve to choose males who provide real benefits (e.g., Thornhill 1976, 1980; Downhower and Brown 1980; Borgia 1982; Pleszczynska 1978). Possibly, selection is so different in these two types of mating systems that one produces adaptive outcomes while the other does not.

The prediction that the runaway process produces nonadaptive outcomes is based on an assumption that male display traits that characterize populations are those that are established first. Thus, models for extreme male displays differ from most other recent evolutionary models in not comparing a full range of competing alternatives and allowing invasion of qualitatively different female preference patterns. If this were done, then the predicted outcomes of runaway models might be very different (but see Kirkpatrick 1982). Competition among alternative female choice patterns should lead to replacement of patterns without beneficial effects for females' offspring by patterns with beneficial effects. A consistent pattern of selection of this type should lead to adaptive patterns of mate choice. This argument implies that our expectations concerning the degree of adaptiveness of female choice will be related to how easy it is for novel, advantageous female choice patterns to replace established, less advantageous alternatives. If replacement is common then female choice will be adaptive, but if replacement is rare then choice will be nonadaptive and arbitrary.

This result suggests a format in which runaway and good genes selection can work together to shape sexual selection. The issue of competition among female preferences is critical in understanding which male display characters evolve and deserves much more attention that can be given here. Competition among female preferences might influence the outcome of sexual selection in several contexts.

1) Preference curves describe the intensity of individual female preference relative to the degree of male character development. Above I suggested that differences in preferences can affect the outcome of sexual selection. However, if some types of female preference curves have consistent advantages in direct competition with others, then the same dominant types of preference curves may characterize female choice in most species. The result might cause widespread evolution of extreme male characters. The recurrent evolution of elaborated male displays in non-resource-based mating systems might reflect this pattern of selection. Of course, if the success of different female preferences is situation-dependent, then a wider array of outcomes might be expected.

2) Male display traits that have sex-limited expression will be favored over those that are fully expressed in both sexes. Even partial expression of male display traits in females can be expensive to them, so selection should favor females who choose male traits that are expressed only in males. This pattern of selection emphasizes the female preference and male trait as coevolving characters. The cost of the display trait as expressed in both daughters and sons will affect the success of these characters in competition with other coevolving female preferences and male display traits.

3) Polygenic display traits lose in competition with similar single gene traits. A substantial substitutional load may occur with rapid changes during the runaway process. Arnold (1985) notes that this load is due to pleiotropic effects of the substitutions of many alleles associated with the evolution of polygenic male display characters and female choice traits. Where replacement of traits can occur female choice traits, which evolve with less load, should have an advantage. Thus, there should exist a bias in favor of male display and female choice traits that are not highly polygenic, thereby creating less of a substitutional load. Load could affect evolution of display in other ways. If display and choice traits were polygenic, perhaps resulting from selection for modifiers once the choice trait was estabished, then the effects of substitutional load may make large qualitative evolutionary changes less likely to occur. These effects may make it more difficult for new choice patterns to invade or to drift along the line of equilibrium (see Lande 1981).

A central issue that appears to separate these models is the extent to which competing alternatives can be replaced. In the "no genetic correlation" (see above) and current runaway models, the persistence of a choice trait favoring extreme males requires that choice traits are not easily replaced by a more advantageous competing alternatives. The opposite is true for models allowing competition between alternatives; extreme characters are maintained precisely because they are the most successful of the competing alternatives. The first case predicts sexual selection for nonadaptive, arbitrary male display characters, whereas in the second, fitness-maximizing traits are expected.

TESTING MODELS

Several distinct models have emerged that give very different predictions about the outcomes of sexual selection. In order to determine which of these alternatives provides the most realistic view of sexual selection we need to determine if there is an important role for invading preference traits, if female choice is characterized by the type of preferences that are initially established, of if some intermediate condition obtains. Modeling can show which of the above-mentioned scenarios provides realistic possibilities for the evolution of exaggerated displays and if the parameters defined in the model are consistent with properties of real organisms. However, we must depend on empirical tests to determine if particular models actually work for living orgnaisms and if they describe the process that caused the evolution of a particular extreme male display. Two general types of tests have been proposed.

The first involves measurement of parameters described in sexual selection models to determine if the conditions stated in the proposed model are met under real conditions. This is a potentially important test of models, but it may be exceedingly difficult to carry out. Species with the most extreme male character development, such as birds-of-paradise, tend to be long-lived and have complex mating behaviors that are difficult to observe in nature. Individual females may mate only once a year in situations that cannot be easily duplicated in the laboratory. This complicates attempts to measure variables such as genetic covariance between males and females and heritability of fitness which must be measured with exceedingly large sample sizes. Deviations from the assumptions of models may require more difficult measurements of parameters. For example, a deviation from the assumptions of mutation-selection balance would require that variance and covariance for choice and display characters be measured not as point estimates but as functions of the intensity and duration of selection. Even with these difficulties attempts should be made to measure factors affecting the models we have discussed including the degree of polygeny of display and choice traits, the amount of load incurred in attempts to select on these characters, the degree of sex-limited expression (and causes for its expression in females), the shape of female preferences curves, and the heritability of fitness.

The second option for testing models for the evolution of extreme male characters is the detailed study of species with male display characters to determine which of the available models best predicts observed patterns of male character evolution and display. Specific versions of the indicator, proximate benefit, and advertisement models make distinctive predictions that can be tested. For example, the male dominance model predicts that male display characters should be good indicators of dominance. The model can be falsified by tests to determine if females choose males with the indicator and if the indicator can be used to predict dominance (see Borgia 1985a,b). Similar predictions can be tested for indicators of age, vigor, or disease resistance. Proximate mechanism models can be directly tested to determine if male displays provide some immediate benefit. This is not so with current versions of runaway models that predict arbitrary or indetermine results. Models that predict all possible outcomes cannot be directly falsified. However, indirect tests can be made to determine if male displays show a consistent pattern in relation to the habitats where they occur. For example, bower of bowerbirds appear to differ in relation to the habitat where a species lives (Borgia et al. 1985). In open habitats, bowers offer more cover to females. This is consistent with a hypothesis that the bower functions to protect visiting females and suggests that the bower display is not arbitrary. However, there are several alternative possibilities.

The interpretation of results and attempts to use them to evaluate models will not doubt be controversial. Consider Arnold's (1983) prediction from Lande's model suggesting that there should be high intraspecific geographical variation in male display characters among polygynous species with extreme male character development. This hypothesis has not been systematically tested for any species with extreme display. However, the most extreme cases known of intraspecific variation in male display traits occurs in monogamous sparrows in which the function of the display song appears to have little to do with female choice (Baptista and Morton 1982). This result emphasizes the need for careful tests and interpretation of results.

In most instances the hypotheses lead to straightforward predictions about the types of male display traits and associated behaviors that are expected to evolve. An exception, however, are the runaway models which, in some versions, can lead to almost any outcome (Lande 1981; Kirkpatrick 1982, 1986). Predicting the outcome of the runaway process is further complicated because Fisher's original version of the model included the elaboration of a male character through female choice for adaptive characters.

Comparisons of male displays in closely related species provide an opportunity to evaluate predictions about the stability of display traits. If display traits are unstable then large differences in male display characters among closely related species should be common, whereas a model of stable sexual selection would predict the species differentiate and develop different display patterns in a bifurcating pattern with few evolutionary reversals. If male display traits are commonly stable (and the conditions for male display are not grossly different), then the degree of differentiation in male traits should follow a pattern very similar to a dendogram based on genetic overlap among species. Shifts in patterns of male displays should be associated with changes in conditions that can qualitatively change the display. The Fisherian mechanism should give a more irregular pattern, often with closely related taxa showing little similarity in their sexual display traits. Frequent shifts in male display patterns that are not associated with any obvious adaptive cause, e.g., a shift in habitat, would support Fisherian runaway models.

The discussion of the evolution of sexually selected traits has taken on a much wider focus than a mere comparison of models. At issue is a whole complex of ideas of how selection works in a complicated social environment. Lande's model for runaway selection required that characters be highly polygenic, that there be a mutation-selection balance, that male display characters are arbitrary and have no impact value to the female or her offspring, that initially established characters be resitant to invasion by new competing types, that selection for male display characters is rapid rather than gradual, that the results of sexual selection be unstable and subject to drift in small populations, and that drift in small populations be responsible for the evolution of species-identifying characters. These assumptions need to be tested.

Not all models for the evolution of extreme characters incorporate Lande's assumptions, and these models may provide more realistic views of how extreme characters evolve. Even so these models still rely on important simplifying assumptions (e.g., haploidy) that may have important and unrecognized effects on their conclusions. Efforts must be made to produce models that are more faithful to the genetic system being described. This may require an emphasis on simulation rather than analytic models. It is critical to explore the significance of competition among qualitatively different female preferences. Arnold (1985) and others have drawn attention to the nonadaptive aspects of sexual selection, but this appears to depend on the assumption that competition between preferences does not occur.

Very little real progress on these issues will be made with only theoretical arguments. We are in need of empirical tests of assumptions of current models so that we can make preliminary tests of hypotheses and eventually build a more refined set of models. As Arnold (1983) points out, 'Models should be viewed as guides for field and laboratory work: only empirical work can produce generalizations about the natural world.' A large body of data on the behavior, ecology, and genetics of mate choice in species with extreme displays is needed to resolve current issues. Collection of such data should be the focus of an intensive effort by behavioral and evolutionary biologists.

Acknowledgements

I thank the many attendees at the Dahlem Workshop for their helpful comments. Additional suggestions came from A. Houde, C. Loffredo, and J. Lauridsen.

REFERENCES

Andersson, M. 1982. Sexual selection, natural selection and quality advertisement. Biol. J. Linn. 17: 375–393.

Andersson, M. 1986. Evolution of condition-dependent sex ornaments and mating preferences: sexual selection based on viability differences. Evolution 40: 804–816.

Arnold, S. 1983. Sexual selection: the interface of theory and empiricism. In: Mate Choice, ed. P. Bateson, pp. 67–107. Cambridge: Cambridge Univ. Press.

Arnold, S. 1985. Quantitative genetic models and sexual selection. Experientia 41: 1296–1310.

Baptista, L.F., and Morton, M.L. 1982. Song dialects and mate selection in montane white-crowned sparrows. Auk 99: 537–547.

Borgia, G. 1979. Sexual selection of the evolution of mating systems. In: Sexual Selection and Reproductive Competition, eds. M.S. Blum and N.A. Blum, pp. 19–80. New York: Academic Press.

Borgia, G. 1982. Experimental changes in availability of oviposition sites: effect on mating success of male *Scatophaga stercoraria*. Evolution 36: 307–312.

Borgia, G. 1985a. Bower destruction and sexual competition in the satin bowerbird (*Ptilonorhynchus violaceus*). Behav. Ecol. Sociobiol. 18: 91–100.

Borgia, G. 1985b. Bower qualtiy, number of decorations, and mating success of male satin bowerbirds (*Ptilonorhynchus violaceus*): an experimental analysis. Anim. Behav. 33: 266–271.

Borgia, G.; Pruett-Jones, S.; and Pruett-Jones, M. 1985. The evolution of bowers as markers of male quality. Zeit. Tier. 67: 225–236.

Davis, J., and O'Donald, P. 1976. Sexual selection for a handicap: a critical analysis of Zahavi's model. J. Theor. Bio. 57: 345–354.

Downhower, J., and Brown, L. 1980. Mate preferences of female mottled sculpins, *Cottus bairdi*. Anim. Behav. 28: 728–734.

Fisher, R.A. 1930. The Genetical Theory of Natural Selection. Oxford: Clarendon Press.

Fisher, R.A. 1958. The Genetical Theory of Natural Selection, 2nd ed. New York: Dover.

Halliday, T. 1978. Sexual selection and mate choice. In: Bahavioral Ecology: and Evolutionary Approach, eds. J.R. Krebs and N.B. Davies, pp. 180–213. Sunderland, MA: Sinauer.

Hamilton, W.D., and Zuk, M. 1982. Heritable true fitness and bright birds: a role for parasites? Science 218: 384–387.

Heisler, I.L. 1984. A quantitative genetic model for the origin of mating preferences. Evolution 38: 1283–1295.

Howard, R.D. 1979. Estimating reproductive success in natural populations. Am. Natural. 114: 221–231.

Kirkpatrick, M. 1982. Sexual selection and the evolution of mate choice. Evolution 36: 1–12.

Kirkpatrick, M. 1986. The handicap mechanism of sexual selection does not work. Am. Natural. 125: 788–810.

Lande, R. 1981. Models of speciation by sexual selection on polygenic traits. Proc. Natl. Acad. Sci. USA 78: 3721–3762.

LeCroy, M.; Kulipi, A.; and Peckover, W. 1980. Goldie's bird of paradise: display, natural history, and traditional relationships of people to the bird. Wilson Bull. 92: 298–301.

Maynard Smith, J. 1956. Fertility, mating behavior, and sexual selection in *Drosphila subobscura*. J. Genet. 54: 261–279.

Maynard Smith, J. 1985. Sexual selection, handicaps and true fitness. J. Theor. Bio. 115: 1–8.

Mayr, E. 1963. Animal Species and Evolution. Cambridge, MA: Harvard Univ. Press.

O'Donald, P. 1983. Sexual selection by female choice. In: Mate Choice, ed. P. Bateson, pp. 53–66. Cambridge: Cambridge Univ. Press.

O'Donald, P., and Majerus, M.E.N. 1985. Sexual selection and the evolution of preferential mating in ladybirds. I. Selection for high and low lines of female preference. Heredity 55: 401–412.

Parker, G.A. 1983. Mate quality and mating decisions. In:Mate Choice, ed. P. Bateson, pp. 141–166. New York: Cambridge Univ. Press.

Partridge, L. 1980. Mate choice increases a component of offsrping fitness in fruit flies. Nature 283: 290–291.

Pleszczynska, W. 1978. Microgeographic prediction of polygyny in the lark bunting. Science 201: 935–937.

Schaeffer, S.W.; Brown, C.J.; and Anderson, W.W. 1984. Does mate choice affect fitness? Genetics 107: 94.

Seger, J. 1985. Unifying genetic models for the evolution of female choice. Evolution 39: 1185–1193.

Thornhill, R. 1976. Sexual selection and feeding behavior in *Bitticas apicalis* (Insecta: Mecoptera). Am. Natural. 110: 529–548.

Thornhill, R. 1980. Competitive, charming males and choosy females: was Darwin correct? FLA Entomol. 63: 5–30.

Trivers, R.L. 1972. Parental investment and sexual selection. In: Sexual Selection and the Descent of Man, ed. B.G. Campbell, pp. 136–179. Chicago: Aldine Press.

Williams, G.C. 1975. Sex and Evolution. Princeton: Princeton Univ. Press.

Zahavi, A. 1975. Mate selection – a selection for a handicap. J. Theor. Bio. 53: 205–214.

Sexual Selection: Testing the Alternatives
eds. J.W. Bradbury and M.B. Andersson, pp. 67–82
John Wiley & Sons Limited

The Evolutionary Forces Acting on Female Mating Preferences in Polygynous Animals

M. Kirkpatrick

**Department of Zoology*
University of Texas
Austin, TX 78712, USA

Abstract. A framework for discussing the forces responsible for the origin and evolution of female mating preferences is developed. Although preferences can arise as adaptations to enable females to discriminate between prospective mates, they may often arise as side effects of biases intrinsic to the sensory system used by females. Once established, these biases may often evolve in response to selection pressures unrelated to mating. This is illustrated by an example from hylid frogs in which ecological factors can cause body size to evolve, which causes female auditory tuning and hence mate preference to change as a biproduct. The forces involved in the evolution of established mating preferences can be organized into three categories: (a) direct selection, in which preferences are selected because of their effects on female survival and fecundity; (b) indirect selection, in which preferences evolve as a correlated response to selection on other traits (such as male secondary sexual characters); and (c) other forces, such as group selection and genetic drift. The qualative effects and relative importance of each of these are discussed. The final section of the paper presents a model for the evolution of preferences when females experience costs while searching for mates.

INTRODUCTION

There is now widespread, though not universal, acceptance of Darwin's (1859, 1871) hypothesis that female mating preferences have been responsible for the evolution of secondary sexual characters in the males of many species of animals. In contrast to this consensus, there is little agreement about what forces are important in the evolution of the preferences themselves.

This paper develops a framework for the discussion of evolutionary forces that give rise to and cause the subsequent evolution of mating preferences. The topic is developed in three sections. First is an examination of how preferences may originate. It is emphasized that observation of the effects of preferences on male characters can mislead the search for what caused their origin. Second, a genetic perspective is used to organize a discussion of several evolutionary forces that may be involved in the evolution of preferences once they are established. The goal here is to categorize the various types of evolutionary forces, hopefully clarifying the question of what kinds of empirical data are needed to discriminate between them. Third, the effects of a specific form of selection on mating preferences, that due to female search costs, is investigated using a quantitative genetic model. This illustrates how our theoretical understanding of preference evolution can be expanded by studying new models.

THE ORIGIN OF PREFERENCES

The factors involved in the origin of mating preferences may often be very different from those involved in the evolution of established preferences. Preferences can originate in at least three different ways. First, preferences may be due to biases that are built into the sensory modality being used and may have no intrinsic adaptive value. Second, preferences may orginate in response to evolutionary forces unrelated to intraspecific mate discrimination and may cause females to bias matings towards certain male phenotypes as a side-effect. Third, preferences can initially spread due to the direct effects they have on mate choice. Although the third possibility has received most of the attention in theoretical discussions of sexual selection, the first two possibilities may be biologically the most important.

Intrinsic biases

Mate preferences have their proximal bases in neurological, morphological, and physiological systems of females that bias how they perceive and react to males. Many such biases are inherent in those systems from the time they first appear in evolution. No sensory modality is without biases: eyes are most sensitive to certain colors, ears are tuned to certain frequencies. Once present, such biases can evolve in one direction or another if there is genetic variation that allows them to do so. Regardless of whether they do subsequently evolve, however, these intrinsic biases can cause females to favor matings with certain male phenotypes and so result in sexual selection. If biases that exist in the primitive evolutionary state of a female sensory system are the basis for a "preference," it would be fruitless to search for the selective forces responsible for its origin.

Side effects of other evolutionary processes

A second possibility is that a female preference arises in response to evolutionary forces unrelated to intraspecific mate discrimination. Selection on females to avoid hybridization with other species may be the most widespread case. Behaviors established to avoid hybridization can secondarily cause females to bias matings towards certain types of males within their own species. Ecological factors unrelated to mating can have similar effects. Imagine a bird that forages for red fruit in the understory of a rainforest. Mutations that give individuals greater visual sensitivity to red will spread by simple natural selection. A pleiotropic effect of these genes might then be to make red males more conspicuous to females than brown males and sexual selection favoring red males is thereby established. These new 'preference' genes are adaptive in the context of foraging but might be nonadaptive or even maladaptive in the context of mate choice.

Origin caused by effects of preferences on mating

A third possibility is that preferences initially spread due to the effects they have on mate choice within the species. Fisher (1958) discussed how a gene causing a female to prefer mating with males carrying an advantageous mutation could spread through a population in which females initially showed no mating biases. Other possibilities include selection for females to mate with males offering the best material resources. These forms of selection on preferences are discussed below in relation to the evolution of established mating preferences.

Cause vs. effect

It is critical to keep clear the distinction between the evolutionary forces acting on the preferences themselves and the evolutionary effects these preferences have on male traits. Selection on mating preferences can be studied directly by documenting variation in the preferences and then showing how this variation affects female fitness, or by using a comparative approach. It is not possible, however, to demonstrate selection on preferences by looking only at the action of preferences on male traits, as many workers have attempted (e.g. Borgia, 1985). For example, a study which shows that females tend to mate unparasitized males more often than highly parasitized males could document the effects of parasitism and sexual selection on male traits but would not demonstrate how selection acts on the mating preferences. If a female preference for a high male display rate is established in a species for any reason, it is likely that males with fewer parasites and a greater aerobic capacity will mate more often. The observation that females tend to mate with the less-diseased males would not, however, indicate that the preference arose or is maintained because of this effect.

The distinction between cause and effect is well illustrated by the hilltopping behavior of many insects. Males and females aggregate at topographic prominences in order to mate. Parker (1978) has described the positive feedback loop that may be responsible for the evolution of this behavior. It begins with the establishment of a tendency for either sex to clump spatially around some landmark, such as hilltops. Once such a convention is initiated for any reason, it selects for increased compliance by both sexes. Females are favored to go to hilltops to minimize the time, energy, and risk of locating a mate. Males aggregate there to maximize the chance of encountering and mating females.

Once established, hilltopping selects for a series of secondary adaptations. Males that are strong fliers are favored because they arrive earliest at the mating site. The female hilltopping behavior thus has the side effect of causing females to mate with the strongest fliers. In this scenario females are selected to aggregate because natural selection favors rapid mating not because the behavior causes them to mate with a particular type of male. The forces that gave rise to female hilltopping cannot be uncovered by examining the influence the behavior has on the mating success of different male phenotypes. Evolution of the preferences can only be studied by examining the preference themselves, as will now be discussed.

THE EVOLUTION OF ESTABLISHED MATING PREFERENCES

Preferences, like any other kind of character, evolve in response to a number of forces, including natural selection. The following is a short survey of some of the evolutionary forces acting on preferences, treating them in three major categories: (a) direct selection on preferences, (b) indirect selection (i.e., correlated response to selection on other characters), and (c) other evolutionary forces (such as group selection and genetic drift). The goal is to provide a general framework for discussion and investigation of forces acting on the preferences. Although the list of specific effects provided here no doubt is incomplete, the categories seem to be inclusive. By organizing our theroretical and empirical knowledge in this framework (or one like it), we can begin to understand under what conditions each of the different forces will be important.

In the following, I will discuss only polygynous mating systems in which females are free to choose a mate from among many males. While sexual selection is not restricted to polygynous mating systems (Darwin 1871; Fisher 1958, O'Donald 1983), polygyny is associated with the most extreme forms of sexual selection and therefore has been the focus of most of the research on sexual selection. As a point of departure I will assume that female preferences affect mating behavior but do not alter female survival or fecundity. This implies that females do not receive any material benefits, including parental care, from their mates (other than gametes).

If female mating preferences do not affect female survival or fecundity, they are evolutionarily neutral characters. The simplest models of sexual selection in fact assume there are no selective forces acting directly on female mating preferences. The neutral stability of the preferences consequently leads to lines or curves of equilibria. These curves relate the average strength of the preference among females in the population to that degree of development of the male character that balances the forces of natural and sexual selection. The neutrally stable curves of equilibria have been extensively discussed previously (Lande 1981; Kirkpatrick 1982; Seger 1985; reviewed in Arnold 1983, 1985).

The assumption that mating preferences have no effect on the survival or fecundity of females provides a null model for the joint evolution of preferences and traits. Even if there are no real populations that meet this assumption, the null model is useful because it shows how evolution would act in the absence of additional complications. Once we understand the evolutionary behavior of the null model we can construct more elaborate scenarios by adding appropriate modifications. The following sections review how various possible departures from this null condition impose evolutionary forces on mating preferences. It is important to keep in mind that these are not mutually exclusive alternatives — any real population probably experiences several of these forces simultaneously. Empirical studies of the evolution of mating preferences must therefore consider the relative importance of each.

The curves of equilibria, predicted from models that assume there are no evolutionary forces acting directly on the female preference, are a useful aid in situations where selection does act on the preference. These curves help to visualize how evolutionary forces affecting the preference determine the joint evolutionary equilibrium (or equilibria) for the preference and the male character. Typically, if one knows the preference equilibria from consideration of the evolutionary forces acting on them, the equilibria for the male character can be found by referring to the curve to determine the corresponding points. I will therefore be referring to the 'curve of equilibria' below in situations where selection on the preferences actually results in a single point rather than a curve of equilibria — by this I mean the curve that would result if the selection on the preference was removed.

Direct selection

Mating preferences, like other kinds of characters, can evolve under the force of natural selection. It is worth remembering, however, that preferences are not directly subject to sexual selection, which is targeted on male traits. This point is sometimes confused, as when Butlin et al. (1985) write of 'the possibility of stabilizing sexual selection on the mate recognition system.'

Within the context of mate choice an obvious selection pressure on female preferences is to avoid mating with other species. Since hybridization usually results in a complete loss of gametes, this is a powerful kind of selection, as emphasized by Mayr (1963), Dobzhansky (1970), and others. Reproductive character displacement is one possible outcome. This causes the mating preferences of two species that hybridize to diverge so that the frequency of hybridization is reduced. A secondary effect of the divergence of the mating preferences may then be the divergence of male sexual characters, which evolve in response to the new regime of sexual selection created by the new preferences. Although this interesting topic has been widely discussed, much of the empirical evidence is equivocal and the theory is incomplete. Perhaps the most convincing field study to date is Levin's (1985) work on plants of the genus *Phlox*. Intensive work has also been done on frogs (Littlejohn 1965; Fouquette 1975).

Mate choice within a species may or may not result in natural selection acting directly on female preferences. One way in which direct selection on preferences can arise is from the costs females incur while searching for mates. Females of some species may be able to exercise a certain preference without expending substantially more energy or incurring more risk than females that show some other preference. This may often be the case in species where males aggregate, and females can survey many males simultaneously. In such cases there are no search costs associated with the preference, and selection does not discriminate between females with different preferences.

In other species, however, selection does act on females as they exercise preference. Consider, for example, what happens if each female has a search image for the male phenotype she prefers to mate, and females search until they encounter an acceptable one. Females seeking rarer male phenotypes will then spend more time and energy searching for mates and so be at a selective disadvantage. In this case, the effect will be to cause the mean of the female search image to evolve towards the mean of the male character. This, in turn, will alter the form of sexual selection experienced by the males. The result of the coevolution of the trait and preference in this situation is examined later in the paper where I present a simple model. In that model, evolution can result either in both the trait and preference equilibrating at a point that maximizes male viability or at some other point, depending on the kind of selection that is acting.

Another kind of direct selection occurs when a female's immediate fecundity is influenced by the phenotype of her mate. Male parental care is a common form of material benefit that females receive from males that affects their fecundity. Other forms of material benefits include providing females with nutrition (such as in a nuptial offering or spermatophore) or a nesting territory or oviposition site (as seen in many birds, frogs, and insects). Differences in male fertility will also affect immediate female fecundity. Freeland (1976) has pointed out that females will be selected to avoid contagion from males that are diseased or parasitized. In each of these cases, the type of male with whom a female mates affects her fecundity and so natural selection acts directly on the preference. The evolutionary equilibrium for the preference is at the point that maximizes immediate female fecundity (Kirkpatrick 1985). If, however, the correlation between a female's preference phenotype and the quality of the material benefit she receives is low, the strength of selection acting on the preference will be weak.

When females base their preferences on heritable characters that affect male viability, the male offspring of some females will be more viable than others because of genes they inherited from their fathers. This does not impose a direct selective pressure on female mating preferences. Variation in offspring viability affects the evolution of mating preferences through indirect selection, as discussed below.

In summary, direct selection can act on female preferences through a variety of mechanisms. In general, it acts as a directional evolutionary force working to maximize female fitness by maximizing female survivorship and immediate fecundity. If there is a single mating preference that maximizes female fitness, this will be the unique evolutionary equilibrium for the preference. The equilibrium for the male character is then expected to be at the point on the curve of equilibria that corresponds to this degree of development of the preference. The male character will be at a compromise between what is favored by the female preferences and what is favored by natural selection.

Indirect selection

Indirect selection on the preferences occurs when some other character that is under natural or sexual selection is gentically correlated with the preference genes. Basic genetic theory (see Falconer 1981, chap. 9) tells us that the preference will then evolve as a 'correlated response,' a sort of evolutionary side effect. Indirect selection is the only evolutionary force affecting the preference in the simplest models of sexual selection (e.g., Lande 1981; Kirkpatrick 1982, 1986; Seger 1985). This is because sexual selection, which acts directly on male traits, affects female preferences only through the genetic correlation between the preferences and the male traits (see below).

Indirect selection causes evolution of the preferences only while the character correlated to the preference is evolving directionally. If the correlated character that is under direct selection reaches an equilibrium, indirect selection on the preference ceases. Because most characters are thought to be near equilibrium for most of their evolutionary history, the effects of indirect selection typically will be transient. Hamilton and Zuk's (1982) complex hypothesis for the cyclic coevolution of pathogens, host resistance, and sexual selection takes exception to this generalization, as it was specifically formulated to keep a population under continuous directional selection.

Genetic correlation is not itself an evolutionary force. Its presence or absence does not affect the existence or even the location of the neutral curve of equilibria. Genetic correlation does, however, have important evolutionary consequences for the preference because it transmits the effects of selection that are acting on other characters to the preference. When the preference is postively genetically correlated with a character, such as the male trait, selection acting to increase the male trait causes the preference to increase also. This is the basis of Fisher's (1958) hypothesis for the origin of preferences: a male trait gene that has net positive fitness will tend to increase and, as a secondary effect, cause an increase of female preferences for it. Fisher (1958), O'Donald (1980), and Heisler (1984) suppose that the initial advantage that the male trait gene enjoys is due to natural selection. Other theoretical work, however, has shown alternatively that a male trait gene can gain a net fitness advantage through sexual selection alone, even when it decreases male viability, if a preference gene becomes sufficiently common by some other mechanism (Kirkpatrick 1982). In this case, both the deletrious male trait gene and the female mating preference for it will become more common. This second possibility is ignored in many of the "good genes" discussions of sexual selection (e.g., Trivers 1972; Zahavi 1975; Borgia 1979, 1985; Thornhill and Alcock 1983).

There are two sources of genetic correlation: pleiotropy and linkage disequilibrium. Mating preference genes, like other kinds of genes (Dobzhansky 1970, pp. 60–64), can be expected to show pleiotropic effects. When genes affecting female mating preferences also affect other characters, and these other characters are under selection, the preference will evolve as a correlated response. While I have included this kind of selection under the heading "indirect selection," because the preferences per se are not the target of selection, the genes influencing preference are themselves under direct selection. This form of selection therefore can be a powerful force in the evolution of preferences.

An interesting example comes from the work of Wilczynski et al. (1984) on the tree frog *Hyla crucifer*. Females respond to vocalizing males when calls excite the basilar papilla of their auditory system. The basilar papilla is thought to act as a resonator whose tuning depends on the size of the individual: a larger body size tunes them to lower frequencies, a smaller body size to higher frequencies. Female auditory tuning is, in effect, a component of mating preference, since it determines the range of male call frequencies that the female can detect. Any ecological selection that causes an evolutionary change in the body size of the females will thus have the secondary effect of shifting the female auditory sensitivity and hence the mating preferences in the population.

The second source of genetic correlation is linkage disequilibrium, which can arise for a number of reasons. The source of linkage disequilibrium that has received the most attention in discussions of sexual selection is the nonrandom mating that female mating preferences cause. Females that prefer to mate with long-tailed males will mate with such males more often than females that prefer short-tailed males. Following mating and genetic recombination, the genes for long-tail preference and the genes for the long tail itself will become correlated: an individual carrying a gene for long tails also will tend to carry a gene for the corresponding preference (Fisher 1958; O'Donald 1980; Lande 1981; Kirkpatrick 1982).

Indirect selection via genetic correlation is the source of a number of interesting evolutionary phenomena, including the "runaway process" of sexual selection (Fisher 1958) in which the correlated response of the preference to the evolution of the male trait causes an unstable feedback that results in the preference and trait rapidly coevolving in a maladaptive way. While this interesting process has been the focus of many recent discussions (e.g., Lande 1981, 1982), it should be emphasized that it is only one of the many mechanisms by which preferences can evolve to extreme states.

Female preferences, then, can evolve in any direction in response to indirect selection. The strength of this force depends on the strength of selection on the correlated character and the magnitude of the genetic correlation. The ultimate equilibrium for the preference depends on its initial state, the strength and duration of the selection on the correlated character, and the magnitude of the genetic correlation. Thus indirect selection will move populations up or down the curve of equilibria but will not generally favor any particular point on the curve. Alternatively, a sufficiently strong genetic correlation can cause indirect selection to become so strong that the curve itself becomes evolutionarily unstable. In this case, populations will evolve rapidly away from the curve in an unpredictable and maladaptive runaway process (Fisher 1958; Lande 1981).

Other evolutionary forces

Individual selection is not the only source of evolutionary forces acting on mating preferences. Here I will briefly discuss two more: group selection and genetic drift. This list is not exhaustive — mutation pressure, for example, is another potential force — but the effects of other forces have not yet been found to have effects special to the context of sexual selection.

Group selection operates on differences between populations in their mean fitness. Sexual selection can provide the variation on which group selection acts if evolution of the preference and male character causes a change in the net reproductive output of a population (Lande 1981; Kirkpatrick 1982; Seger and Trivers 1986). Sexual selection may therefore be an important agency for creating the variation in mean population fitness on which group selection acts. Standard forms of group selection would cause preferences for the more viable male phenotypes to spread at the expense of preferences for less viable ones, even when selection within a population does not discriminate between preferences. This effect is the basis for the directional evolution of preferences in a model of sexual selection proposed by Seger and Trivers (1986). At a higher level, the analogous phenomenon of species selection could likewise favor species that persist longer because sexual selection has not resulted in radical reductions in their mean fitness. (Sexual selection, however, also can enhance speciation rates [Fisher 1958; Lande 1981, 1982; West-Eberhard 1983; Ryan 1986] thereby counteracting the effects of extinction.) It is an open empirical question as to whether group selection has had measurable impact on the evolution of female mating preferences. Note that group selection imposes a predictable directionality on the evolution of the preference: it causes preferences that increase mean population fitness to spread. Group selection therefore drives populations towards the point on the curve of equilibria that maximizes population fitness. Whether this is a strong or weak evolutionary force depends on the population structure.

Genetic drift is absent only from theoretical populations of infinite size — it is constantly at work on all heritable characters in all real populations. What makes genetic drift of particular interest in the context of sexual selection is that the genetic correlation between the preference and male trait can greatly magnify the effect of drift in causing random evolution in the preference (Lande 1981). Thus drift is expected to be more important in the evolution of female mating preferences than it is for many other kinds of characters. Drift, of course, causes the preferences to evolve in an unpredictable direction along the curve of equilibria.

Determining what forces are important in the evolution of mate preferences

It should be clear that there is substantial opportunity for further theoretical and empirical work. Additional models are needed to fill out a menu of the possible evolutionary effects that various ecological and behavioral mechanisms have on the preferences. Empirical studies of the forces acting on mating preferences are badly needed. These studies will have to: (a) identify variation in female mating behavior, and (b) quantify selection acting on that variation. The first of these tasks is by far the most difficult. O'Donald and Majerus's (1985) work on ladybirds provides a rare example of demonstrated genetic variation in preferences within a species but has not yet identified the male attribute used by females in their discrimination or shown how selection acts on preferences in a natural population. Once variation can be measured, direct experimental manipulation (e.g., Thornhill 1981; Gwynne 1984) and/or statistical inference (Lande and Arnold 1983) can be used to determine if and how selection is acting on the preferences.

Having framed the subject in the broadest terms, I will now investigate a particular form of direct selection acting on female mating preferences: the costs associated with searching for mates. This will provide an illustration of the way in which additional forces can be incorporated into models for the evolution of preferences, as well as indicating some interesting evolutionary consequences of seach costs.

A MODEL FOR THE EVOLUTION OF PREFERENCES WITH SEARCH COSTS

As discussed earlier, natural selection may favor female behaviors that minimize the time, energy, and risk involved in mating. Intuitition suggests that the outcome of this will be to maximize female fitness. If, for example, females have a search image of the male phenotype with which they would most prefer to mate, we expect that evolution of the preference would result in the female search image coming to coincide with the most common male phenotype in the population, thereby minimizing the time females spend searching. It is less obvious, however, what the outcome of evolution will be when the coevolution of the male character is considered. I therefore constructed a quantitative genetic model to study this question.

The model is of a polygynous species in which there is no male parental care or other direct contribution of males to their mates other than sperm. Females choose mates on the basis of a male secondary sexual character. In the interests of space, derivation of the model will be kept very brief here; details have been discussed elsewhere (Lande 1981; Kirkpatrick 1985).

The value of the trait for a particular male (the length of his tail, for example) is written z, and is normally distributed with mean $\bar{z}$ and variance σ^2. The male character is assumed to be under stabilizing natural selection, with a fitness function that is proportional to a Gaussian function with an optimum at $z = \theta_m$ and a variance of ω^2.

The preference phenotype of a female, written y, specifies the value of the male character with which the female would most prefer to mate. The strength of her preference for males deviating from this value falls off in proportion to a Gaussian function with variance ν^2. This mating preference function (or 'search image') is therefore described by the function:

$$\psi\,(z|y) = exp\,\{\,-\,(z\,-\,y)^2/2\upsilon^2\}. \qquad (1)$$

The preference phenotype y is normally distributed among females with mean $\bar{y}$ and variance τ^2. Up to this point, the model is identical to the 'absolute preference' model of sexual selection developed by Lande (1981).

To incorporate a cost for females searching, I simply assume that the fitness of a female is proportional to the degree to which the distribution of available males overlaps with her preference function (or 'search image'). That is,

$$W(y) \propto \int \psi\,(z|y)\,p^*\,(z)\,dz, \qquad (2)$$

where the $p^*\,(z)$ is the distribution of the male character after natural selection has acted. The mechanism that imposes the search cost could involve decreased fecundity caused by wasted time and energy, or decreased female survival caused by predation or other risk. This model for the cost assumes that there is a linear relationship between the time spent searching and the fitness of a female. This may not be the case for many species in which females can survey several males at negligible cost but take on proportionally greater costs as the number of males they assess increases. While I have not modeled this more realistic situation, I believe it produces the same evolutionary equilibrium as the model under discussion. A second possible objection to this formulation is that females with no preference at all ($\nu^2 = \infty$) always have higher fitness than females with some preference. This model will assume that truly indiscriminant females are not present, because sensory constraints or selective forces (e.g., avoidance of hybridization) cause some preference in all females.

A much more important assumption implicit in Eq. (2) is that female fitness depends only on the extent of overlap between her preference function and the distribution of available males.

This assumption might be appropriate, for example, for a female frog trying to locate calling males in an acoustically simple environment. If, however, some types of calls attenuate more rapidly or are more difficult to localize than others, there will be an additional fitness function imposed on females favoring those that prefer the most apparent or conspicuous male phenotypes. This possibility is not included in the present model but its possible ramifications will be discussed below.

The fitness function of Eq. (2) can be used to calculate the selection differential for the female mating preference:

$$s_y \approx \frac{(\bar{z}^* - \bar{y})\ \tau^2}{\nu^2 + \tau^2} \tag{3}$$

where $\bar{z}^* \approx \bar{z} + (\theta_m - \bar{z})\ \sigma^2 / \omega^2$ is the mean of the male trait following natural selection. (In this and following equations the algebra has been simplified by assuming that the strength of stabilizing selection acting on the male character is weak ($\omega^2 >> \sigma^2$) and that the accuracy of the female preference is poor ($\nu^2 >> \sigma^2$). These assumptions do not alter the qualitative conclusions.)

Natural and sexual selection produce a selection differential in the male trait of:

$$s_z \approx \bar{z}^* - \bar{z} + (\bar{y}^* - \bar{z}^*)\, \sigma^2 / \nu^2 \tag{4}$$

where $\bar{y}^* = \bar{y} + s_y$ is the mean of the female preference following selection acting through the searching costs.

These selection differentials are then used to calculate the per-generation rate of evolutionary change in the mean of the male trait and female preference using the equations:

$$\Delta\bar{z} = \frac{1}{2}\left(\frac{g_z}{\sigma^2}\ s_z + \frac{g_{zy}}{\tau^2}\ s_y\right) \tag{5}$$

$$\Delta\bar{y} = \frac{1}{2}\left(\frac{g_y}{\tau^2}\, s_y + \frac{g_{zy}}{\sigma^2}\, s_z\right) \tag{6}$$

where g_z is the additive genetic variance for the male trait, g_y is the additive genetic variance for the female preference, and g_{zy} is the additive genetic covariance between the trait and preference (Kirkpatrick 1985).

The evolutionary equilibrium for this model can be determined by setting Eqns. (5) and (6) to zero and solving for the values of $\bar{z}$ and $\bar{y}$. Doing so shows that there is a single equilibrium point: that at which the means of the male trait and the female mating preference both coincide with θ_m, the optimum for male survival. This result can be visualized intuitively as follows.

In each generation, search costs act to drive the female's search images to coincide with the mean of the male character at $\bar{z}$. This follows from the assumption that females whose search image deviates in either direction from the mean of the males spend excess time searching for a mate. Simultaneously, the male character is driven by the combined forces of natural and sexual selection to a compromise intermediate between the survival optimum at θ_m and the sexual selection optimum at $\bar{y}$. This coevolutionary process—the mean female preference evolving towards the mean of the male character, while the mean male character evolves towards a point between the mean of the females and the survival optimum—ultimately ends when both the preference and male character lie at θ_m.

This result depends critically on the assumption that a female's fitness depends only on the overlap between her preference function and the distribution of males and is completely independent of its actual value. The physical properites of the environment and the biophysics of sensory systems may often invalidate this assumption. For example, a male frog calling at 3 kH may be more apparent to females tuned to 3 kH than is a male calling at 4 kH to females tuned to 4 kH, simply because 3 kH calls propagate better through that environment. The model described above can be extended to these situations by including an additional fitness function that reflects the absolute apparency of males as a function of their phenotype z. Results show that two types of outcomes are possible. One is an equilibrium in which the preference and trait differ from the survival optimum at θ_m and from each other. At this equilibrium, female mating cost is minimized. In the alternative outcome there is no equilibrium, and the preference and trait evolve in an unstable process away from θ_m until gentic variation is exhausted. The general conclusion is that search costs can be very important in the evolution of the preference and trait. The exact evolutionary outcome, however, depends on details of how the costs affect females.

I have not answered two basic questions concerning this model. First, I have only studied a single type of mating preference function. To develop an analogous model for open-ended preferences, for example, requires assumptions about how the steepness of the preference function affects search costs. What specific assumptions are plausible is less obvious for this case than for the absolute preferences studied above, since Eq. (2) seems inappropriate. Second, I have not determined the stability properties of the equilibrium. It seems likely, however, that a sufficiently high genetic covariance between the preference and trait could cause an equilibrium point to become unstable, in which case both characters would rapidly evolve away from the equilibrium in a maladaptive process that would cause the female fitness and total reproductive output of the population (mean population fitness) to decline.

CONCLUSION

The greatest challenge now confronting evolutionary biologists interested in sexual selection is to understand the evolution of female mating preferences. Recent theoretical developments have illuminated the evolutionary consequences of a number of selection scenarios, but it is clear that much more work remains to be done. Empirical studies, from both lab and field, have not yet even begun to quantify the various evolutionary forces acting on preferences. Doing so will be difficult, requiring the clever choice of experimental systems in which the relevant measurements and manipulations can be made. Progress in our understanding depends on success in this venture.

Acknowledgements

I am very grateful to T. Price and M.J. Ryan for numerous talks that provided ideas for this paper. I thank D.K. Hews, R. Lande, P. O'Donald, S. Sakaluk, B.K. Sullivan, R. Thornhill, and the participants of the Dahlem Conference for discussions.

REFERENCES

Arnold, S.J. 1983. Sexual selection: the interface of theory and empiricism. In: Mate Choice, ed. P. Bateson, pp 67–108. Cambridge: Cambridge Univ. Press.

Arnold, S.J. 1985. Quantitative genetic models of sexual selection. Experientia 41: 1296–1310.

Borgia, G. 1979. Sexual selection and the evolution of mating systems. In: Sexual Selection and Reproductive Competition in Insects, eds. M.S. Blum and N.A. Blum, pp. 19–80. New York: Academic Press.

Borgia, G. 1985. Bower quality, number of decorations, and mating success of male satin bower birds (*Ptilonorhynchus violaceus*): an experimental study. Anim. Behav. 33: 266–271.

Butlin, R.K.; Hewitt, G.M.; and Webb, S.F. 1985. Sexual selection for intermediate optimum in *Chorthippus brunneus* (Orthoptera, Acrididae). Anim. Behav. 33: 1281–1292.

Darwin, C. 1859. On the Origin of Species. London: John Murray.

Darwin, C. 1871. The Descent of Man, and Selection in Relation to Sex. London: J. Murray.

Dobzhansky, T. 1970. Genetics of the Evolutionary Process. New York: Columbia Univ. Press.

Falconer, D.S. 1981. Introduction to Quantitative Genetics, 2nd ed. London: Longman.

Fisher, R.A. 1958. The Genetical Theory of Natural Selection, 2nd ed. New York: Dover.

Fouquette, M.J. 1975. Speciation in chorus frogs. I. Reproductive character displacement in the *Pseudacris nigrita* complex. Syst. Zool. 24: 16–23.

Freeland, W.J. 1976. Pathogens and the evolution of primate sociality. Biotropica 8: 12–24.

Gwynne, D.T. 1984. Courtship feeding increases female reproductive success in bushcrickets. Nature 307: 361–363.

Hamilton, W.D., and Zuk, M. 1982. Heritable true fitness and bright birds: a role for parasites? Science 218: 384–387.

Heisler, I.L. 1984. A quantitative genetic model for the origin of mating preferences. Evolution 38: 1283–1295.

Kirkpatrick, M. 1982. Sexual selection and the evolution of female choice. Evolution 36: 1–12.

Kirkpatrick, M. 1985. Evolution of female choice and male parental investment in polygynous species: the demise of the "sexy son". Am. Natural. 125: 788–810.

Kirkpatrick, M. 1986. The handicap mechanism of sexual selection does not work. Am. Natural. 127: 222–240.

Lande, R. 1981. Models of speciation by sexual selection on polygenic traits. Proc. Natl. Acad. Sci. USA 78: 3721–3725.

Lande, R. 1982. Rapid origin of sexual isolation and character divergence in a cline. Evolution 36: 213–223.

Lande, R., and Arnold, S.J. 1983. The measurement of selection on correlated characters. Evolution 37: 1210–1226.

Levin, D.A. 1985. Reproductive character displacement in *Phlox*. Evolution 39: 1275–1281.

Littlejohn, M.J. 1965. Premating isolation in the *Hyla ewingi* complex (Anura: Hylidae). Evolution 19: 234–243.

Mayr, E. 1963. Animal Species and Evolution. Cambridge, MA: Belknap.

O'Donald, P. 1980. Genetic Models of Sexual Selection. Cambridge: Cambridge Univ. Press.

O'Donald, P. 1983. The Arctic Skua: A Study of the Ecology and Evolution of a Seabird. Cambridge: Cambridge Univ. Press.

O'Donald, P., and Majerus, M.E.N. 1985. Sexual selection and the evolution of preferential mating in ladybirds. I. Selection for high and low lines of female preference. Heredity 55: 401–412.

Parker, G.A. 1978. Evolution of competitive mate searching. Ann. R. Entom. 23: 173–196.

Ryan, M.J. 1986. Neuroanatomy influences speciation rates among anurans. Proc. Natl. Acad. Sci. USA 83: 1379–1382.

Seger, J. 1985. Unifying genetic models for the evolution of female choice. Evolution 39: 1185–1193.

Seger, J., and Trivers, R. 1986. Assymmetry in the evolution of female mating preferences. Nature 319: 771–773.

Thornhill, R. 1981. *Panorpa* (Mecoptera: Panorpidae) scorpionflies: systems for understanding resource-defense polygyny and alternative male reproductive efforts. Ann. Rev. Ecol. Syst. 12: 355–386.

Thornhill, R., and Alcock, J. 1983. The Evolution of Insect Mating Systems. Cambridge, MA: Harvard Univ. Press.

Trivers, R. 1972. Parental investment and sexual selection. In: Sexual Selection and the Descent of Man 1871–1971, ed. B. Campbell, pp. 136–179. Chicago: Aldine.

West-Eberhard, M.J. 1983. Sexual selection, social competition, and speciation. Q. Rev. Biol. 58: 155–183.

Wilczynski, W.; Zakon, H.H.; and Brenowitz, E.A. 1984. Acoustic communication in spring peepers: call characteristics and neurophysiological aspects. J. Comp. Phys. A 155: 577–584.

Zahavi, A. 1975. Mate selection — a selection for a handicap. J. Theor. Bio. 53: 205–214.

Sexual Selection: Testing the Alternatives
eds. J.W. Bradbury and M.B. Andersson, pp. 83–94
John Wiley & Sons Limited

Genetic Correlations between the Sexes in the Evolution of Sexual Dimorphism and Mating Preferences

R. Lande

Department of Biology
University of Chicago
Chicago, IL 60637, USA

Abstract. Genetic correlations between the sexes constrain the evolution of sexual dimorphism and promote the joint evolution of mating preferences and secondary sexual characters in ways that may be nonadaptive or maladaptive. Quantitative genetic models have elucidated their role in the evolutionary dynamics of sexually selected traits and help to explain two of Darwin's major generalizations from comparative studies: a) in species where males have exaggerated characters, females often express homologous traits in a rudimentary form that appears useless or deleterious, and b) closely related species of higher animals often differ most in male secondary sexual characters.

INTRODUCTION

Salient features of sexual selection and the evolution of sexual dimorphism can be understood only through the study of genetic mechanisms. Of special significance in the evolutionary dynamics of sexual dimorphism and mating preferences are genetic correlations between the sexes, that is, the extent to which male and female characters are influenced by the same genes or segregating factors. Genetic correlations between the sexes, due to pleiotropy (manifold effects of particular genes) and/or linkage disequilibrium (nonrandom association between alleles at loci influencing different characters), imply that male and female characters can not evolve independently; natural or sexual selection on one sex produces not only a direct evolutionary response in the characters of that sex but also an indirect or correlated response in characters of the opposite sex. The results of this process are manifest in the morphology and behavior of sexually selected species.

The present paper reviews genetic mechanisms that help to explain two of the major generalizations made by Darwin (1874). The first is that in many species in which males have evolved exaggerated secondary sexual characters serving to increase success in inter-male competition or the acquisition of mates, homologous characters are expressed in rudimentary form in females without any apparent advantage to them. Darwin termed this phenomenon the 'transference' of the character betwen the sexes. The second generalization is that in many taxa of arthropods and vertebrates closely related species differ most in secondary sexual characters of males, which are therefore among the most raidly evolving and taxonomically important morphological traits in higher organisms.

Genetic mechanisms underlying the evolution of sexual dimorphism and mating preferences were first outlined by Fisher (1930, 1958, chap. 6), who also clearly described the connection between sexual selection and speciation by divergent evolution of mating preferences between populations. Fisher filled a significant gap in Darwin's theory of sexual selection by proposing genetic mechanisms for the evolution of female mating preferences for exaggerated male traits in polygamous species where males invest nothing but gametes in their offspring. Until recently, scant attention has been devoted to detailed anlaysis of genetic mechanisms involved in the process of sexual selection described by Darwin and Fisher. Here I summarize some of the current theory of sexual selection on quantative characters and review a number of errors and misunderstandings that have appeared in the recent literature; for more extensive discussion see Arnold (1985).

SEXUAL DIMORPHISM

Homologous characters of the sexes, that have similar expression in males and females in a population (e.g., body size) usually show a very high genetic correlation between the sexes due primarily to pleiotropy: similar sets of genes are expressed in a similar fashion in both sexes (Lande 1980). This places a strong constraint on the joint evolution of male and female characters. For simplicity, consider the joint evolution of one quantitative (polygenic) character in each sex, under stabilizing natural selection toward an intermediate optimum phenotype that may differ between males and females. If on some scale of measurement (e.g., logarithmic) homologous male and female characters are equally variable and equally heritable, and if they are subject to different selective forces, as in polygamous species where sexual selection acts only on males, the evolutionary dynamics of the mean phenotypes toward an equilibrium occurs in two phases. Initially there is rapid, almost parallel evolution of male and female characters until the net selective forces on the two sexes become nearly equal in magnitude but opposite in sign. There follows a slow evolution of sexual dimorphism in which male and female traits change in opposite directions, as described by Fisher (1958). The timescales for these two phases are proportional respectively to $1/(1 + \gamma)$ and $1/(1 - \gamma)$, where γ is the

additive genetic correlation betw
to 1.0, the evolution of sexual dir
and characters that have been sel
be 'transferred,' at least temporar
are useless or even deletrious for
Leutenegger and Cheverud (19
also be produced by sex differences
when the force of directional sel
asymmetry in genetic or phenotyp
dimorphism to proceed on a timesca
those described above, so that when
high, the evolution of sexual dim
followed by a slow phase.
When the population finally achie
time, the mean female phenotype is
whereas the mean male phenotype i
opposing forces of natural selection fo
success. Darwin emphasized that the
sexual selection, and genetic models o
mean fitness of males can be greatest
weak natural selection toward an inte
is a complete genetic correlation betwe
of females can also be permanently di
(Lande 1980).
The inheritance of sexual dimorphis
mainly on autosomal genes with differe
The amount of sexual dimorphism in
sex-linked genes is approximately propor
some(s) (the X-chromosome), which gen
genome (Bird and Schaffer 1972; Coyne
species with heteromorphic sex chromos
the heterogametic sex (the Y-chromosome
the W-chromosome in female birds and lep
has little or no effect on morphological o
For polygenic characters with extreme se
with sex limited effects have evolved, the ge
should be expected to be quite low (γ n
sex-limited, its rudimentary expression in
neutral, and if, as is common with vestigial ch
a low (additive genetic) correlation with oth
geologically long time. As with genetic cor
pressed in the same individual, genetic cor
be measured from artificial selection exper

resemblance bet
Lande 1980).

MATING PREF

In contrast to i
advantage of ma
species (with pr
fertilizing their
extreme charac
Fisher suggest
increase in freq
selection, but
comes a select
character beyo
natural selecti
additive genet
character. Eve
nervous mech
male morpho
genetic correl
mating genera
genetic varian
due to assort
between the
correlation b
their mates, i
choose mates
to the averag
contribute to
a male chara
Once esta
self-reinforc
male chara
genetic corr
female mat
process thu
male chara
exponentia
against ext
their diffic
that, in spe
mates, the

Sexual Selection: Testing the Alternatives
eds. J.W. Bradbury and M.B. Andersson, pp. 83–94
John Wiley & Sons Limited

Genetic Correlations between the Sexes in the Evolution of Sexual Dimorphism and Mating Preferences

R. Lande

Department of Biology
University of Chicago
Chicago, IL 60637, USA

Abstract. Genetic correlations between the sexes constrain the evolution of sexual dimorphism and promote the joint evolution of mating preferences and secondary sexual characters in ways that may be nonadaptive or maladaptive. Quantitative genetic models have elucidated their role in the evolutionary dynamics of sexually selected traits and help to explain two of Darwin's major generalizations from comparative studies: a) in species where males have exaggerated characters, females often express homologous traits in a rudimentary form that appears useless or deleterious, and b) closely related species of higher animals often differ most in male secondary sexual characters.

INTRODUCTION

Salient features of sexual selection and the evolution of sexual dimorphism can be understood only through the study of genetic mechanisms. Of special significance in the evolutionary dynamics of sexual dimorphism and mating preferences are genetic correlations between the sexes, that is, the extent to which male and female characters are influenced by the same genes or segregating factors. Genetic correlations between the sexes, due to pleiotropy (manifold effects of particular genes) and/or linkage disequilibrium (nonrandom association between alleles at loci influencing different characters), imply that male and female characters can not evolve independently; natural or sexual selection on one sex produces not only a direct evolutionary response in the characters of that sex but also an indirect or correlated response in characters of the opposite sex. The results of this process are manifest in the morphology and behavior of sexually selected species.

The present paper reviews genetic mechanisms that help to explain two of the major generalizations made by Darwin (1874). The first is that in many species in which males have evolved exaggerated secondary sexual characters serving to increase success in inter-male competition or the acquisition of mates, homologous characters are expressed in rudimentary form in females without any apparent advantage to them. Darwin termed this phenomenon the 'transference' of the character betwen the sexes. The second generalization is that in many taxa of arthropods and vertebrates closely related species differ most in secondary sexual characters of males, which are therefore among the most raidly evolving and taxonomically important morphological traits in higher organisms.

Genetic mechanisms underlying the evolution of sexual dimorphism and mating preferences were first outlined by Fisher (1930, 1958, chap. 6), who also clearly described the connection between sexual selection and speciation by divergent evolution of mating preferences between populations. Fisher filled a significant gap in Darwin's theory of sexual selection by proposing genetic mechanisms for the evolution of female mating preferences for exaggerated male traits in polygamous species where males invest nothing but gametes in their offspring. Until recently, scant attention has been devoted to detailed anlaysis of genetic mechanisms involved in the process of sexual selection described by Darwin and Fisher. Here I summarize some of the current theory of sexual selection on quantative characters and review a number of errors and misunderstandings that have appeared in the recent literature; for more extensive discussion see Arnold (1985).

SEXUAL DIMORPHISM

Homologous characters of the sexes, that have similar expression in males and females in a population (e.g., body size) usually show a very high genetic correlation between the sexes due primarily to pleiotropy: similar sets of genes are expressed in a similar fashion in both sexes (Lande 1980). This places a strong constraint on the joint evolution of male and female characters. For simplicity, consider the joint evolution of one quantitative (polygenic) character in each sex, under stabilizing natural selection toward an intermediate optimum phenotype that may differ between males and females. If on some scale of measurement (e.g., logarithmic) homologous male and female characters are equally variable and equally heritable, and if they are subject to different selective forces, as in polygamous species where sexual selection acts only on males, the evolutionary dynamics of the mean phenotypes toward an equilibrium occurs in two phases. Initially there is rapid, almost parallel evolution of male and female characters until the net selective forces on the two sexes become nearly equal in magnitude but opposite in sign. There follows a slow evolution of sexual dimorphism in which male and female traits change in opposite directions, as described by Fisher (1958). The timescales for these two phases are proportional respectively to $1/(1 + \gamma)$ and $1/(1 - \gamma)$, where γ is the additive genetic correlation between the sexes (Lande 1980). Thus if γ is close to 1.0, the evolution of sexual dimorphism may be an exceedingly slow process, and characters that have been selected to an extreme expression in males can be 'transferred,' at least temporarily, to a lesser degree in females in which they are useless or even deletrious for survival, as suggested by Darwin.

Leutenegger and Cheverud (1985) have noted that sexual dimorphism can also be produced by sex differences in heritability or phenotypic variance, even when the force of directional selection is the same on both sexes. Sexual asymmetry in genetic or phenotypic variance causes the evolution of sexual dimorphism to proceed on a timescale that is (roughly) a linear combination of those described above, so that when the genetic correlaion between the sexes is high, the evolution of sexual dimorphism itself consists of a rapid phase followed by a slow phase.

When the population finally achieves equilibrium, after an indefinitely long time, the mean female phenotype is at its optimum under natural selection, whereas the mean male phenotype is determined by a balance between the opposing forces of natural selection for viability and sexual selection for mating success. Darwin emphasized that the viability of males may be diminished by sexual selection, and genetic models of the process reveal that the decrease in mean fitness of males can be greatest for characters that are initially under weak natural selection toward an intermediate optimum phenotype. If there is a complete genetic correlation between the sexes ($\gamma = 1.0$) the mean fitness of females can also be permanently diminished by sexual selection on males (Lande 1980).

The inheritance of sexual dimorphism in quantitative characters is based mainly on autosomal genes with different expression in males and females. The amount of sexual dimorphism in polygenic characters attributable to sex-linked genes is approximately proportional to the length of the sex chromosome(s) (the X-chromosome), which generally is a small fraction of the entire genome (Bird and Schaffer 1972; Coyne, 1983; Carson and Lande 1984). In species with heteromorphic sex chromosomes, the chromosome confined to the heterogametic sex (the Y-chromosome in male mammals and most insects; the W-chromosome in female birds and lepidoptera) usually is degenerate and has little or no effect on morphological or behavioral traits (Bull 1983).

For polygenic characters with extreme sexual dimorphism, after many genes with sex limited effects have evolved, the genetic correlation between the sexes should be expected to be quite low (γ near zero). When a trait is nearly sex-limited, its rudimentary expression in one sex may be almost selectively neutral, and if, as is common with vestigial characters, the rudimentary trait has a low (additive genetic) correlation with other characters it could persist for a geologically long time. As with genetic correlations between characters expressed in the same individual, genetic correlations between the sexes can be measured from artificial selection experiments or from the phenotypic resemblance between relatives of opposite sex (Falconer 1981; and refs. in Lande 1980).

MATING PREFERENCE

In contrast to inter-male combat in which the winners achieve an obvious advantage of mating success, Darwin was unable to explain why in polygamous species (with promiscuous males that expend little or no parental effort beyond fertilizing their mates) females should prefer to mate with males bearing extreme characters that are seemingly useless or detrimental for their survival. Fisher suggested that genes for female mating preference could initially increase in frequency by their association with a male character under natural selection, but that once established in a population, mating preference becomes a selective force in its own right which can further elaborate the male character beyond the point where it has ceased to have an advantage under natural selection. In modern terms, the critical feature of this process is the additive genetic correlation between female mating preference and the male character. Even when female mating preferences are mediated by sensory and nervous mechanisms influenced by a set of genes distinct from those affecting a male morphological or behavioral trait subject to mate choice, a positive genetic correlation between them will be created and maintained by assortative mating generated by genetic variance in mating preference, as long as additive genetic variance in both characters persists. In this case, linkage disequilibrium due to assortative mating is responsible for the additive genetic correlation between the sexes. Here assortative mating means that there is a positive correlation between the mating preferences of females and the phenotype of their mates, i.e., females with extreme preferences (in either direction) tend to choose mates with extreme characters (in the corresponding direction) relative to the average mate chosen. Of course, in some situations pleiotropy may also contribute to the genetic correlation between female mating preference and a male character.

Once established in a population, female mating preference can become self-reinforcing: insofar as mating preference causes directional change in a male character it also indirectly augments its own expression through the genetic correlation between the sexes, because of the correlated response in female mating preference to selection directly on the male character. The process thus has the potential to produce runaway evolution in which both the male character and female mating preference evolve at geometrically or exponentially increasing rates until they are halted by severe natural selection against extreme males, or against females with extreme preferences due to their difficulty in finding a mate (Fisher 1930, 1958). Fisher (1915) also noted that, in species where females have limited time or ability to judge prospective mates, the expression of intense mating preference for one trait may limit the opportunity for sexual selection on other traits. The evolution of new secondary sexual characters and associated mating preferences may therefore contribute to the decline of old ones.

Quantitative genetic models, in which additive genetic variability in a male character and in female mating preference is continually replenished by mutation, have confirmed Fisher's description of the evolution of mating preference (Lande 1981). The feasibility of this mechanism has also been confirmed by simplier two-locus models with two or three alleles per locus (O'Donald 1980; Kirkpatrick 1982), although in those models the genetic system greatly restricts the possible range of evolution of the characters.

In polygamous species where females have many prospective mates to choose among, and every female is inseminated, sexual preference is a selectively neutral character, assuming that the numbers of offspring produced by females are independent of their mate choices. (The common practice of assigning the genetic quality of the offspring as part of the mother's fitness, rather than that of the offspring themselves, confounds selection with heredity and has led many workers to the erroneous conclusions that mate choice must benefit females). As a selectively neutral trait, female mating preference evolves deterministically only as a result of natural and sexual selection on the male character and the additive genetic correlation between the sexes. Despite stabilizing natural selection toward an intermediate value of the male character, there is then a line of possible equilibria for the mean phenotypes of the two characters. No matter how maladaptive the male character is with respect to survival (short of lethality for all males), there is an intensity of sexual selection by females that will exactly counteract and balance the natural selection, and this represents an equilibrium for the population because there is no selection acting directly on mating preference.

The existence of a line of equilibria implies that evolution is indeterminate because displacements of the population along the line will cause no further evolution. The line of equilibria as a whole may be stable or unstable depending on the genetic parameters of the population and the form and intensity of natural and sexual selection. Instability, corresponding to Fisher's runaway process, is promoted by weak natural selection on the male character and abundant genetic variance (with strong expression of choosiness) in female mating preferences. The possibility of runaway sexual selection (toward increased or decreased values of the characters) occurs regardless of whether the mating preferences of individual females are unimodal functions of male phenotypes with the most preferred males determined either absolutely (e.g., inborn preference) or relative to the population mean, or monotonic (open-ended) functions. Even when the line of equilibria is stable, rapid diversification between geographically isolated populations can occur by random genetic drift interacting with natural and sexual selection, because populations perturbed from the line of equilibria generally will not return to the same point (Lande 1981).

Rapid indeterminate evolution of male secondary sexual characters and female mating preferences, which may function as reproductive isolating barriers between populations, helps to explain Darwin's observation that closely related species of higher animals often differ in male secondary sexual characters. Indeed, the process of sexual selection via evolution of mating preferences and sexual dimorphism may be a major factor promoting rapid speciation in many taxa, as suggested by Fisher (1930, 1958), Carson (1978), and others. Quantitative genetic models have also confirmed Fisher's view that sexual selection can promote speciation along a cline, across a broad geographic region, in the absence of geographic isolation (Lande 1982).

Many of the above conclusions regarding sexual selection in polygamous species are rather robust since they can be derived from diverse assumptions about mechanisms of inheritance and systems of mating. In particular, lines of equilibria and indeterminate evolution of male traits and female mating preferences appear in simple two-locus models with different rules of mate choice (Taylor and Williams 1981; Kirkpatrick 1982; Seger 1985), as well as in three-character quantitative genetic models, with an additional male or female character under natural selection (Lande and Arnold 1985; Heisler 1985; Kirkpatrick 1986b).

The potential for maladaptive evolution still exists when natural selection acts directly on female mating preference by increasing the number of progeny produced, through variation in male parental effort, mate guarding or provisioning, or when there is a cost to female mate choice through delayed mating or increased risk of predation. In this situation there is an equilibrium point where the mean fitness of females is maximized but that of males is not, as suggested by Lande (1981) and demonstrated in detailed models by Kirkpatrick (1985). Even then there is the possibility of evolutionary instability of the equilibrium, and runaway sexual selection, depending on the pattern of selection and genetic parameters of the population.

The demonstration of direct selection on some aspects of mate choice does not preclude the existence of lines (or hyperplanes) of equilibria involving other aspects of mating preference. Direct costs to female mate choice (or the opportunity for choice) versus no choice or indiscriminate mating, may be offset by direct benefits such as avoiding interspecific hybridization, maximizing resources provided by mates, or seeking male aggregations to ensure insemination. Nevertheless, the particular choice among males with different secondary sexual characters may be selectively neutral. Adaptive and maladaptive female mating preferences may therefore operate simultaneously on different male characters within a species.

Although a great potential for maladaptive and indeterminate evolution exists in sexual selection, especially with respect to male secondary sexual characters, female mate choice may often be adaptive (if not optimal) in the sense of acting in the same direction as natural selection on a character. This is necessarily the case during the origin of mating preferences by genetic correlation with a male character under natural selection, before its optimum expression has been overshot by the self-reinforcing evolution of mating preference (Fisher 1958; Heisler 1984, 1985). If females can assess the overall viability of males by the vigor of courtship behavior (as suggested by Maynard Smith, this volume), adaptive mate choice may become established for high-viability males, with heritable variation in viability maintained by migration and/or deleterious mutation (Charlesworth, this volume); overshoot of an optimum is then impossible, but self-reinforcing mating preferences for certain low-viability traits might still arise within this context. Maladaptive evolution by individual selection could be counteracted to some extent by intergroup selection favouring populations with higher mean fitness.

Empirical tests of whether sexual selection and natural selection (on male characters or on an average of homologous male and female characters) reinforce or oppose each other would require the application of methods for measuring phenotypic selection, ideally in a multivariate context (Lande and Arnold 1983). Tests of 'goods genes' hypotheses of mate choice require demonstration that the sexual attractiveness of males is correlated with their viability, or that of their progeny (assuming some heritable component to variation in viability) (Partridge 1983; Boake 1985; Heisler et al., this volume).

Experimental estimation of an additive genetic correlation between female mating preference and the male characters(s) subject to mate choice may be extremely difficult, for several reasons. Substantial effort may be necessary to separate the effects of inter-male competition from female mate choice and to identify the male character(s) mediating mate choice. Problems are also likely to arise in the analysis of behavioral characters as complex as mating preference, which may depend on past experience and present environment and which can be studied only through the actual outcomes of mate choice. Finally, if the genetic correlation between female mating preference and the male character(s) is caused by linkage disequilibrium, this may be broken down by recombination if the opportunity for the expression of female mating preferences in laboratory stock populations is reduced in comparison to that in natural populations (reducing the intensity of assortive mating), unless the genes influencing the two traits are very tightly linked.

ERRORS AND OVERSIGHTS IN THE LITERATURE

Many evolutionists have espoused the philosophy that mate choice must generally be adaptive in reinforcing natural selection (e.g., Mayr 1972; Trivers 1972; Borgia 1979; Thornhill and Alcock 1983; Hamilton 1983). This view is not supported by any very compelling logical arguments or genetic models, and it disagrees with Darwin's original conception of sexual selection. Maynard Smith (1982, pp. 131–137) has pointed out that strategic reasoning, such as embodied in evolutionary game theory, is inadequate to describe the joint evolution of mating preferences and secondary sexual characters, for which explicit genetic models are necessary.

Mayr (1972) believed that the rudimentary expression in females of characters developed to an extreme degree in males, such as the faint coloration of some female birds, function as species recognition signals and prevent interspecific matings that might produce hybrid offspring with reduced viability or fertility. This argument ignores the observation that in polygamous species it is generally the females that discriminate in mate choice, whereas males are notoriously indiscriminate and will often court females of other species.

The hypothesis of Hamilton and Zuk (1982), that exaggerated secondary sexual characters of males (i.e, the brightly colored plumage and patches of skin in many birds) are adaptive in allowing females to assess the genetic susceptibility of their mates to parasites, fails to account for Darwin's observation that closely related species often differ most in male secondary sexual characters. If parasites were the primary reasons for the evolution of secondary sexual characters, it would seem sufficient for a group of closely related species to share one or a few types of such characters, e.g., for all male birds to display feathers or skin patches of a single color. Three-locus models of Hamilton's mechanism for polygamous populations demonstrate that the mate choice trait can not become established unless it is initially above a threshold frequency, and that the results are not necessarily adaptive (Kirkpatrick 1986a). This and other types of 'conditional' handicap mechanisms of sexual selection and evolution of mating preferences (reviewed by Maynard Smith 1985, this volume; Andersson 1986) by themselves have difficulty in explaining the origin of sexual dimorphism (see Heisler et al., this volume).

Some authors have confused sex linkage of genes with sex-limited, or sexually dimorphic, expression of characters (e.g., Trivers 1972; Bonner and May 1981). This neglects the phenomenon of dosage compensation in species with heteromorphic sex chromosomes, for which the expression of sex-linked genes in the heterogametic sex usually is adjusted to nearly equal that in the homogametic sex (Bull 1983, pp. 229–234). It also overlooks the fact that the proximal cause of sexual dimorphism in quantitative characters is, mainly, autosomal genes with different expression in males and females (refs. above).

There have been numerous misunderstanding's concerning the nature of genetic correlations between the sexes and their role in the evolution of mating preferences and sexual dimorphism. Often, such genetic correlations are simply ignored (equivalent to assuming zero genetic correlation) or the possibility that sexual dimorphism can evolve is overlooked (amounting to an assumption of perfect genetic correlation). Errors and misunderstandings abound even among authors with an awareness of the inheritance of sexual dimorphism and the evolutionary role of genetic correlations between the sexes. For example, in experimental hybridization between species the finding that F1 hybrid females prefer F1 males more than the males of either parental species has been interpreted to imply that the same genes influence the difference between the species in male characters and female receptivity (Kyriacou and Hall 1986 and refs. therein). However, assuming that the females of both species most prefer conspecific males, even if these characters are influenced by distinct sets of autosomal genes, the same result would be predicted from a model of additive gene action (regardless of the linkage relations of the loci), because both male and female characters would then be intermediate in F1 hybrids. A low genetic correlation between male and female traits in the F2 or backcross populations (or later generations such as F3) would indicate that the differences between the species in male and female characters are controlled by largely distinct sets of loosely linked (or unlinked) loci. But a high correlation between characters in the F2 population still would not imply that the same genes influence both characters, because of the difficulty of distinguishing variation in pleiotropic genes from the joint segregation of tightly linked genes affecting different traits.

O'Donald (1983) stated that quantative genetic models of Fisher's process assume an infinite number of loci and that the genetic correlation between female mating preference and the male character is assumed to be constant because linkage disequilibria can not be calculated in polygenic models. Neither of these assumptions was made in my model (Lande 1981). Instead, I assumed that the characters are influenced by an arbitrary (finite) number of autosomal loci, and based on a submodel of mutation with many alleles at each locus and an arbitrary linkage map, I showed that under certain conditions the genetic variance-covariance pattern achieves stability during the evolution of the mean phenotypes. An explicit formula was derived for the critical genetic parameter that determines the stability properties of the line of equilibria, in terms of the relative mutibility of the loci influencing the two characters.

Engen and Saether (1985) repeated O'Donald's criticism and attempted to generalize my model of absolute mating preference to account for limited availability of potential mates, with the goal of ascertaining whether genetic variation can be maintained without mutation. Unfortunately, their dynamic model for variance and covariance of the characters does not permit change in the heritablities of the traits, nor does it correctly account for changes in linkage disequilibria between loci created by selection and assortative mating, both of which are crucial in calculating the amount of genetic variation maintained (see Felsenstein 1981; Kirkpatrick 1985). The conclusion that stable genetic variance can be maintained without mutation is not justified by their model. Their result that variation in female mating preferences contributes to the maintenance of genetic variation, and in certain situations can cause it to increase in an unstable manner, is correct, as pointed out previously in Lande (1981).

Giddings and Templeton (1983) defended a verbal model proposed by Kaneshiro, involving speciation by founder effect and random genetic drift in (temporarily) small populations. They suppose that, during the period of small population size, elements of male courtship are lost (or altered) by random genetic drift, so that in laboratory tests the females of both species prefer to mate with ancestral males that retain the full repetoire of courtship elements. Giddings and Templeton claim that this form of asymmetric reproductive isolation is an inevitable outcome of Fisher's process of sexual selection. However, detailed consideration of the different models of mate choice in Lande (1981) reveals that divergence of a derived population from a large ancestral one during a period of small population size can violate Kaneshiro's prediction in a variety of ways. For example, during the elaboration of novel male secondary sexual characters and associated female mating preferences it is possible for derived females to most prefer mating with conspecific males.

Acknowledgements

I thank C. Boake, B. Charlesworth, L. Heisler, M. Kirkpatrick, P. O'Donald, and T. Price for helpful discussions. This work was supported by U.S. Public Health Service grant No. GM27120.

REFERENCES

Andersson, M. 1986. Evolution of condition-dependent sex ornaments and mating preferences: sexual selection based on viability differences. Evolution 40: 804–816.

Arnold, S.J. 1985. Quantative genetic models of sexual selection. Experientia 41: 1296–1310.

Bird, M.A., and Schaffer, H.E. 1972. A study of the genetic basis of the sexual dimorphism for wing length in *Drosophila melanogaster*. Genetics 72: 475–487.

Boake, C.R.B. 1985. Genetic consequences of mate choice: a quantitaive genetic method for testing sexual selection theory. Science 227: 1061–1063.

Bonner, J.T., and May, R.M. 1981. Introduction. In: The Descent of Man, and Selection in Relation to Sex, by C. Darwin (reprint of the 1st ed.), pp. vii-xli. Princeton: Princeton Univ. Press.

Borgia, G. 1979. Sexual selection and the evolution of mating systems. In: Sexual Selection and Reproductive Competition in Insects, ed. M.S. Blum and N.A. Blum, pp. 19–80. New York: Academic Press.

Bull, J.J. 1983. Evolution of Sex Determining Mechanisms. Menlo Park: Benjamin/Cummings.

Carson, H.L. 1978. Speciation and sexual selection in Hawaiian *Drosophila*. In: Ecological Genetics: The Interface, ed. P.F. Brussard, pp. 93–107. New York: Springer-Verlag.

Carson, H.L. and Lande, R. 1984. Inheritance of secondary sexual character in *Drosophila silvestris*. Proc. Natl. Acad. Sci, USA 81: 6904–6907.

Coyne, J.A. 1983. Genetic basis of genital differences among three sibling species of *Drosophila*. Evolution 37: 1101–1118.
Darwin, C. 1874. The Descent of Man, and Selection in Relation to Sex (2nd ed.). London: John Murrary.
Engen, S., and Saether, B.E. 1985. The evolutionary significance of sexual selection. J. Theor. Bio. 117: 227–289.
Falconer, D.S. 1981. Introduction to Quantitative Genetics, 2nd edn. London: Longman.
Felsenstein, J. 1981. Continuous-genotype models and assortative mating. Theor. Pop. Biol. 19: 341–357.
Fisher, R.A. 1915. The evolution of sexual preference. Eugenics Rev. 7: 184–192.
Fisher, R.A. 1930. The Genetical Theory of Natural Selection. Oxford: Clarendon.
Fisher, R.A. 1958. The Genetical Theory of Natural Selection, 2nd ed. New York; Dover.
Giddings, L.C. and Templeton, A.R 1983. Behavioral phylogenies and the direction of evolution. Science 220: 372–378.
Hamilton, W.D. 1983. Methods in March-hare madness. Nature 304: 563–564.
Hamilton, W.D., and Zuk, M. 1982. Heritable true fitness and bright birds: a role for parasites? Science 218: 384–387.
Heisler, I.L. 1984. A quantitative genetic model for the origin of mating preferences. Evolution 38: 1283–1295.
Heisler, I.L. 1985. Quantitative genetic models of female choice based on 'arbitrary' male characters. Heredity 55: 187–198.
Kirkpatrick, M. 1982. Sexual selection and the evolution of female choice. Evolution 36: 1–12.
Kirkpatrick, M. 1985. Evolution of female choice and male parental investment in polygynous species: the demise of the 'sexy son.' Am. Natural. 125: 788–810.
Kirkpatrick, M. 1986a. Sexual selection of cycling parasites: a simulation study of Hamilton's hypothesis. J. Theor. Bio. 119: 263–271.
Kirkpatrick, M. 1986b. The handicap mechanism of sexual selection does not work. Am. Natural. 127: 222–240.
Kyriacou, C.P., and Hall, J.C. 1986. Interspecific genetic control of courtship song production and reception in *Drosophila*. Science 232: 494–497.
Lande, R. 1980. Sexual dimorphism, sexual selection, and adaptation in polygenic characters. Evolution 34: 292–305.
Lande, R. 1981. Models of speciation by sexual selection on polygenic traits. Proc. Natl. Acad. Sci. USA 78:3721–3725.
Lande, R. 1982. Rapid origin of sexual isolation and character divergence in a cline. Evolution 36: 213–223.
Lande, R., and Arnold, S.J. 1983. The measurement of selection on correlated characters. Evolution 37: 1210–1226.
Lande, R., and Arnold, S.J. 1985. Evolution of mating preference and sexual dimorphism. J. Theor. Bio. 117: 651–664.
Leutenegger, W. and Cheverud, J.M. 1985. Sexual dimorphism in primates: the effects of size. In: Size and Scaling in Primate Biology, ed. W.L. Jungers, pp. 33–50. New York: Plenum Press.
Maynard Smith, J. 1982. Evolution and the Theory of Games. Cambridge: Cambridge Univ. Press.
Maynard Smith, J. 1985. Sexual selection, handicaps and true fitness. J. Theor. Bio. 115: 1–8.
Mayr, E. 1972. Sexual selection and natural selection. In: Sexual Selection and the Descent of Man, ed. B. Campbell, pp. 87–104. Chicago: Aldine Publishing Company.

O'Donald, P. 1980. Genetic Models of Sexual Selection. Cambridge: Cambridge Univ. Press.

O'Donald, P. 1983. Sexual selection by female choice. In: Mate Choice, ed. P. Bateson, pp. 53–66. Cambridge: Cambridge Univ. Press.

Partridge, L. 1983. Non-random mating and offspring fitness. In: Mate Choice, ed. P. Bateson, pp. 227–255. Cambridge: Cambridge Univ. Press.

Seger, J. 1985. Unifying models for the evolution of female choice. Evolution 39: 1185–1193.

Taylor, P.D., and Williams, G.C. 1981. On the modelling of sexual selection. Q. Rev. Biol. 56: 305–313.

Thornhill, R., and Alcock, J. 1983. The Evolution of Insect Mating Systems. Cambridge, MA: Harvard Univ. Press.

Trivers, R.L. 1972. Parental investment and sexual selection. In: Sexual Selection and the Descent of Man, ed. B. Campbell, pp. 136–179. Chicago: Aldine Publishing Company.

Standing, left to right:
Russ Lande, Glenn Hausfater, Gerry Borgia, Steve Arnold, Franz Weissing

Seated (center), left to right:
Lorraine Heisler, Chris Boake, John Maynard Smith, Peter O'Donald

Seated (front), left to right:
Malte Andersson, Mark Kirkpatrick, Randy Thornhill

Sexual Selection: Testing the Alternatives
eds. J.W. Bradbury and M.B. Andersson, pp. 96–118
John Wiley & Sons Limited

The Evolution of Mating Preferences and Sexually Selected Traits
Group Report

L. Heisler, Rapporteur
M.B. Andersson
S.J. Arnold
C.R. Boake
G. Borgia
G. Hausfater
M. Kirkpatrick
R. Lande
J. Maynard Smith
P. O'Donald
A.R. Thornhill
F.J. Weissing

Abstract. This report covers three topics: a) the meaning of some important concepts in sexual selection theory, b) a critical evaluation of models for the evolution of female mating preferences, and c) suggested topics and methods of approach for empirical research. Female choice results from the expression of mating preferences, which are phenotypic traits of females that can cause sexual selection in males. Female mating preferences may evolve as a result of direct selection if females differ in viability or fecundity in relation to their mating preferences. They may also evolve as a result of selection of other traits with which the preferences are genetically correlated (indirect selection). Genetic models for the evolution of mating preferences differ in the degree to which the viability selection or female choice is the primary cause of indirect selection for a preference. We evaluate two particular sets of models: a) the "Fisherian models" of O'Donald, Lande, and Kirkpatrick in which female choice can lead to a self-reinforcing increase in the preference, and b) the Hamilton-Zuk and Andersson "viability indicator" models in which sustained viability selection can promote the evolution of mating preferences. Resolution of open questions about these mechanisms depends on three types of empirical research: a) assessing the pattern of phenotypic selection on attractive traits, b) measuring the genetic variation and correlation of male traits and female mating preferences, and c) measuring the genetic relationships between attractive male characters and components of fitness. We discuss the relevance of such studies to testing the different models. In two appendices, we suggest empirical tests and some topics in need of further theoretical investigation.

INTRODUCTION

Mechanisms for the evolution of extreme male displays and the female mating preferences for them are probably the most controversial subjects in sexual selection. A variety of biological processes have been proposed, and a number of these have now been formalized into explicit mathematical models. An intense debate has arisen over two classes of models — the viability indicator (sometimes called "good genes") and "Fisherian" models — that have been framed as alternative explanations for the evolution of mating preferences for extreme displays. The primary aim of our group was to evaluate the logical scope, biological merits, and empirical implications of models of these two general types and to attempt to generate a set of testable hypotheses that could be used to evaluate the relative importance of the different processes in nature.

This report begins with a discussion of some terminological problems that have hindered the development of a common language for sexual selection workers. We follow this with a critical discussion of some important current models. We then review the kinds of questions subsumed under the idea of "critical tests" of different models and present some suggestions for such tests. Our report ends with a review of some methodological approaches to testing sexual selection theory.

Definitions of fitness

Behavioral ecologists and population geneticists traditionally use quite different languages in discussing evolutionary processes. Members of both disciplines are actively involved in research on sexual selection and terminological differences have often obscured common concerns. Especially troublesome in this regard has been the use of the term "fitness."

Within population genetics, the goal of making precise predictions of evolutionary change has not been met through the use of any single definition of fitness. However, in many cases fitness can usefully be defined as the lifetime progeny count of an individual. For populations having discrete generations, this is equal to the product, $V \times F$, of viability (the probability of survival from conception to reproductive age) times fecundity (the number of zygotes contributed by surviving adults). Lifetime progeny counts also may provide an *approximate* definition of fitness in populations having overlapping generations (Charlesworth 1980). Population geneticists normally define individual fitness over the time period from zygote-to-zygote, for this separates processes of inheritance from those of selection and renders mathematical analysis more tractable. Unfortunately, empirical measurement of fitness defined in this way is difficult, perhaps impossible. Populations of zygotes are not amenable to censussing and even when censussing is done at a later age, components of lifetime fitness may be the most that can be measured effectively. Empiricists, then, must generally use alternative and indirect measures of fitness components from which to infer total fitness.

Defining fitness becomes especially difficult when there are direct parental effects on offspring phenotypes other than through the genes that are passed. Parental care is the most obvious and perhaps the most biologically important example of this type of effect. Male nuptial feeding is another example of special relevance to discussions of mate choice. In such instances, the post-zygotic survival of progeny is directly affected by variation in parental reproductive traits, and the zygote-to-zygote definition of fitness is therefore inappropriate. We believe that direct parental effects are very important in the evolution of mating preferences. However, we excluded them from our discussion both because of the lack of a well-developed theoretical framework for dealing with this class of characters and because of our group's focus on the evolution of elaborate male traits that are apparently unrelated to parental care. Unless otherwise noted, the following discussion is therefore restricted to cases where parental care or other forms of direct investment are either absent of are independent of the male traits and mating preferences being considered. [It has been suggested that the problem of direct parental effects might be addressed using inclusive fitness formulations, by considering the component of offspring survival accounted for by the parental character as a fitness component due to nepotism (Grafen 1985).]

Another source of terminological confusion has been the use of the terms "direct" versus "indirect" selection in discussing the Fisherian and viability indicator models. Both classes of models assume that male parental care and resource contributions are absent; rather, mating preferences evolve as a result of differences in fitness between the progeny of females that possess genotypically different mating preferences. The process that creates these differences in progeny fitness is selection on the traits of the offspring themselves and not selection directly on the female mating preference. We therefore refer to differential effects of mating preferences on progeny viability or mating success as cases of "indirect selection" on female mating preferences. "Direct selection" on mating preferences occurs only when differences in preference lead to differential viabilities of the females themselves (e.g., through predation) or to differences in numbers of progeny born (differences in fecundity). We recommend this usage both for the sake of terminological consistency and because it retains our view of natural selection arising from the actual biological relationships between the traits of organisms and their survival and reproductive success.

Definitions of "choice" and "preference"

The words "preference" and "choice" have often been used as if they were synonymous in the population genetics and ethological literature. Yet in ordinary language "choice" and "preference" denote different types of phenomena. Choice denotes an action, whereas preference denotes a disposition or propensity that an individual possesses whether or not it is exercised. In practice, female mating preferences can only be observed by observing female choice under conditions where the object of the preference is separated from other factors that may influence its expression. For example, Andersson (1982) was able to demonstrate a female preference for long tails in widowbirds by manipulating tail length while simultaneously controlling for other factors, such as territory quality, that might confound the expression of the preference of interest with other responses of females.

With respect to definitions of "female choice" per se, we favor Halliday's definition of female choice as differential mating by females as a result of their possession of mating preferences (Halliday, this volume). Note that this definition requires that female mating preferences simultaneously be both ordinary phenotypic traits and potential agents of sexual selection on males. We disagree with Maynard Smith's suggestion (1985) to reserve the term "female choice" only for those cases in which we consider the preference to have evolved as a result of the selective forces associated with it. Historical inferences about the evolutionary causes of mating preferences are difficult to make and frequently contentious; reserving the term for cases in which particular inferences have been made would probably render it useless. Halliday's definition, on the other hand, lends itself to reasonably straightforward operational definitions that allow us to detect female choice using currently observable processes.

Female mating preferences are most readily detected if individual females differ in their preferences, and these differences can be observed as differences in their choices between alternative male types. In such a case at least some variation in the relative mating probabilities of different male types will be attributable to individual differences between females, and mating preferences in Halliday's sense can thus be inferred. Variation among females in their mating preferences may or may not be genetic; showing that female mating preferences are heritable requires explicit genetic investigations (discussed below). If females are all alike in their responses to different male types, they may still possess preferences that are genotypically invariant.

Given that a mating preference is found, there are three general types of preferences that differ with respect to their evolutionary history and behavioral expression in females. First are what could be called *direct preferences*, where the male character we identify with the preference is indeed the trait to which females differentially respond and through which evolution of the mating preference has been mediated. This class of preferences corresponds to Maynard Smith's definition (1985). The second class of preferences might be called *indirect preferences*. These differ from direct preferences in that the male character we identify with the preference is not the trait to which females actually respond but is phenotypically correlated with it. Identifying whether a particular male trait is the object of a direct or an indirect mating preference requires either experimental manipulation, such as that used by Andersson

(1982) to infer preference for tail length in widowbirds, or a multivariate analysis of the male characters associated with attractiveness (see below and Wilkinson et al., this volume, for a discussion of one such method). The third class of mating preferences could be called *pleiotropic preferences*. These are mating preferences that arise as a result of pleiotropic expression of genes that determine some other female behavior, such as food choice. Like indirect preferences, pleiotropic preferences lead to apparent female choice for male characters that are actually incidental to the ecological function of the preference. Indirect preferences arise from correlations between traits in males, whereas pleiotropic preferences arise from correlated effects of genes in females. Pleiotropic preferences may not therefore be distinguishable from direct preferences by experimental or statistical manipulation; the distinction instead will depend on the identification of genetic correlations between different female behaviors, combined with studies of the relative importance of the different observable preferences in nature.

CRITIQUE AND DISCUSSION OF CURRENT THEORY

Two main schools of thought exist as to whether or not mating preferences for extreme male characters will generally be adaptive. A variety of different mechanisms of sexual selection have been proposed in support of alternative views, and the feasibility of several of these has been investigated theoretically using population genetic models. These models make diverse assumptions about mechanisms of inheritance (e.g., two or three loci versus polygenic inheritance) and use different methods of analysis (e.g., analytic versus computer-simulated results); such differences in modelling techniques complicate the task of comparing their empirical implications. Furthermore, models that differ substantially in their basic assumptions may nevertheless lead to very similar outcomes, making it difficult to identify empirical predictions that discriminate between them. A considerable part of our group's discussion was therefore aimed at identifying the major differences between models, the assumptions most vulnerable to criticism on biological grounds, and common features that might lead to qualitatively indistinguishable results.

Fisherian models

The joint evolution of female mating preferences and male secondary sexual characters in polygamous species has been analysed by O'Donald (1967, 1980), Lande (1981), Kirkpatrick (1982), and Seger (1985). These have confirmed that Fisher's (1930, 1958) basic description of the process is indeed feasible and can decrease male viability through the evolution of exaggerated male traits. During this process, natural and sexual selection may initially reinforce each other, but at equilibrium they act in opposite directions on the male character. In this sense, the evolution of the sexually selected trait may be maladaptive in that female mating preferences lower the mean viability of males.

A primary point of agreement among the various Fisherian models is the existence of a neutral line of equilibria (Lande 1981; Kirkpatrick 1982). The possibility that the line itself can become unstable when the genetic correlation between the female mating preference and male trait is sufficiently large was first demonstrated for polygenic models (Lande 1981), and similar behavior has subsequently been found in other models. Two-locus models like those of O'Donald (1967), Kirkpatrick (1982), and Seger (1985) permit only fixation of alternative alleles and thus do not explicitly model continued and exaggerated evolution of secondary sexual traits. Nevertheless, the requirements for rapid fixation of the male trait in the two-locus models are similar to those for instability in the polygenic models in that fixation occurs when the correlated evolutionary response of the female mating preference is sufficiently large relative to the direct response of the male character to sexual selection.

The realism of the Fisherian models, or at least their applicability to many biological situations, has been questioned on several grounds. Most important is the assumption that female mating preferences are not subject to direct selection, by which we mean that females have different viabilities or produce different numbers of progeny as a consequence of differences in mating preference. In many species, there may be direct costs to females associated with mate choice. Examples include risks associated with searching for mates, increased fecundity as a result of male parental care, and infertility caused by species hybridization. However, the widespread existence of costs to mate choice per se must be distinguished from the subset of cases in which these costs create direct selection on a female mating preference of interest. For example, females who fail to recognize a species-specific character may have reduced fecundity as a result of hybridization. Given that females recognize the trait, however, there may be no differential fecundity as a result of discriminating among conspecific males that differ in the degree of its elaboration. The relative importance of different kinds of costs to mating preferences is virtually unknown and is a critical topic for future research. Note also that the existence of direct selection on female mating preferences applies as a criticism not only to the Fisherian models but also to the various viability indicator models discussed below.

If costs associated with mate choice lead to direct selection on mating preferences, general theoretical considerations show that this will eliminate the neutral line of equilibria. The preference can only equilibrate at a point that balances the forces of selection acting directly on females. Kirkpatrick has studied two explicit models that provide detailed predictions for special cases. The first examines the situation in which males provide females with material benefits, such as a resource (Kirkpatrick 1985). In this situation, females evolve to maximize the quality or quantity of resource they receive. The second model examines the effects of search costs that arise, for example, from predation risks or energetic costs associated with differences in mating preferences (Kirkpatrick, this volume). Under some ecological conditions this model produces evolution of the male trait and female preference toward the viability optimum for the male character. Under other conditions, however, the trait and preference both equilibrate at values that deviate from the viability optimum. In general, both models indicate that direct selection on female mating preferences leads to evolutionary equilibria that are adaptive from the standpoint of females, in that mean female fitness is maximized, but that may be maladaptive with respect to mean male viability. These models still retain the theoretical possibility of an unstable runaway process in which preference and trait evolve in an unpredictable and overall maladaptive way.

A second criticism of the biological applicability of the Fisherian models is that the mating preference rules considered are too simple, and the introduction of more complex behavioral mechanisms of mate choice can strongly affect the models' results. The most dramatic effect is on the stability of the line of equilibria. In the simplest models of female choice studied by O'Donald (1980), Lande (1981), and Kirkpatrick (1982), females are assumed to have access to all males in the population and to evaluate potential mates according to fixed preferences that are unaffected by the frequency with which they encounter males of different types. In Lande's model, the line of equilibria becomes unstable only when the genetic covariance between the preference and trait becomes sufficiently large, as discussed previously. More detailed models of how females choose their mates, however, show that the behavioral mechanism of mate choice can itself affect the stability of the equilibria.

For example, Seger (1985) has extended Kirkpatrick's (1982) model to consider situations in which females choose the most preferred male from a sample of size N, such as might occur when females visit males at a lek. In these models the line of equilibria can become unstable as a direct consequence of this particular behavioral mechanism. O'Donald and Karlin have studied encounter models that assume that females have a variable threshold that determines the number of encounters required before they will mate with nonpreferred males (O'Donald 1980). These models give rise to frequency-dependent expression of mating preferences that can stabilize equilibria that would be unstable under simpler behavioral models. Furthermore, these effects arise even though there is no genetic variance for the female mating preference, and hence no genetic covariance between trait and preference that could contribute to the stability conditions. The above examples show that the mechanism of expression of mating preferences can strongly influence the range of possible evolutionary outcomes of sexual selection. In order to make detailed empirical predictions from theory, we therefore may require knowledge not only of the genetic system but also of the behavioral process by which mate choice occurs.

A third source of criticism of the Fisherian models is that the mating preferences they assume may be vulnerable to invasion and eventual replacement by alternative preferences (Fisher 1915; Borgia, this volume). The equilibria in the Fisherian models are intrinsically immune to invasion by *quantitatively* different mating preferences directed toward more viable values for the male character. However, there is a need for theoretical investigation of the possibility that mating preferences for maladaptive male traits can be replaced by *qualitatively* different preferences involving either different behavioral mechanisms or different male characters.

The argument that preferences for maladaptive traits will be replaced by preferences for adaptive traits is based on the widespread observation of apparently adaptive mate choice in resource-based systems. Existing models of choice in such systems suggest that female mating preferences are adaptive in that they maximize mean female fitness (but not necessarily male fitness). It is proposed that adaptive mating preferences arise in resource-based systems because of selection among alternative mating preferences, wherein preferences for male traits that enhance female fitness are more likely to evolve than preferences that do not have this effect. By analogy, it has been argued that in nonresource-based mating systems, mating preferences for male traits that enhance viability will outcompete mating preferences for male traits that have no effect or that decrease viability (Borgia, this volume). It is feasible that traits that increase male viability could bias the evolution of novel mating preferences in their favor yet still lead ultimately to the Fisherian trade-off between sexual and viability selection. The possibility that competition among alternative mating preferences could lead to the prevalence of preferences for adaptive male traits is an area in which explicit genetic models are needed.

Viability indicator models

Several mechanisms have been proposed that might lead to the evolution of predominantly adaptive mating preferences. By "adaptive mating preferences," we mean preferences that cause females to mate with males carrying genes that increase the viability of their progeny. The following is a list of several ways in which a mating preference could have this effect.

(A) Females choose mates directly on the basis of viability (e.g., if courtship vigor were perfectly correlated with viability; see Maynard Smith and Charlesworth, both this volume).

(B) Females choose mates using a character that indicates viability.

- (B1) The indicator itself increases viability (e.g., Kodric-Brown and Brown 1984).
- (B2) The indicator itself decreases viability (i.e., is a handicap) but is positively correlated with total viability through other traits (models reviewed by Maynard Smith 1985; Andersson 1986).

The following discussion is restricted to mechanisms of type (B2) above, because they are the most pertinent as explanations for the evolution of mating preferences for exaggerated male characters that must typically entail some cost to their possessors.

The first of the viability indicator models to be analyzed genetically was Zahavi's (1975, 1977) handicap mechanism. In its original form, this model was criticized by Maynard Smith (1976), who showed that under realistic conditions the survival of males with an unconditional handicap could not serve as an indicator of overall genetic quality. The male handicap and the female preference for it would then be lost from the population. Following this a variety of mechanisms involving "facultative," "conditional," or "revealing" handicaps have been suggested (Hamilton and Zuk 1982; West-Eberhard 1979; Dominey 1983; and others) in which a slightly deleterious indicator trait is fully expressed only in the most viable males. These models represent extensions of the basic Fisherian model to include genetic variance for viability in one or more characters in addition to the male trait. They are reviewed and analyzed by Maynard Smith (1985) who concluded that a male viability indicator and a female mating preference for it can spread if they reach a threshold frequency, but neither will spread if initially rare.

A version of the viability indicator hypothesis suggested by Hamilton and Zuk (1982) has received much attention. They proposed that the coevolution of species and their parasites can produce cyclical selection on genes in the host that confer resistance to particular species of parasites. When the parasite load of an individual affects the expression of a male secondary sexual trait, parasite selection will drive the evolution of a mating preference for the parasite indicator, thus giving resistant males a further advantage under sexual selection. It is postulated that population mean fitness increases by this process and that the mechanism represents a major cause for the evolution of exaggerated male traits in nature.

The first genetic models proposed by Hamilton have been studied by Kirkpatrick (1986) and Pomiankowski (unpublished manuscript). These models have found that the fluctuating selection produced by the parasites leads to different evolutionary equilibria than do the basic Fisherian models or those viability indicator models that assume constant viabilities. The neutral line of equilibria is lost, and the trait and preference either both spread or are both lost. In the simple haploid models studied thus far, the trait and preference will not spread when rare but require some other process to raise the preference allele to some critical threshold frequency needed to initiate effects from the parasite mechanism. Two important criticisms of the parasite hypothesis have been raised on the basis of these results. First, evolution need not be unidirectional. If external forces can initiate the process, they may also reverse it by causing the preference to drop below the threshold frequency. Which outcome will be more prevalent depends on the parameters of the model; it is a question that can only be resolved by empirical measurement of those quantities. For example, the mechanism as originally formulated depends on special quantitative relationships between host and parasite life spans and the periodicity of fluctuating selection; the frequency of these conditions in nature is not known. A second criticism is that the current models do not lead to an increase in mean population fitness as proposed by Hamilton. Thus, although the parasite mechanism can influence the direction of sexual selection and can have strong effects on its outcome, current models have neither shown that it can initiate the process nor that it will lead ultimately to an adaptive outcome. There is a need for more theoretical and empirical work before the importance of the parasite hypothesis can be assessed.

Recently, Andersson (1986) analyzed a model of a conditional handicap in which a male indicator trait has no effects on viability or mating success except when associated with a high viability allele at another locus. To assess whether the mechanism suffices for the spread of trait and preference when there is no Fisherian mating advantage, the model assumed a monogamous mating system without variation in female fecundity. (In principle, indicator mechanisms can also operate in polygynous systems although their importance relative to the Fisherian mating advantage mechanism should decrease as the level of polygyny increases.) Andersson's model showed that under these conditions a male facultative handicap and female mating preference for it can evolve solely by association with genes for high viability. In this model there is neither sexual selection, in the sense of differences between males in the numbers or fecundities of mates, nor is there direct selection on female mating preferences, which evolve only as a correlated response to viability selection. Andersson's model accommodates Borgia's (1979) idea of a female preference for a trait that indicates dominance. It also covers, as a special case, variation in male viability caused by differences in resistance to parasites. This type of indicator model seems even more favorable for the evolution of trait and preference than does the Hamilton-Zuk host-parasite model. For these reasons, studies of indicator processes should not be restricted to parasite cycles as a cause of heritable viability differences but should consider also recurrent mutations and other environmental or genetic processes that might maintain additive genetic variance for fitness (e.g., Charlesworth, this volume).

An important criticism of the conditional indicator mechanism as a widespread explanation for mating preferences is that, operating alone in monogamous mating systems, it cannot account for the frequent occurrence of sexual dimorphism in monogamous species. This is because in the absence of any initial asymmetry in the expression of the characters in males and females there is no intrinsic force of selection for sexual dimorphism. This is easily seen where a conditional handicap character reveals genetic resistance to parasites by being fully expressed only in healthy individuals that are parasite-free. In such a case, with monogamous matings it would equally benefit either sex to choose healthy mates, either by increasing their own fitness if parasites are transmitted during copulation (direct effect) or the fitness of their progeny (indirect effect). Thus, to the extent that the conditional handicap mechanism occurs in natural populations, it must operate in conjunction with other mechanisms that are capable of creating the sexual asymmetry required for the evolution of sexual dimorphism.

An alternative mechanism that *can* explain the existence of sexual dimorphism in monogamous species was proposed originally by Darwin (1871) and elaborated later by Fisher (1958). This postulates nonheritable differences in female fecundity wherein the most fecund females breed earlier in the season and have first choice of mates. This creates a force of sex-limited sexual selection on male traits that can lead to evolution of both the mating preference and sexual dimorphism via the classic Fisherian mechanism. Evidence for sexual selection of this sort has been found in arctic skuas by O'Donald (1983) and in kittiwakes by Coulson and Thomas (1983), and genetic models of its operation have been studied by O'Donald (1973, 1980). Although the Darwin-Fisher mechanism by itself can account for mating preferences and sexual dimorphism in monogamous populations, it could also operate in conjunction with a facultative handicap mechanism so as to bias evolution in favor of male traits that indicate viability.

The criticism that conditional handicaps cannot explain the evolution of sexual dimorphism also applies to polygynous mating systems. Here, however, a force of sexual selection will arise once mating preferences have begun to evolve, and it is probable that the Fisherian process, once initiated, will be a stronger force than the viability mechanism. Empirical tests of conditional handicap models should therefore be designed not only to indicate whether these mechanisms are operating but also to assess their importance relative to other processes that are expected to be operating simultaneously.

Commonalities of the models

Although the different models make strikingly different assumptions about underlying genetic mechanisms and the relationship between male traits and viability, they all share a common requirement for a mechanism that initiates a female mating preference from a prior state in which females show no bias in their mating. Either a mechanism is built into the models to create a force of viability selection that will lead to increase of an initially rare mating preference (as in Andersson's model) or alternative mechanisms, extrinsic to the models themselves, are required to bring mating preferences to a threshold frequency needed to initiate sexual selection. Several such extrinsic mechanisms have been suggested by Kirkpatrick (this volume). The ability of viability mechanisms to lead to the origin of mating preferences for quantitative characters has been examined in models by Heisler (1984, 1985). However, theoretical work is needed to investigate other mechanisms for the origin of mating preferences and, especially, to examine the period of transition from a primarily viability selected to a sexually selected phase of evolution.

A second feature common to both the pure Fisherian models and the viability indicator models that also include Fisherian mating advantages is that as long as there is no direct force of selection on mating preferences and viabilities remain constant over time, there exists a neutral line of equilibria. In some models the equilibria are always unstable and therefore play no role in the evolutionary outcome of sexual selection. In either instance, however, there appears to exist a region of genotypic or phenotypic space in which initial conditions will lead to maladaptation on the part of the male character. Viability mechanisms reduce the set of initial conditions that will lead to maladaptation, but the Fisherian process appears always to be able to induce maladaptive evolution under at least some conditions.

ARE THERE DISCRIMINATING TESTS OF ALTERNATIVE MODELS?

Kinds of tests

Any discussion of the nature of critical tests of sexual selection theory must draw distinctions between the explanatory aims of different types of tests. Most fundamental is the difference between tests of historical hypotheses about the long-term processes that gave rise to the types of traits we observe across a range of taxa and tests of hypotheses about processes currently in operation within specific populations. Workers on sexual selection are interested in both kinds of questions, and failure to identify differences in perspective has sometimes hindered communication within the field.

Tests of hypotheses about current selective processes can be used to refute the existence of particular processes in individual instances and to strengthen evidence for the presence of alternatives. In aggregate many such tests can be used to make inferences about the relative frequency of different processes in other species and in the past. Models for how selection is currently affecting mating preferences and male characters can be tested using either critical assumptions that differ between the models or distinctive predictions. Andersson (this volume) has argued that tests of assumptions are especially desirable, so long as the assumptions are responsible for critical differences between alternative models. An example of one such critical assumption is the absence of direct costs or benefits that create direct selection on female mating preferences, an assumption required for the neutral line of equilibria. Likewise, discriminating tests of predictions about ongoing processes can be used in much the same way as tests of assumptions. An example of one such prediciton is the existence of a positive correlation between the attractiveness of males and the average viability of their daughters, predicted by some of the viability indicator models (Boake 1986).

Tests of hypotheses about current processes, however, provide only indirect evidence about the relative importance of different processes in the past. To address questions about the historical origin and elaboration of different classes of characters, predictions are needed for the long-term results of different evolutionary mechanisms. Such tests do not require that the assumptions of different models currently be met in populations; they use long-term extrapolations from processes postulated within species to test predicted patterns of character distribution across species. An example of one such prediction is that male characters that reveal parasite loads should be more highly developed in species that are subject to high parasitism than in species where parasites are less common. The explanatory scope of historical tests is much larger than that of tests of current processes. However, it often may be difficult to obtain the kinds of predictions desired, in that models that have quite different assumptions may nevertheless be indistinguishable in their long-term results. Furthermore, the results of historical tests will typically be more subject to dispute than will tests of current processes, since the former tests depend on a larger number of biological assumptions, such as long-term constancy in the pattern of selection on traits.

The relative roles of genetic and phenotypic studies

An integrated approach to studying evolution requires both genetic and ecological investigations. The Fisherian and viability indicator models differ in specific statements about the genetic relationship between female mating preferences, male characters, and components of total fitness. These genetic relationships can be measured as genetic correlations, and such correlations may be of different magnitude and even different sign than the corresponding correlations based on phenotypic observations (Falconer 1981). Testing hypotheses that involve statements about such relationships necessarily requires explicit genetic analyses. Consequently, it has been questioned whether the different models can ever be distinguished without conducting arduous genetic investigations, an impossible task for many of the species whose traits are of most interest. Although genetic studies are indispensible for answering many important questions about the evolutionary consequences of mate choice, we believe that at least some discriminating tests can be done without explicit genetic information, using purely phenotypic observations.

Phenotypic studies

One approach to investigating sexual selection is to measure the selective pressures acting in a contemporary population. The idea is to exploit natural variation in traits that are presumed to be sexually selected and to analyze associations with measured components of fitness. This approach yields estimates of the strength of selection that are analogous to the selection

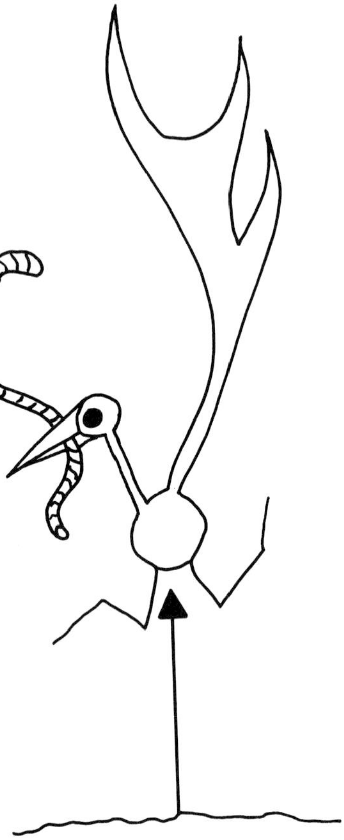

FIG. 1: Arnold's Bird-of-Paradise, illustrating three kinds of sexually selected traits. Tail size is sexually dimorphic and extravagant in males. The vigor of the male's courtship dance, measured as the average height of leaps, is a potential viability indicator. The average size of worms presented to a female at the end of the dance has potential effects on many male fitness components and on the fitness of his progeny.

coefficients used in equations for evolutionary change (Lande and Arnold 1983; Arnold and Wade 1984). A limitation of the method is that it cannot by itself address questions about ongoing evolutionary change, for which information about the pattern of genetic variation is required. The phenotypic approach treats selection as a process that can be studied separately from the analysis of inheritance. Even in the absence of genetic information, it permits a number of important questions to be explored. Some examples of these are illustrated in Figures 1 and 2. Wilkinson et al. (this volume) lists the assumptions implicit in the particular statistical approach discussed here.

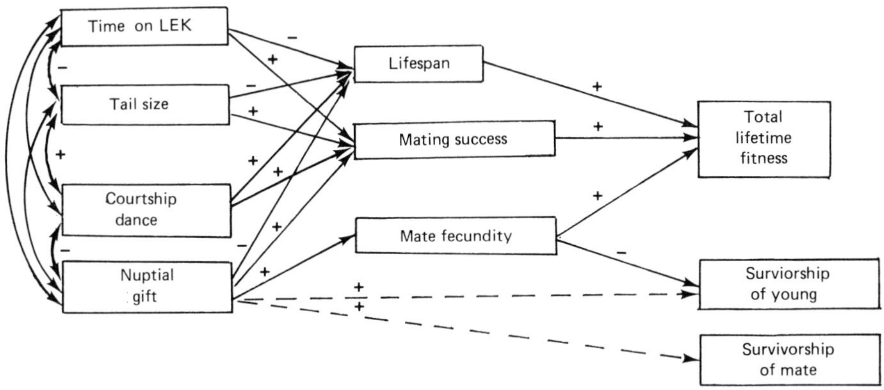

FIG. 2: An example of sexually selected traits that are phenotypically correlated with various components of fitness. Double-headed arrows represent phenotypic correlations between traits. Solid single-headed arrows represent partial regression coefficients of the fitness components on the individual characters. Dashed single-headed arrows represent maternal and paternal phenotypic effects on the survivorship of offspring. Lifespan is measured in units of time, mating success in numbers of mates per unit time, and male fecundity in units of average number of offspring per mate. The product of these three fitness components is total lifetime fitness in units of numbers of offspring (see text for discussion).

The hypothetical male shown in Figure 1 is a member of a lek-breeding species in which the male displays his extravagant tail to the female while displaying in front of her and offering her a nutritious nuptial gift. Assume that we have measured four phenotypic traits in males: time spent on the lek, tail size, vigor of the courtship dance, and the average size of nuptial gifts. Assume also that we have measures of three components of individual fitness — lifespan, mating success, and mate fecundity — that completely account for total lifetime fitness. Because we measure total lifetime fitness as a simple zygotic progeny count, we also score the survivorship of each male's progeny in order to assess maternal and paternal phenotypic effects on offspring quality (mediated, e.g., through the nuptial gift).

The hypothetical male traits are chosen to illustrate major categories of sexually selected traits that are presumed to exist under different selective regimes. Thus, tail size is the kind of trait commonly considered in Fisherian models (e.g., Lande 1981), and for it we have the expectation that opposing forces of sexual and viability selection balance in an equilibrium population. From the Darwin-Fisher hypothesis we might therefore predict that tail length would be positively correlated with mating success, negatively correlated with lifespan, and uncorrelated with total fitness.

However, to illustrate the kind of complications that arise due to character correlation suppose that tail size is negatively correlated with time on the lek (perhaps because males with the largest tails exhaust quickly and retire sooner from the lek each day), but that time on the lek is a major phenotypic determinant of mating success. In this case the simple correlation coefficient between tail length and mating success might be negative or zero, giving the misleading picture of a mating preference for short tails or of no preference at all. A multiple regression anlaysis, however, would provide partial regression coefficients of lifespan and mating success on tail length, and these would control for the correlations between tail length and time on the lek and the other characters. Experimental manipulations of tail size could subsequnetly be used to assess directly the effect of tail length on mating success.

The vigor of the male's courtship dance illustrates how we might test expectations derived from a viability indicator model. We hypothesize that dance vigor is an indicator of male viability and, accordingly, we expect both lifespan and mating success to have positive partial regression coeficients associated with this trait. As before, experimental manipulations may be necessary to verify our interpretation of the statistical results.

Finally, the size of the male's nuptial gift illustrates the case of a male trait that is presumed to have complicated effects both on the fitness of his mates and on the fitness of his progeny. For example, suppose we wish to test the expectations that the male's nuptial gift has positive effects on his own mating success, on his mate's fecundity, on his mate's survivorship, and on the survivorship of his young. In addition, we expect a negative effect on the male's lifespan (because large gifts increase his risk of predation) as well as a negative correlation with dance vigor (because a large gift lowers his agility). As before, we could use partial regression analysis to sort out direct effects of the nuptial gift on these different components of fitness. In a situation this complex, however, experimental research would surely be needed to ascertain the actual biological processes occurring.

Genetic studies

Some types of genetic investigations will be essential to evaluate the applicability of different models. However desirable, it will not normally be feasible to perform analyses of both selection and inheritance on the same species, although in exceptional cases reasonable estimates of genetic parameters may be possible using field-collected data. Generally, however, genetic studies require species which can be reared in the laboratory in large numbers. Furthermore, if laboratory-based estimates are to be used to infer heritabilities in nature, the species must behave normally in an experimental environment.

Perhaps the most crucial information needed at present concerns the genetic basis and magnitude of heritable variation for female mating preferences. Artificial selection is a direct but as yet little-used approach to investigating this question. Unfortunately, selection on mating preferences is more difficult than selection on simpler behavioral or morphological traits, owing to difficulties both in measuring the character itself and in separating its selection from selection on other traits that influence mating probabilities. Preferences under investigation may be directed at discrete male characters or at quantitative traits, and methods of selection will differ somewhat in the two cases. In general, the procedure is to assess female choice and retain those females that mated with "preferred" males as mothers of the next generation. As the preference increases in response to selection, the population will naturally come to consist more and more of the preferred phenotype. A complication arises, unless special measures are taken, in that this will alter both the intensity of selection on the preference and the level of frequency-dependence in its expression. In selecting on preferences for discrete traits, one can artificially start each generation with the same proportion of preferred and nonpreferred males. For example, Majerus et al. (1982) used a ratio of 3 melanic: 7 nonmelanic beetles in each generation of selection for a mating preference in ladybirds. In selecting for a preference directed at a quantitative character, the intensity of selection can be held constant by selecting some fixed proportion of females that mated with the most extreme males. Because the distribution of the male trait changes over time, however, measuring the response to selection requires that females must periodically be tested using males from a control line having a standard distribution of phenotypes.

During the course of selection for a mating preference, male competition may cause evolutionary change in male characters other than the one under investigation, and this may confound the response to selection for the mating preference with unknown genetic changes. This problem can be avoided if it is possible to measure female preference using males from an unselected line and subsequently to breed the selected females at random with males from within their population. An alternative and less time-consuming procedure is to test males and females from the selected and unselected lines in various combinations, so as to assess the degree to which male characters have evolved (Majerus et al. 1982).

Several other types of genetic studies may be important in evaluating hypotheses about sexual selection. These include estimation of the number of genetic loci influencing male characters and/or female mating preferences and measurement of the genetic covariance (linkage disequilibrium) between male traits and female preferences. A particularly crucial class of measurements is the genetic correlations between sexually selected male characters and other male traits, most importantly components of male fitness.

To estimate genetic covariances for a set of characters, an efficient procedure is to use the regression between parental and offspring traits. The technique is to measure a set of behavioral and/or morphological traits in males, breed the males, and measure the same traits in their sons. The details of such designs are covered in several texts (e.g., Falconer 1981). An important consideration is that each design has its own limitations and advantages. One such limitation is that genetic analyses often require large sample sizes if reasonable standard errors for the estimates are to be obtained (Bulmer 1980). An alternative procedure for estimating genetic covariances is artificial selection, but this method is hazardous because genetic correlations can be generated through unintentional assortative mating.

Although some of the viability indicator models are falsifiable using solely phenotypic observations, such as demonstrating that male parasite loads are not negatively correlated with attractiveness, other models require explicit genetic analyses of the correlations between male attractiveness and components of offspring fitness. The general procedure for obtaining such estimates is to evaluate male attractiveness and subsequently measure the viability of their progeny (Boake 1986). A correlation between fathers' attractiveness and daughters' viability will often avoid the problem of nongenetic parental effects on progeny, though this may not be the case if males contribute resources or care to their mates or offspring.

Genetic analyses should only be undertaken after careful consideration of their utility in answering the questions of interest for a particular species. Since the labor involved in genetic studies is considerable, experimental designs require careful development to ensure that the results will yield reasonably precise estimates. A variety of pilot procedures can provide information important in designing successful experiments. For example, measures of the repeatability (correlation between subsequent measures on the same individual) of behavioral traits should always precede heritability tests, since behavioral measures often have large components of random variation that place an upper limit on the magnitude of the heritability that can be detected.

CONCLUSION

This report summarizes a discussion that has not ended. A number of specific predictions were proposed that, if tested empirically, should be capable of falsifying particular models as explanations in specific instances. Some suggestions for such tests are listed in Appendix A. The list is not complete — there may well exist tests that are either more decisive or more readily conducted than those suggested. It is also possible that future research will show that some proposed tests are not in fact decisive. Nevertheless, the list includes enough important empirical questions to keep field and laboratory workers occupied for some time to come. Meanwhile, more theoretical work is needed. For example, some of the equilibrium phenotypic or genetic correlations between either mating preferences or male indicator traits and the various components of fitness are expected to differ among the Fisherian and various indicator models. These potentially distinguishing relationships have yet to be detailed in full for any single model. Not only do we need new theory to address empirically reasonable proposals that as yet lack formal treatment, but perhaps more importantly, we need to examine more fully the empirical implications of the theory we already possess.

APPENDIX A: PROPOSED TESTS OF SOME MODELS OF SEXUAL SELECTION

Below we list some empirical predictions that could be used to test different proposed mechanisms of sexual selection. Some of these tests involve measuring relationships within particular species while others are comparative tests across a range of species.

Fisherian models

The pure Fisherian models of O'Donald (1967, 1980), Lande (1981), and Kirkpatrick (1982) make the following equilibrium predictions. Showing that any of these fails to hold demonstrates that the Fisherian mechanism cannot solely account for the pattern observed (i.e., is a partial falsification of the model).

1) Across a range of species, attractive male traits should be "arbitrary" in the sense that they should not be demonstrable predictors of male or offspring viability or total fitness.
2) Within populations assumed to be in evolutionary equilibrium, viability and sexual selection should act in opposing directions on attractive male traits.
3) Female mate choice should not positively influence the viability or total fitness of her progeny.

Viability indicator models

Several kinds of male characters have been suggested as possible indicators of the ability of males to sire viable progeny. Four of these are listed below with specific predictions that, should they fail to hold, would eliminate the mechanism as the sole explanation of the trait in question.

1) *Indicators of Male Age*. This model postulates that females prefer traits that accurately indicate male age because older males have demonstrated viability. The specific predictions are: (a) females preferentially mate with older males; (b) the male trait in question is an accurate indicator of age; and (c) females use the proposed indicator during mate choice.
(2) *Indicators of Male Dominance*. This model postulates that males that win dominance contests produce progeny having high viability or total fitness. The

specific predictions are: (a) females prefer to mate with dominant males; (b) the proposed indicator is correlated with dominance; and (c) females use the proposed indicator during mate choice.

3) *Parasite Indicators*. This is the model of Hamilton and Zuk (1982), described in detail in the text. Four specific predictions are: (a) females within a given species prefer males in inverse relation to the number of their parasites; (b) individual display quality (e.g., plumage brightness) declines with increasing parasite loads; (c) display quality is favored by females during mate choice; and (d) in comparisons between related species, species that are heavily parasitized should possess more elaborate displays. (This last test requires careful control for confounding effects of mating systems and the degree of aggregation, which also should affect parasitism).

4) *Indicators of Male Vigor*. This model postulates that vigorous males carry genes for high viability or total fitness and indicate their vigor with elaborate traits. Three specific predictions are: (a) male vigor is positively correlated with mating success; (b) the proposed indicator is positively correlated with male vigor; and (c) females use the indicator during mate choice.

APPENDIX B: SUGGESTIONS FOR FUTURE THEORETICAL WORK

A. Exploration of Existing Models
   1. Dynamical properties of models — trajectories of approach to equilibrium, rates of evolution in different areas of phenotypic space
   2. Further examination of fluctuating selection (e.g., the parasite hypothesis)
   3. Equilibrium correlations between traits and fitness components in various models

B. Addition of New Factors to Existing Models
   1. Explicit models of female choice behavior – further examination
   2. Effects of limited polygyny
   3. Further examination of costs to female mating preferences

C. New Models
   1. Mate choice and sexual selection in hermaphrodites
   2. Effects of learned or culturally inherited mating preferences.
   3. Interactions between male dominance/combat and female choice
   4. Elaboration of specific 'good genes' models
      a. Mate choice based on genome-wide viability differences generated by mutation-selection balance.
      b. Mate choice based on male age as an indicator trait
      c. Effects of heterozygote advantage for male characters

SUGGESTED READINGS

Arnold, S.J. 1983. Sexual selection: the interface of theory and empiricism. In: Mate Choice, ed. P. Bateson, pp. 67–107. Cambridge: Cambridge Univ. Press.
Maynard Smith, J. 1985. Mini review: sexual selection, handicaps and true fitness. J. Theor. Bio. 115: 1–8.
O'Donald, P. 1980. Genetic Models of Sexual Selection. Cambridge: Cambridge Univ. Press.

REFERENCES

Andersson, M. 1982. Female choice selects for extreme tail length in a widowbird. Nature 299: 818–820.
Andersson, M. 1986. Evolution of condition-dependent sex ornaments and mating preferences: sexual selection based on viability differences. Evolution 40: 804–816.
Arnold, S.J., and Wade, M.J. 1984. On the measurement of natural and sexual selection: theory. Evolution 38: 709–719.
Boake, C.R.B. 1986. A method for testing adaptive hypotheses of mate choice. Am. Natural. 127: 654–666.
Borgia, G. 1979. Sexual selection and the evolution of mating systems. In: Sexual Selection and Reproductive Competition in Insects, eds. M.S. Blum and N.A. Blum, pp. 19–80. New York: Academic Press.
Bulmer, M.G. 1980. The Mathematical Theory of Quantitative Genetics. Oxford: Clarendon Press.
Charlesworth, B. 1980. Evolution in Age-Structured Populations. Cambridge: Cambridge Univ. Press.
Coulson, J.C., and Thomas, C.S. 1983. Mate choice in the Kittiwake Gull. In: Mate Chice, ed. P. Bateson, pp. 361–376. Cambridge: Cambridge Univ. Press.
Darwin, C. 1871. The Descent of Man and Selection in Relation to Sex. London: Murray.
Dominey, W.J. 1983. Sexual selection, additive genetic variance and the 'phenotypic handicap.' J. Theor. Bio. 101: 495–502.
Falconcr, D.S. 1981. Introduction to Quantitative Genetics, 2nd ed. London: Longman.
Fisher, R.A. 1915. The evolution of sexual preference. Eugenics Rev. 7: 184–192.
Fisher, R.A. 1930. The Genetical Theory of Natural Selection. Oxford: Clarendon Press.
Fisher, R.A. 1958. The Genetical Theory of Natural Selection, 2nd ed. New York: Dover.
Grafen, A. 1985. A geometric view of relatedness. Oxford Surv. Evol. Biol. 2: 28–90.
Hamilton, W.D., and Zuk, M. 1982. Heritable true fitness and bright birds: a role for parasites? Science 218: 384–387.
Heisler, I.L. 1984. A quantitative genetic model for the origin of mating preferences. Evolution 38: 1283–1295.
Heisler, I.L. 1985. Quantitative genetic models of female choice based on "arbitrary" male characters. Heredity 55: 187–198.
Kirkpatrick, M. 1982. Sexual selection and the evolution of female choice. Evolution 36: 1–12.
Kirkpatrick, M. 1985. Evolution of female choice and male parental investment in polygynous species: the demise of the "sexy son." Am. Natural. 125: 788: 788–810.

Kirkpatrick, M. 1986. Sexual selection and cycling parasites: a simulation study of Hamilton's hypothesis. J. Theor. Bio. 119: 263–271.

Kodric-Brown, A., and Brown, J.H. 1984. Truth in advertising: the kinds of traits favored by sexual selection. Am. Natural. 124: 309–323.

Lande, R. 1981. Models of speciation by sexual selection on polygenic traits. Proc. Natl. Acad. Sci. USA 78: 3721–3725.

Lande, R., and Arnold, S.J. 1983. The measurement of selection on correlated characters. Evolution 37: 1210–1226.

Majerus, M.E.N.; O'Donald, P.; and Weir, J. 1982. Female mating preference is genetic. Nature 300: 521–523.

Maynard Smith, J. 1976. Sexual selection and the handicap principle. J. Theor. Bio. 57: 239–242.

Maynard Smith, J. 1985. Mini review: sexual selection, handicaps and true fitness. J. Theor. Bio. 115: 1–8.

O'Donald, P. 1967. A general model of sexual and natural selection. Heredity 22: 499–518.

O'Donald, P. 1973. Frequency-dependent sexual selection as a result of variations in fitness at breeding time. Heredity 30: 351–368.

O'Donald, P. 1980. Genetic Models of Sexual Selection. Cambridge: Cambridge Univ. Press.

O'Donald, P. 1983. The Arctic Skua. A Study of the Ecology and Evolution of a Seabird. Cambridge: Cambridge Univ. Press.

Seger, J. 1985. Unifying genetic models for the evolution of female choice. Evolution 39: 1185–1193.

West-Eberhard, M.J. Sexual selection, social competition, and evolution. Proc. Am. Phil. Soc. 123: 222–234.

Zahavi, A. 1975. Mate selection — a selection for a handicap. J. Theor. Bio. 53: 205–214.

Zahavi, A. 1977. The cost of honesty (further remarks on the handicap principle). J. Theor. Bio. 67: 603–605.

Sexual Selection: Testing the Alternatives
eds. J.W. Bradbury and M.B. Andersson, pp. 119–142
John Wiley & Sons Limited

Sexual Selection: Games between the Sexes

P. Hammerstein* and G.A. Parker**

*Max-Planck-Institut für Verhaltensphysiologie
Abteilung Wickler
8131 Seewiesen, F.R. Germany
**Department of Zoology, P.O. Box 147
University of Liverpool
Liverpool L69 3BX, England

Abstract. Mating involves several conflicts between the sexes. These conflicts are discussed from a game theoretic point of view. In order to mate, the sexes must first meet. If one sex exerts a high search effort, individuals of the other sex can profit by countering this pattern with low mobility; in view of the costs of moving, each sex would be better off if the other took on the high search effort. The problem of who moves thus shows the characteristic of a game between the sexes. We present a new model, the mobility game, in order to deal with the following question: why is the mobility pattern so typically linked with sex and why do females so often win this conflict? It may seem intuitively obvious that males should be the more mobile sex if females invest far more in the offspring and represent a limiting resource for males. This would mean that the sex who "loses" the parental investment game should "win" the mobility game. Surprisingly this intuitive logic turns out to be fragile; in our model both the "roaming male strategy" and the "roaming female strategy" are alternative ESSs for the same game. One sex may thus represent the loser of both the conflict over parental investment and the conflict over mate searching! We make a first attempt to show why the case of losing both these games seems less likely to occur in evolution than the case of winning one and losing the other. Another problem area is more directly related to sexual selection: even when the sexes meet they may not want to mate. A game-like situation arises in this context if it pays one sex to mate but not the other. Conflicts over mating decisions concerning incest, superfluous matings, mate quality, and hybridization are discussed. A previous model by Parker on disagreement between the sexes with regard to incestuous matings is modified. Conditions for the existence of this conflict are discussed. Interestingly, this discussion is related to that of the evolutionary maintenance of sexual reproduction. From the incest model we also conclude that incest avoidance does not seem to be the major principle which would explain sex-specific animal dispersal patterns. The final part of this paper deals with parental investment games. Once again, the problem of multiple ESSs for the same game occurs. This problem seems typical for many models of conflict between the sexes.

INTRODUCTION

A famous paradigm in classical game theory is called the "battle of the sexes" (Luce and Raiffa 1957). The background story of this paradigm relates to an old-fashioned cultural stereotype and reads as follows. A man and a woman (the two players in the game) have to decide about their evening's entertainment. There are two alternatives as to where to spend the day's precious spare time, namely at a prize fight or at a ballet. It is needless to explain which entertainment the old-fashioned players each would prefer if the other joined him or her for the evening. However, it would also cause both players a lot of grief if they ended up spending the evening separately.

This is a rather uninteresting story but it gives rise to an interesting two-person game (see Luce and Raiffa 1957 for the payoff matrix). Mathematicians in the social sciences were puzzled by the battle of the sexes because this simple model seems to have three alternative game theoretic solutions instead of a single one. Two of these solutions are obvious: (a) both players go to the ballet or (b) both spend their evening at the prize fight. The third solution (c) is one in which both individuals play mixed strategies, i.e. the man chooses "prize fight" and "ballet" with appropriate probabilities p and 1-p, the woman makes her choices independently with probablities q and 1-q. Technically speaking, (a), (b), and (c) are alternative Nash equilibria for the sexes. A Nash equilibrium for a two person game is defined as a pair of strategies, each of which is optimal against the other. It therefore does not pay either sex to deviate from a Nash equilibrium if he/she expects the other sex to stick to that solution.

As a theory of rational decision making, classical game theory almost finds itself trapped in a dilemma here. Rational players should play strategies that belong to a Nash equilibrium but they could not easily tell which of the alternative Nash equilibria to choose. In the battle of the sexes the first two solutions would each strongly favor one sex. The third solution seems more fair but it would imply for each player an equilibrium payoff which is even lower than the benefit from seeing the less preferred entertainment together.

Problems with multiple solutions of games have hampered classical game theorists in various contexts due to the obsession that there must be a single rational answer to a given strategic conflict. In contrast, evolutionary game theorists do not share this rigid philosophy and have fewer conceptual problems; it is a generally held view that natural selection may resolve the same conflict in different ways. However, in some of the cases where biological games do have multiple solutions (here solution = evolutionary stable strategy or ESS), major theoretical questions may remain unanswered by a simple ESS analysis. For example, imagine that the process of evolution (instead of the process of rational thinking) would resolve the above conflict between the sexes over evening entertainments. Quite like in classical game theory, one would not be able to predict the winner of this conflict on the basis of elementary ESS theory. This is so because both (a) and (b) are ESSs.

Taken literally, this example does not seem to have relevant biological content. However, the mathematical structure found in the battle of the sexes can be reinterpreted biologically as a parental investment game. Under many circumstances in nature a mate would have an advantage from imposing all or most of the parental tasks on the other sex. However, if the offspring ended up left alone by both parents, this would involve a great loss in fitness to both sexes. Here, the strategy of "looking after the baby" corresponds to the strategy of "accepting the less preferred entertainment." As far as payoffs are concerned, the negative consequences of bilateral desertion from parental care correspond to the negative aspects of bilateral desertion with regard to the evening's entertainment.

Maynard Smith (1977) did the pioneer work on evolutionary games in relation to parental investment. His models bear some resemblance to the simpler battle of the sexes. For a wide range of parameters his results do indeed confirm those gained from the latter model but also share with it the problem of leaving the major theoretical question open: depending on which of two alternative ESSs (for the same numerical payoff matrix) animals play, either males or females may have to carry the burden of caring for the offspring. The alternative ESSs are called the "stickleback" and the "duck" strategy (there are also parameter conditions for which only one of these two strategies is evolutionarily stable; furthermore, "both guard" or "both desert" can be the only ESS).

These results do not quite support an earlier suggestion by Trivers (1972) that a simple logical connection can be constructed between unequal investment in gametes (anisogamy) and unequal investment in offspring. It is not easy to decide whether Trivers committed the so-called Concorde fallacy (Dawkins and Carlisle 1976) or not, but his idea should at least be handled with great care. In the present paper we shall discuss, in particular, further game theoretic arguments which do not quite support the suggestion by Trivers. These arguments are based on the sequential structure of parental investment games and can lead to a unique ESS.

Sexual conflict over parental investment is, like many other games between the sexes, an example of an asymmetric contest. Hammerstein (1981) has shown, for two-person games, how the original definition of an ESS for symmetric conflicts relates to such models in which the two players are "labelled" and "know" about their different labels. ESSs for asymmetric contests often share the following two properties. First, escalation (e.g., bilateral mate desertion) is avoided by attaching "role-specific" behavior to the different labels (e.g., sexes). This often precludes a "fair" outcome so that one player more or less "wins" the game. Second, alternative ESSs exist in which alternative labels are used to determine the "conventional" winner. Therefore, the theory of asymmetric contests sheds some light on the general problem of explaining behavioral differences between the sexes.

In addition to the problem of parental investment, there are two other major areas of evolutionary struggle between the sexes: conflict over mate searching and mating decisions (Parker 1979, 1983b). We shall investigate a new model, the mobility game, in order to discuss mate searching strategies. Some game theoretic insights can be gained as to why one sex should disperse more than the other and why males rather than females represent so often the more mobile sex. This model is a population game with many players. In order to derive its solutions, more complicated techniques than those for solving simple asymmetric contests are required. These techniques will only be outlined here. We find that the problem of alternative ESSs arises in the mobility game, as in the simple battle of the sexes. However, this problem can be overcome to some extent by looking at the differential strength of the selective forces which stabalize the alternative ESSs. It will also be shown that the problem of inbreeding avoidance does not seem to give deep functional insights into sex-specific dispersal patterns. Concerning mating decisions we shall discuss the following game-like situations: conflict between the sexes over incestuous matings, superfluous matings, mate quality, and hybridization.

The present paper is in no way intended to be a complete review of conflicts between the sexes. In particular, we do not cover direct competition between males which can often affect female interests very adversely. In dungflies, for example, paired females can be smeared with the liquid dung or badly damaged during battles between males. Similar effects can occur in toads (Davies and Halliday 1979). A related conflict between the sexes concerns infanticide exhibited by males. None of these topics will be considered here.

SEARCHING FOR MATES

In order to mate, the sexes must first meet. The mechanisms by which this is achieved in nature are amazingly diverse. However, if we search for generality in the animal kingdom we find most often that males exert a much greater search effort than females. Why should this be so?

Intuitively it seems advantageous to individuals of one sex if members of the other sex do a lot of searching; mobility would be expected to be costly. The idea seems appealing that if one sex exerts a high search effort, the other sex can profit by countering this pattern with low mobility. Each sex would be better off if the other took on the high search effort. The problem of who moves thus shows all the characteristics of a game between the sexes. There seem to be two obvious questions to answer. First, why is the mobility pattern so typically linked with sex? Second, why do females so often win this conflict? We seek to answer these questions by looking at a very simple model that we call the mobility game.

The mobility game

In this game all adult members of a given population are involved as players.

The size of the adult population is $2N$, and we assume an unbiased adult sex ratio. There are thus N male players and N female players. We focus attention on a player's time investments with regard to various activities. An adult individual goes through several reproductive cycles, each of which consists of three phases: (a) a mate searching period, (b) a period involving courtship, copulations, parental efforts, as well as gamete production, and (c) a period of reproductive inactivity during which the animal "pays" for previous expenses of movement. The order of these phases is unimportant in our model.

Period (c) can be interpreted in various ways. Time may be needed for resource replenishment (such as filling the tank of a car) or for recovering from injury and exhaustion (such as repairing a heavily used car). Perhaps most important, the time costs of moving, i.e; the duration of phase (c), may partly be thought of as a reduction of an individual's expected lifetime (the car wears out faster if it is used a lot). For the sake of mathematical simplicity, this reduction in lifetime will be accounted for in an indirect way, namely by artificially lengthening a player's reproductive cycle instead of reducing the time he participates in the game. Therefore, with mortality costs of mobility in mind, phase (c) of a reproductive cycle contains not only components for recovery and replenishment but also a "virtual" component which stands for this cycle's "share" of the lifetime-reduction caused by mobility.

The average duration of each phase of a reproductive cycle will be denoted as follows for a female:

s_f = time spent (actively or passively) searching for a mate
g_f = time spent courting a mate and caring for the offspring
r_f = time cost of moving (replenishment, recovery, mortality).

The expected duration T_f of an individual female's reproductive cycle then is $T_f = s_f + g_f + r_f$. The corresponding expected value for a male will be expressed analogously as $T_m = s_m + g_m + r_m$. We assume a 1:1 population sex ratio at the beginning of adulthood and no inherent differences between the sexes in adult survival. As pointed out above the effect of differential mobility on survival is included in r_f and r_m. Finally, we assume the number of surviving offspring per reproductive cycle to be constant and independent of cycle length.

These assumptions have the following important consequence. In a diploid species with a 1:1 sex ratio and no phases of parthenogenetic reproduction, the average cycle length for the male subpopulation, $T_m{}^*$, should not differ from that of the female subpopulation: $T_m{}^* = T_f{}^*$ (including the virtual components). In order to justify this equality imagine that one sex had a lower value for T. This sex then would go through more reproductive cycles during his adult life than the other. With the assumptions made this would mean that the members of the sex with a lower value for T would on the average produce more offspring than the members of the opposite sex. However, such sex-dependent reproductive success cannot occur if each individual offspring has exactly one father and one mother. We thus have shown that the equality in question must hold. It can be restated as follows:

$$s_f^* + g_f^* + r_f^* = s_m^* + g_m^* + r_m^*. \tag{1}$$

At this point it must be emphasized that equation (1) relates to those values s, g, and r that are averaged over the entire subpopulations of members of the same sex (the asterisk denotes such averages). Clearly, a single individual may have reproductive cycles that differ in expected duration from the mean subpopulation. In the present crude model we simply take the length of an individual's reproductive cycle as an inverse measure of its reproductive success: the shorter the cycle, the more offspring will be expected. An evolutionary stable pair of movement strategies for the two sex classes of players is then mainly characterized by the following game theoretic equilibrium condition: if the population plays these strategies, no mutant strategist of either sex can expect a shorter reproductive cycle in this population.

The chessboard paradigm

With all these preliminaries in mind, we now define strategies and payoffs. Imagine that our population's mate-searching game takes place on a large chessboard, with n squares on which individuals can be located like chessmen. Whenever a female and a male meet on the same square, both are removed from the board for an amount of time that corresponds to the length of their two periods during which they are mating, caring, and reproductively inactive. This means, for example, that after meeting a male, a female spends $g_f + r_f$ time units outside the board. At the end of this off-board phase individuals are reintroduced onto randomly chosen empty squares.

Thus at any instant we find those members of the population on the chessboard who are ready for mating, but are still searching for a mate. Suppose that at equilibrium there are F such females and M such males (out of a total of N members of each sex). Time is modelled in a discrete fashion. At each time step, a "player" (here every chessman corresponds to a player) can either stay on his present square or move. If he/she moves, he/she will be transported to another randomly assigned square which is not occupied by a member of the same sex. For mathematical convenience, no intermediate squares are visited during such a movement. An individual's movement may result in meeting a member of the opposite sex, in which case both disappear from the board until their next mate-searching period begins.

In our mobility game we consider only one strategic parameter for each sex, namely the level of mobility. This parameter, v, is defined as the probability of moving to another square at any time step (only one such displacement to another square is possible for an individual per time step). It indicates what proportion of the search time an individual spends moving as opposed to passive waiting. For example, $v_f = 1$ means that a female moves all the time until she finds a mate. Conversely, $v_f = 0$ means that a female sits and waits. It must be emphasized at this point that we do not make the level of time investement in gametes, courtship, and offspring a subject of strategic analysis. The values of g_m and g_f are fixed. We assume in the following that $g_m < g_g$. Furthermore, the time loss with respect to replenishment and mortality is assumed to be proportional to an individual's mobility level and search time:

$$r_m = c_m v_m s_m, \text{ and } r_f = c_f v_f s_f \tag{2}$$

where c_m, c_f are positive constants which correspond to the cost of one unit time spent in mobility, expressed in time taken for "replenishment."

The calculation of how search time depends on mobility levels of the sexes is an essential part in constructing the relation between these strategic parameters and the length of an individual's reproductive cycle. Let $v_f{}^*$ and $v_m{}^*$ denote the mean levels of mobility exhibited by males and females in the given population. Consider single "mutant individuals" of both sexes who play deviant values v_f and v_m. These individuals will meet a mating partner with the following approximate probabilities p_f, p_m respectively:

$$p_f = [v_f + (1 - v_f)v_m{}^*]\,\frac{M}{n}, \tag{3}$$

$$p_m = [v_m + (1 - v_m)v_f{}^*]\,\frac{F}{n}, \tag{4}$$

where n is the number of squares on the chessboard. With respect to these probabilities for success, the mate-searching process of an individual can be thought of as a sequence of so-called Bernoulli trials. Therefore, the expected search times for males and females are $s_f = 1/p_f$ and $s_m = 1/p_m$. Note, however, that the numbers F and M of females and males on the "mating market" are endogenous parameters, i.e., they depend on the strategies played in the population. Expected search time depends on these parameters F and M but, conversely, they depend on expected sex-specific search time of the population too:

$$F = \frac{s_f{}^* N}{s_f{}^* + g_f + r_f{}^*}, \text{ and } M = \frac{s_m{}^* N}{s_m{}^* + g_m + r_m{}^*} \tag{5}$$

We are thus faced with an implicit definition of the mean population search times $s_f{}^*$ and $s_m{}^*$. Sweeping under the carpet the mathematical problem of proving existence and uniqueness of these implicitly characterized search times, we are now able to determine the length of an individual's reproductive cycle. Let $T_f(v_f, v_f{}^*, v_m)$ denote the length of a mutant female's reproductive cycle if her level of mobility is v_f, in a population playing $v_f{}^*$ and $v_m{}^*$. We know that this length is given by the sum $s_f + g_f + r_f$. Therefore,

$$T_f(v_f, v_f{}^*, v_m{}^*) = s_f + g_f + c_f v_f s_f, \tag{6}$$

where g_f and c_f are independent of mobility levels. The analogous formula holds for the length of a male's reproductive cycle.

Analysis of the mobility game

A complete mathematical analysis of this mobility game would be beyond the scope of the present paper and will be given elsewhere. Interestingly, there exist two alternative evolutionarily stable pairs of strategies for the sexes: one in which the male mobility level is 1 and the female mobility level is 0, and the other vice versa. We shall call these two alternatives the "roaming male ESS" and the "roaming female ESS."

Why is the "roaming female strategy" evolutionarily stable? Consider a population in which the males sit and wait ($v_m{}^* = 0$) and the females move "restlessly" ($v_f{}^* = 1$) during their mate-searching period. At any time step a female then finds a mate with probability $p_f{}^* = M/n$, where M is the equilibrium number of males on the mating market and n is the number of squares on the chessboard. Here expected search time is $s_f{}^* = 1/p_f{}^* = n/M$. Now let us draw attention to a mutant female whose mobility level v_f is smaller than that of "wild-type females." This mutant's probability p_f of finding a mate is $p_f = v_f M/n = v_f p_f{}^*$. Thus the search times of mutant and wild type are related to one another by $s_f = s_f{}^*/v_f$. An even simpler relation holds for the time of reproductive inactivity: $r_f = c_f v_f s_f = c_f s_f = r_f{}^*$, i.e., the time cost of moving does not differ between mutant and wild type. Therefore the mutant female ($v_f < 1$) has a reproductive cycle which is longer than that of a wild-type female ($v_f{}^* = 1$). The difference in length between these cycles is

$$T_f - T_f' = \left(\frac{1}{v_f} - 1\right) s_f{}^* . \tag{7}$$

We omit the trivial proof that in our model it does not pay male mutants to break the mobility rules set by the roaming female ESS. These mutants would incur the costs involved in active searching but would not benefit from moving.

Let us now look at the roaming male ESS, which is characterized by the mobility levels $v_m = 1$ and $v_f{}^* = 0$. Here a "lazy" male mutant individual ($v_m < 1$) would pay for his laziness with the following increase in length of his reproductive cycle:

$$T_m - T_m{}^* = \left(\frac{1}{v_m} - 1\right) s_m{}^* . \tag{8}$$

Comparing equations (7) and (8) we are now able to point out an important difference between the two alternatives ESSs. Imagine two populations, each playing one of the alternative stable strategies. In both populations the less mobile mutants of the roaming sex find themselves under negative selective pressure, as indicated by equations (7) and (8). However, if males show lower time investment in offspring and if the time cost of moving is not too high, the strength of this stabilizing selective force differs considerably between the two alternative ESSs. Male mutants in the population of roaming males pay a higher evolutionary price for their lower mobility than female mutants do in the other population; i.e., if we assume that $g_f > g_m$, the expression $(T_m - T_m^*)/T_m^*$ derived from equation (7) is greater than the expression $(T_f - T_f^*)/T_f^*$ derived from equation (8). This is because the following inequality holds if moving is equally costly to both sexes ($c_f = c_m = c$) and if the cost-factor c is not excessive:

$$\frac{s_m^*}{T_m^*} \text{ (roaming male ESS)} > \frac{s_f^*}{T_f^*} \text{ (roaming female ESS)} . \tag{9}$$

Although this important inequality may seem obvious in view of the higher female time investment in offspring, its proof turns out to be more complicated than one would expect.

Let us first show that equation (9) holds if $c = 0$. Note with regard to the right-hand side of (9) that $s_f^*/T_f^* = F_{rf}/N$, where F_{rf} is the equilibrium number of females on the chessboard in a population playing the roaming female ESS. Similarly, we find for the left-hand side in equation (9) that $s_m^*/T_m^* = M_{rm}/\mathrm{N}$, but here M_{rm} is the equilibrium number of males in a different population playing the roaming male strategy. This makes it difficult to compare, the left-hand and right-hand side in equation (9). However, for the case $c = 0$ both ESSs lead to the same equilibrium numbers of males and females on the chessboard. In order to show this let us look at the population with roaming males first. At any time step the expected number of males (or of females) who are leaving the board (the mating market) equals $F_{rm}M_{rm}/n$. Conversely, males and females re-enter the board at time rates, M_{rm}'/g_m and F_{rm}'/g_f, where $M_{rm}'=N-F_{rm}$ denotes the number of off-board males and $F_{rf}'=N-F_{rf}$ denotes the number of off-board females. At equilibrium the rates of leaving the board must equal those of entering the board for both sexes. This gives us the following two balance equations:

$$F_{rm}M_{rm}/n = F_{rm}'/g_f \text{ and } F_{rm}M_{rm}/n = M_{rm}'/g_m. \tag{10}$$

From these balance equations it follows that $F_{rm}'=N-F_{rm}$ must be a solution of the following quadratic equation (with the constraint that $0 < F_{rm}' < N$):

$$(N - F_{rm}')(N - \frac{g_m}{g_f} F_{rm}') = \frac{n}{g_f} F_{rm}' . \tag{11}$$

Equation (11) characterizes F_{rm}', F_{rm}, respectively. An analogous equation can be derived for M_{rm}', M_{rm}, respectively. Furthermore, the same arguments that have led to equation (11) also hold if we study the roaming female ESS and ask for the equilibrium number of females in a population playing this ESS; therefore both F_{rf} and F_{rm} are characterized by the same quadratic equation

(11), so that both alternatives ESSs give rise to the same equilibrium numbers of individuals on the board: $M_{rm}=M_{rf}$ and $F_{rm}=F_{rf}$. Now, since we know from the inequality $g_m<g_f$ and from equation (1) that $M_{rm}>F_{rm}$, the latter inequality must also hold between ESSs: $M_{rm}>F_{rf}$.

At this point we have established the crucial inequality (9) for the case $c = 0$. It is easy to see that the introduction of small values of c should not change this result. The discussion of large values of c will be given elsewhere.

Conclusions from the mobility game

As we have seen from the mobility game, we can quite readily explain why one sex should expend a far greater search and movement effort than the other. It is less clear *which* of the sexes should search and move. Even in species where males have little investment with their mates and offspring, females could be far more mobile than males; there exist two alternative ESSs like in many other games between the sexes. However, the "roaming male ESS" seems more likely than the "roaming female ESS" for the following reason: the selective forces which stabilize a high level of male mobility are stronger than those which stabilize a high level of female mobility. Ideally, we would like to take this comparison a step further and determine for both ESSs their "range of attraction" in an explicitly specified dynamic model of the selection process. However, this seems a difficult task which is well beyond the scope of the present paper.

Mobility and inbreeding avoidance

As an alternative to our model, avoidance of inbreeding has been a widely accepted interpretation of the movement patterns of the sexes. Here the argument is that inbreeding depression can be prevented if one sex moves away from the natal area (e.g., Greenwood and Harvey 1976; Greenwood et al. 1978; Bischof 1985; Harvey and Ralls 1986). We expect that a dispersal game which allows for inbreeding depression, but where there is no difficulty to find mates, would lead to an ESS in which one sex moves far more than the other. Therefore, at first glance there seems to be game theoretic support for the suggestion that inbreeding avoidance may structure the dispersal pattern.

We see, however, the following problems in justifying this popular idea. In species showing little male investment with mates and offspring, females would have much more to gain from avoiding inbreeding than males; indeed, it may commonly pay males to engage in incestuous matings (see next section on *MATING DECISIONS*). Hence with this dispersal game the probability of achieving an ESS in which the female moves out of the natal area seems much greater than the possibility of an alternative ESS where males move out. This is clearly contrary to the predominant dispersal patterns found in nature, though "roaming female" patterns are found in many birds.

As an alternative to females moving out, the need to avoid incest could result in females being reluctant to mate with males who grew up with them in their natal area. In this case, if males can be effectively rejected, then at least one sex (but not necessarily the males) would be forced to search elsewhere for mates. Hence, we are faced essentially with the logic of the mobility game once again. Our result would be that for most species, males do the major part of the moving. But we stress that this explanation of dispersal patterns depends more crucially on the rationale of the mobility game than on the argument of inbreeding avoidance.

Could the latter argument nevertheless prove more substantial if males cannot effectively be rejected? In this case, females would have strong reasons to avoid incest by displacing at least some distance from their natal area, unless the costs of moving outweigh those of inbreeding. Thus we reach the same problem that in isolation the argument of incest avoidance more readily predicts high female dispersal than the general biological pattern of high male mobility.

Our arguments above all rely on the assumption that there will be sexual conflict over incest, i.e., selection favors males mating with close relatives but favors females avoiding such matings. Over a wide range of inbreeding costs this will indeed be the case. There are, however, two other cases: (a) both profit from incest; (b) both lose from incest. Clearly, in (a) the inbreeding avoidance argument cannot immediately explain dispersal pattern because inbreeding should not be avoided. In case (b), the inbreeding argument can explain a displacement from the natal area by one sex. But where females invest more, selection pressure on females to avoid incest is likely to be higher than on males.

Mobility and contests

Fighting has been used to explain mobility patterns in mammals (see Greenwood 1980). Most important, in species where sexual selection has favored aggressive guarding of females by males, this obviously has the side effect that dominant males will eject younger rivals from their immediate natal area. Even after being ejected these losers may expect many other encounters with guarding males before they eventually find a mating partner available to them. Therefore, intrasexual contests can be an important factor contributing to male dispersal.

MATING DECISIONS

Even when the sexes meet, they may not want to mate. A game-like situation arises if it pays one sex to mate but not the other (Parker 1979, 1983b). The four most obvious mating conflicts concern incest, superflous matings, mate quality, and hybridization.

Incest

Although it may seem intuitively appealing that incest avoidance between close relatives should be mutually beneficial to both sexes, this need not be so. Selection can often result in a sexual conflict such that it pays the male to mate, but not the female. Let us consider the question whether conflict will arise between potential mates whose coefficient of relatedness is 1/2 (i.e., full sibs, mother-son, father-daughter). It is easy to extend this analysis to cover other forms of inbreeding. The model is modified from Parker (1979); parts of the analysis were derived independently by Bengtsson (1978) and Packer (1979).

Imagine that the decision to mate lies solely under the control of the male. Consider two heritable traits: "avoid incest" and "attempt incest." Suppose that a male who avoids incest would produce m outbred offspring. In contrast, a male who attempts incest would lose a proportion p of these offspring, but gains f offspring via the incestuous mating. Here, f is the number of offspring the female produces from a mating, whereas n is the total number of a female's offspring. Now we suppose that the survivorship of an inbred offspring is $(1\text{-}d)$ times that of an outbred offspring; therefore d represents the reduction in survivorship due to inbreeding. We calculate the inclusive fitness of the two alternative traits as follows:

$$W_m(\text{avoid}) = m \cdot 1/2 + n \cdot 1/4, \tag{12}$$

$$W_m(\text{attempt}) = (1\text{-p})\text{m} \cdot 1/2 + (n\text{-}f) \cdot 1/4 + (1\text{-}d)\text{f} \cdot 3/4. \tag{13}$$

The fractions 1/2, 1/4, and 3/4 represent respectively the degree of relatedness between the male and (a) his own outbred offspring, (b) his sister's outbred offspring, and (c) incestuously produced offspring. Attempting incest will be better than avoiding if W_m(attempt) is greater than W_m(avoid), i.e., if

$$mp/f < 1 - d \cdot 3/2 \tag{14}$$

The world looks different from the female's point of view. We consider the analogous two heritable traits and calculate their inclusive fitness:

$$W_f(\text{avoid}) = n \cdot 1/2 + m \cdot 1/4, \tag{15}$$

$$W_f(\text{attempt}) = (n\text{-}f) \cdot 1/2 + (1\text{-}d)f \cdot 3/4 + (1\text{-}p)m \cdot 1/4. \tag{16}$$

It is easy to see that the female should avoid incest if the following inequality holds:

$$mp/f > 1 - 3d. \tag{17}$$

Conclusions from the incest model

From the inequalities (14) and (17) we can draw the interesting conclusion that there is a range of parameter values for the model that leads to sexual conflict in the following sense: males should attempt to engage in incestuous mating, but females should avoid incest. This range is given by

$$1 - 3d < \frac{mp}{f} < 1 - \frac{3}{2}\text{d}. \tag{18}$$

In order to interpret condition (18) we remind the reader that d is the inbreeding depression, f the brood size, and mp the number of outbred offspring a male sacrifices by opting for the incestuous mating. Suppose now that incest has virtually no opportunity costs for the male, so that $p = 0$. This would apply if a male's instantaneous mating decision had little influence on his future mating prospects. Condition (18) then looks as follows:

$$1/3 < d < 2/3. \tag{19}$$

The range of sexual conflict varies with the number of outbred offspring a male misses through choosing incest. This number (mp) can be related to other biological characteristics of the mating system. Let g be the reproductive time investment due to mating (time taken to copulate, to replenish the gametes expended, to supply parental care, etc.); we have g_f for the female, and g_m for the male. For simplicity, we assume that females do not have to wait for mates, so that g_f is also the time taken for her to produce the typical brood of f offspring. If a male spends time g_m on each mating, then his probability of finding a mate in this time is $g_m g_f^{-1} s^{-1}$, where s is the adult sex ratio (no. of males/no. of females). Therefore, the number pm of outbred offspring lost due to opting for an incestuous mating will be $f g_m g_f^{-1} s^{-1}$.

Calling $\alpha = g_m g_f^{-1} s^{-1}$, we then find a new way to express condition (18) for the range of sexual conflict over incest:

$$(1\text{-}\alpha)/3 < d < (1\text{-}\alpha)\cdot 2/3. \tag{20}$$

What do we gain by this reformulation of condition (18)? It enables us to interpret the sexual conflict range in terms of the adult sex ratio, the time investment ratio, and the inbreeding depression. Interestingly, we are able to relate α to a measure for the operational sex ratio (sensu Emlen and Oring 1977). We would suggest the following definition for an operational sex ratio (OSR):

$$\text{OSR} = \frac{\text{potential mating rate for adult male population}}{\text{potential mating rate for adult female population}} \tag{21}$$

Here, potential mating rate means the maximum number of matings per time unit all the individuals of one sex could perform if mating partners were in unlimited supply, the only limit being the time investment g. Therefore

$$\mathrm{OSR} = sg_f/g_m = \alpha^{-1} \qquad (22)$$

We hope that an interpretation of OSR in this sort of way may both facilitate theoretical developments in sexual selection models and yet simultaneously help to sharpen the empirical measurement of such a quantity (see also Sutherland 1985).

The biological consequences of condition (20) can now be stated as follows. If the OSR is extremely high, condition (20) becomes the same as condition (19). This means that there is a wide range of degrees of inbreeding depression for which there is conflict between the sexes over incest. Our model assumptions are such that the minimum value for OSR is 1, because for simplicity we have not included any time costs for females to find males. This minimum OSR could, for example, exist with a unity adult sex ratio and equal time investment for the sexes. In this case, conflict between the sexes over incest cannot occur; it will here pay both sexes to avoid incest, whatever the inbreeding depression.

One obvious conclusion from our model is that viewed across species, we would expect that attempts at incest would become more common as the operational sex ratio increases. A word of caution here would be that our model merely states when conflict is expected to occur; it does not tell us how the conflict will be resolved. The latter problem would require a game theoretic treatment and would lead into the conceptual difficulties posed by games between relatives. One could no longer rely on the inclusive fitness approach (Grafen 1979) and one would also have to face the fact that both the degree of inbreeding depression and the coefficient of relatedness will depend on what strategies the sexes adopt.

Contrary to the above model conventional wisdom has tended to be that it would pay both sexes to avoid incest, though the consequences of much greater need for avoidance by females have often been discussed (e.g., Clutton-Brock and Harvey 1976). We stresss again that incest can also be (a) favorable to both sexes or alternatively (b) advantageous to males while disadvantageous to females (see also Bengtsson 1978; Packer 1979; Parker 1979). This may explain why males in some species attempt incestuous matings. In some instances females show no reluctance to mate with fathers or brothers (e.g., many Hymenoptera). From condition (20) we would expect that the inbreeding cost d would be smaller than 1/3, and that if male time investment is high, d must be very small indeed. However, the lack of observable reluctance need not always be explained by the absence of conflict: females may be in an unfavorable strategic situation, where their means of defence are strongly limited. In some cases females do show a marked reluctance to mate incestuously (Bischof 1985; Harvey and Ralls 1986), indicating that sexual conflict over incest can occur.

Our model does not allow an operational sex ratio less than 1. This, however, seems biologically possible if, for example, there is sex role reversal so that male time investment exceeds that of females. In order to sudy such a case, it is possible to reinterpret the model by assuming that males now never have to wait for females and that the other assumptions about males and females are also reversed. Surprisingly, females should then try to impose incestuous matings on reluctant males if the cost of inbreeding lies between the appropriate threshold values for sexual conflicts. Practical difficulties arise in this case, however.

Finally, there is a sense in which inbreeding can be said to reduce conflict over mating decisions. So far we have analyzed the case of incest with a coefficient of relatedness $r = 1/2$, where d is the corresponding inbreeding depression. Now consider a meeting between two unrelated individuals of opposite sex ($r = 0$), but where there is nevertheless a mating disadvantage d' so that offspring have survivorship ($1\text{-}d'$) relative to the usual expectation of 1. Following our previous logic, if males invest nothing ($p = 0$) we would expect the male to attempt to mate and the female to avoid mating if $0 < d' < 1$. Clearly, the range of conflict is now *greater* than that suggested by condition (19). We thank Gerry Wilkinson for drawing this to our attention.

The reasons for this effect are as follows. In our model the male loses nothing from a mating if $p = 0$ (copulation costs nothing). He should therefore always mate if there are any prospects greater than zero. In contrast, for the female, there are no costs associated with rejecting a male; it is assumed that she can easily mate with an alternative male offering better offspring survivorship. Hence she should always avoid matings where survivorship of offspring would be less than 1. It is thus clear that for a given level of mating disadvantage, inbreeding reduces the probability of conflict because of kin selection. It can "pay" a sister to "help out" a brother by allowing him to mate, since he has a probability of 1/2 of sharing a gene which contributes to this "decision."

By similar reasoning, it seems obvious that if there is no cost of inbreeding, close relatives should actively prefer each other as mating partners. This problem quickly converges to the question of why sexual reproduction is not abandoned in favor of asexuality. The evidence is that at least in many birds and mammals (see Harvey and Ralls, 1986) females do attempt to avoid mating with closely related males. This implies that d lies within the conflict range proposed in condition (19).

Superfluous matings

A major source of sexual conflict arises when a male encounters a female that has recently mated and so contains enough sperm to fertilize all her eggs. Such conflicts occur throughout the animal kingdom, but are especially prevalent in groups such as insects where sperm can be stored for long periods of time. The conflict is created by the fact that male gains from achieving these matings are often high, whereas they typically impose greater costs than benefits for the female. The male gains are fairly obvious, though one has to take into account the devaluation due to sperm competition (Parker 1970; Smith 1984). Female gains are likely to be much more restricted. Clear cases of female benefit from repeated mating include nuptial gifts, nutritional use of sperm, protection from sexual harrassment, access to resources, etc.

More contentious candidates for female benefit have to do with properties of the genetic material transmitted, e.g., replacing sperm of low genetic quality with higher quality sperm or increasing the genetic variance of the stored sperm (Walker 1980). We discuss decisions over mate quality in the next section. The genetic argument suffers from the difficulty that in the simplest model the expected fitness of a multiply mating female is the same as that of a singly mating female (Parker 1984). More complex models generate conclusions that can differ from this result (Knowlton and Greenwell 1984), though they require analytical investigation before their quantitative importance can be assessed, especially if multiple mating is more costly to females than single mating.

In contrast, the costs to the female of superflous matings can be substantial. At the very least there is the time loss of a copulation and the risk of parasitization or infection. Sometimes females can be seriously damaged through male-male contests during copulation. In dungflies, struggles during copulation can lead to the female being torn or smothered in liquid dung (Hammer 1941). Female toads can be drowned during struggles between males but fertilization is external and so superfluous mating cannot occur (Davies and Halliday 1979). In the dungflies, most females do not need sperm (there is plenty left from previous matings) and so the costly matings are usually superfluous, though the female may benefit by having a mate to guard her during ovipositon (Parker 1970).

Superfluous matings can therefore often constitute cases of sexual conflict in which it pays the male to mate but pays the female to avoid matings. However, the level of conflict may depend upon the effects of male-male competition (Parker 1984). We give an illustration by the example of the copulatory plug, which has commonly evolved through sperm competition as a mechanism to reduce the probability that future males can transfer sperm and share the offspring (i.e., as a "chastity belt"). Plugs can be quite large and hence quite expensive for males to produce. They probably also have some small immediate costs to the female, which must therefore also be felt by the male donor.

Suppose that before any plug evolves superflous matings are either selectively neutral or marginally beneficial to females, so that females are fully receptive to such matings. There would then be high selection pressure on males to produce plugs, and expenditure on the plug would probably proceed to some optimal level for males (maximum prevention of second matings can never occur, since this would make it not worthwhile attempting to mate with plugged females; this would in turn cause expenditure on plugs to be reduced).

However, if this "'male-optimized" plug encumbers females to the extent that superflous matings are now a direct disadvantage to female interests, we expect that females should begin to develop unreceptivity after their first mating. This unreceptivity now makes the current level of expenditure on the plug super-optimal for males; the system will thus have very complex dynamics, having both inter- and intrasexual frequency dependence. Parker (1984) has suggested that the outcome of such games may be a Nash equilibrium in which male and female strategies stabilize at competitive optima. Cycles also seem possible. A full game theoretic analysis of such systems might prove interesting.

Mate quality

This area of sexual conflict was treated by Parker (1983b) in a preliminary and prospective analysis. Conflicting over mating arises if at least one sex is discriminative with regard to mating partners. If variance in mate quality is similar for each sex, then the discriminate sex will be the sex with the higher reproductive time investment. If, on the other hand, reproductive investment is similar for each sex, then it will be the sex with the least variance in mate quality that will be discriminative. In some cases both sexes will be discriminative.

Hybridization

A sexual conflict may commonly occur when a male meets a female of a different race, ecotype, or sibling species (Parker 1979). The resolution of this conflict will have an important influence on sympatric speciation and species/race isolation.

Consider a case in which meetings between the sexes of two sibling species (A and B) are rare, so that hybrid offspring are infrequent and do not influence the following calculations by altering search times, etc. Let the mean fitness of offspring from conspecific matings be 1; 1-d is the fitness of hybrid offspring. Once again we express the search times for each sex as s_m and s_f and the reproductive time investments as g_m and g_f. For simplicity, assume initially that the hybrid disadvantage d and the time investment patterns of the two species are similar and that the sex ratio is unity in both species.

For an A-male, selection favors mating with a B-female that has just been encountered if the gain rate from the A-B mating exceeds the expected rate due to ignoring the B-female and searching for an A-female. Thus it will pay to mate if

$$(1-d_m)/g_m > 1/(g_m+s_m). \tag{23}$$

This follows from the logic of MacArthur and Pianka (1986) modified for mate searching. The decision to mate is as good as the decision not to mate at the threshold

$$d_m = s_m/(g_m + s_m) \tag{24}$$

for the male, and (by the same reasoning)

$$d_f = s_f/(g_f + s_f) \tag{25}$$

for the female. At a 1:1 sex ratio we have $g_m + s_m = g_f + s_f$, and the difference between the thresholds is

$$d_m - d_f = (s_m - s_f)/(g_f + s_f), \text{ or} \tag{26}$$

$$d_m - d_f = (g_f - g_m/(g_f + s_f). \tag{27}$$

This difference gives a measure of the probability of conflict since the more disparate the thresholds, the more likely that a given, actual level d of hybrid disadvantage will fall between them.

It is clear from (27) that the greatest difference between thresholds will occur when $g_f >> g_m$, i.e., when female parental time investment vastly exceeds that of the male. Another intuitively obvious conclusion is that the easier it is for the female to find a mate (s_f small), the higher the probability of conflict. The smaller the value of s_f, the less it costs the B-female to reject a present A-male in favor of searching for a future B-male. In the normal case, where female reproductive investment exceeds that of the male ($g_f > g_m$), the conflict will be such that it pays the male to attempt to mate, but pays the female to avoid mating. There is much biological evidence for male persistence and female reluctance in meetings between closely related species. For example, Bateman (1948) pointed out that in *Drosophila*, females tend to be reluctant to mate across species, but males are typically the reverse. In contrast, where there is sex role reversal ($g_m > g_f$), the contest will be such that it pays the female to mate, but not the male.

We can measure the intensity of selection on each sex faced by a conflict of this sort. Suppose that offspring fitness in an A-B mating is 1-d_x. Selective pressure acting on the male to favor mating will be proportional to the difference in payoff between the alternative strategies (mating versus withdrawing). This difference D_m equals the difference between left- and right-hand side of equation (24). Similarly, selective pressure on the female favoring withdrawing or rejecting the male (rather than mating) will be proportional to the difference D_f between right- and left-hand side of equation 25. The intensity of selection on males to persist will be greater than that on females to reject if $D_m > D_f$, i.e., if

$$d_x < 1 - \frac{2g_m g_f}{(g_m + g_f)(g_f + s_f)}. \tag{28}$$

If the inverse inequality holds, selection is greater on the female to reject than it is on the male to persist.

In fact, it can be shown (Parker 1979) that the value of d_x at which selection intensity (favoring the opposite outcome in the two sexes) is equal always lies between d_m and d_f, but closer to d_m than to d_f if female investment (g_f) exceeds male investment (g_m). When g_m is trivially small (approaching zero), selection is always greater on male persistence.

This offers a contrast to the classical view that the females are under the most intense selection to avoid cross mating and that males are under similar but much weaker selection to avoid cross mating.

A more realistic model must take into account differences between *A* and *B* in their "aptitude for encounter," which will determine the search times for each species and in the ratio of parental investment, g_m/g_f. There are two quite separate conditions to consider if these parameters differ. These are (a) male *A* × female *B*, and (b) male *B* × female *A*. It would be theoretically possible, say, for conflict to exist in case (a) but not in case (b). This could arise if *A* has much longer search times, s_f, s_m, than *B*; the long search time s_f could make it favorable for *A* females to accept *B* males. Gene flow could thus be restricted in one "mating direction," but not in the other. A further possibility is that if male investment g_m becomes large, a conflict could occur in which the *B* female should "want" to mate, but the *A* male should "want" to withdraw. Though *B* type males may not be in danger of being raped by females, it is possible that they could be "cheated" in evolutionary terms (at least until counterselection operates) if selection favors *A*-type females that mimic the *B*-type females.

The implications of these sorts of asymmetries between the sexes and between ecotypes for the probability of sympatric speciation, or for the maintenance or breakdown of separate gene pools, would seem to be a rich source for further study. In general, where the biology of the two exotypes or races is not radically different the original model should serve as a good approximation. Especially where male investment is relatively low, sexual conflict is highly likely and selection is likely to be more intense on the male to persist than on the female to reject. Thus where there is disruptive selection acting on a species, selection will often favor females that reject males which are unlike themselves. Unchecked, this adaptation could theoretically lead to assortative mating and maintenance of separate gene pools. However, this is to assume that the female can always win the sexual conflict. If the male interests prevail, gene flow will be maintained and sympatric speciation prevented.

Resolution of sexual conflict over mating

Resolution of sexual conflict games where it pays the male to mate but not the female is a complex topic that will repay further study and development. One type of model (Parker 1979; Knowlton and Greenwell 1984) consists of a game in which each sex has two discrete strategies: males can either be active in their attempts to mate or can give up quickly if a female is uncooperative; similarly, females can either attempt to reject male advances or submit to them without appreciable rejection effort. Such games can generate complex dynamics (see Knowlton and Greenwell 1984) in which outcomes depend on both the starting frequencies of strategies and on the payoffs. These games are rather similar to Dawkins' (1976) "battle of the sexes."

Another type of model (Parker 1979, 1983b) discusses resolution of sexual conflict over mating decisions when strategies can vary continuously for each sex. The two games proposed are given below.

Sexual war of attrition: the persistence game

Mating is impossible unless the female allows the male to mate. Each sex can choose a persistence time to "hold out" against the other. During a "battle," contest costs accrue unequally for the two opponents. Mating occurs if the female gives up being unreceptive before the male gives up courtship. The benefit of winning differs for each sex.

This game seems analogous to the asymmetric "war of attribution," sensu Hammerstein and Parker (1982), provided that there is sufficient variation *within* each sex in the payoff relevant asymmetries (benefits of winning, contest costs) to "override" the obvious sex difference between the two players. The reason that this restriction is important cannot be explained fully here, but it has to do with the following. Hammerstein and Parker (1982) showed that an ESS for the asymmetric war of attrition can occur only if the individuals make occasional mistakes about the payoff-relevant asymmetries in the contest. For the sexual war of attrition it would make no sense to assume that the players could make mistakes about their sex, but mistakes could be made over payoff-relevant asymmetries prodiving these do not correlate perfectly with the sexual asymmetry.

The ESS for the asymmetric war of attrition is such that the player with the higher value for

$$\frac{\text{value of winning the contest}}{\text{rate at which contest costs accrue}} \tag{29}$$

typically chooses a high persistence time, and the player with the lower value for the same index typically chooses a low persistence time (Hammerstein and Parker 1982). For the sexual war of attrition this may mean that one sex commonly wins the contest, but not always.

Sexual arms race

Contest costs are trivial. The outcome depends on each opponent's level of armament (weaponry, body size) so that a male can either enforce mating or not. However, increasing the arms level results in increased costs (e.g., reduced survivorship); the best strategy therefore depends on the opponent's current level of armament.

Arms races have been discussed in general terms by Dawkins and Krebs (1979) and modelled by Parker (1979, 1983a). The sexual arms race is a version of the "opponent-independent costs" game (Parker 1979, 1983a), so named because the costs do not depend on the actual contest with the opponent but depend on prior morphological "choice" of arms level. Providing that "choosing" a given arms investment level specifies only a probability distribution of armament actually attained, an ESS can consist of a pure strategy — a competitive optimum for arms level. For the sexual arms race an ESS may consist of a pair of arms levels, say x^* for the male and y^* for the female, from which it would pay neither to deviate unilaterally (Parker 1983a, b). Because of asymmetries in the sexes in the value of winning and in the cost of overcoming a unit increase in the opponent's arms level, the ESS may be such that again one sex usually wins the conflict, but not always.

PARENTAL INVESTMENT GAMES

There are two kinds of parental investment games. Desertion games, on the one hand, bring into focus the problem of whether an individual of a given sex should participate at all in the effort of parental care. On the other hand, continuous allocation games focus on the problem of how much the sexes should each contribute to raising the offspring in the case of biparental investment.

Desertion games

The simplest desertion game is the so-called "battle of the sexes" from classical game theory. Dawkins (1976) introduced the term "battle of the sexes" in the biological context but did not refer to this classical paradigm. Maynard Smith (1977) studied more complex but similar games in which each sex may either "guard" or "desert." Often there are two alternative ESSs: in the "stickleback" ESS the male guards and the female deserts; in the "duck" ESS the sex roles are reversed. There are also conditions where either "stickleback" or "duck" is the only ESS.

Interestingly, the confidence of paternity does not seem to have a major effect on male parental investment in modified versions of this game. In contrast, Trivers (1972) has argued that the more uncertainty exists on paternity, the less there would be an ultimate cause for males to engage in parental care. However, this simple argument involves logical problems because changes in "opportunity costs" may be ignored. Admittedly, the gains to a male from a partner's offspring decrease if confidence of paternity decreases, but opportunity costs decrease as well if females are more promiscuous (see Grafen [1980] for the concept of opportunity costs).

The existence of alternative ESSs in the model by Maynard Smith (1977) shows that there is no simple logical link between anisogamy and a female's role as the major parental investor. More complicated desertion models can be considered in which the sequential structure of the game is taken into account. It is possible to show mathematically that the sex who is "left with the zygote" has an important strategic disadvantage (work in progress using games in "extensive form" and the so-called "trembling hand approach"). This idea was originally suggested by Maynard Smith (1978). He presents comparative data about the relation between parental care and internal versus external fertilization in taxonomic families of bony fishes. These data support the view that external fertilization can favor an ESS in which the male guards and the female deserts.

Continuous-allocation games

When both parents care for the offspring, each parent's strategy could consist of a continuously varying amount of investment from zero up to some maximum. It is easy to model a sexual conflict in which each parent can increase or decrease its investment into a particular offspring against the current level of allocation provided by the other parent. The simplest such game is that of Maynard Smith (1982), who derived an ESS amount of provisioning for gametes in an isogamous species in which the zygote's provisioning is the sum of provisioning provided by the two parents. An analogous model has been proposed independently by Parker (1985) for a parental investment game. Here investment is regarded as the allocation of a fixed resource for each parent. Portions of the resource are therefore allocated to each offspring very much in the same way that slices are allocated from a cake. The solution depends on the mating system. If there is "true" monogamy in the sense that there is no chance to remate after one partner dies, the total investment per offspring (sum of the two parents) should be the same as when one parent invests unaided. In contrast, if one parent can remate after the death of its partner, then paradoxically the *total* investment per offspring should be *less* than that for the one parent case.

In such games, the important consideration is the extent to which self's reproductive success is affected by the expenditure of the mate's investment. With "true" monogamy, self cannot reproduce after the death of the mate and the ESS is thus the same as for uniparental investment; it is the solution that maximizes the product of offspring viability and offspring number. But when a mate's death does not have any effect on self's reproductive success, a mutant parent which allocates slightly less investment per offspring can profit by surviving to produce more offspring, even though each offspring has marginally poorer survivorship. Because of high investment by the mate the drop in survivorship is not as costly as the benefit of higher offspring numbers. In this game the investment per offspring should gradually reduce towards the new level, which is the competitive optimum.

A feature which would act to destabilize biparental investment is that of differential response to the offsprings' solicitation (Parker 1985). Consider a system in which, although sex roles may typically be different, both parents are nevertheless sensitive to the extent or intensity of the offsprings solicitation for parental investment. Suppose that if one parent ignores high solicitation, the other parent in some way meets the offspring's demands for investment. Such circumstances could also lead to insensitivity in one sex and ultimately to uniparental investment by the other sex.

Continuous parental allocation games (e.g., games of incubation duration, vigilance versus feeding, etc.) seem a rich source for further study.

REFERENCES

Bateman, A.J. 1948. Intra-sexual selection in *Drosophila*. Heredity 2: 349–368.

Bengtsson, B.O. 1978. Avoiding inbreeding: at what cost? J. Theor. Biol. 73: 439–444.

Bischof, N. 1985. Das Rätsel Ödipus. München: Piper.

Clutton-Brock, T.H., and Harvey, P.H. 1976. Evolutionary rules and primate socieites. In: Growing Points in Ethology, eds. P.P.G. Bateson and R.A. Hinde, pp. 195–237. Cambridge: Cambridge Univ. Press.

Davies, N.B., and Halliday, T.M. 1979. Competitive mate searching in male common toads, *Bufo bufo*. Anim. Behav. 27: 1253–1267.

Dawkins, R. 1976. The Selfish Gene. Oxford: Oxford Univ. Press.

Dawkins, R., and Carlisle, T.R. 1976. Parental investment, mate desertion and a fallacy. Nature 262: 131–133.

Dawkins, R., and Krebs, J.R. 1979. Arms races between and within and among species. Proc. Roy. Soc. Lon. 205: 489–511.

Emlen, S.T., and Oring, L.W. 1977. Ecology, sexual selection, and the evolution of mating systems. Science 197: 215–223.

Grafen, A. 1979. The hawk-dove game played between relatives. Anim. Behav. 27: 905–907.

Grafen, A. 1980. Opportunity costs, benefit and degree of relatedness. Anim. Behav. 28: 967–968.

Greenwood, P.J. 1980. Mating systems, philopatry and dispersal in birds and mammals. Anim. Behav. 28: 1140–1162.

Greenwood, P.J., and Harvey, P.H. 1976. The adaptive significance of breeding area fidelity of the blackbird (*Turdus merula* L.). J. Anim. Ecol. 45: 887–898.

Greenwood, P.J.; Harvey, P.H., and Perrins, C.M. 1978. Inbreeding and dispersal in the great tit. Nature 271: 52–54.

Hammer, O. 1941. Biological and ecological investigations on flies associated with pasturing cattle and their excrement. Vidensk. Medd. Dan. Naturhist. Foren. 105: 1–257.

Hammerstein, P. 1981. The role of asymmetries in animal contests. Anim. Behav. 29: 193–205.

Hammerstein, P., and Parker, G.A. 1982. The asymmetric war of attrition. J. Theor. Biol. 96: 647–682.

Harvey, P.H., and Ralls, K. 1986. Do animals avoid incest? Nature 320: 575–576.

Knowlton, N., and Greenwell, S.R. 1984. Male sperm competition avoidance mechanisms: the influence of female interests. In: Sperm Competition and the Evolution of Animal Mating Systems, ed. R.L. Smith, pp. 61–84. New York: Academic Press.

Luce, R.D., and Raiffa, H. 1957. Games and Decisions. New York: Wiley.

MacArthur, R.H., and Pianka, E.R. 1966. On optimal use of a patchy environment. Am. Natural. 100: 603–609.

Maynard Smith, J. 1977. Parental investment: a prospective analysis. Anim. Behav. 25: 1–9.

Maynard Smith, J. 1978. The Evolution of Sex. Cambridge: Cambridge Univ. Press.

Maynard Smith, J. 1982. Evolution and the Theory of Games. Cambridge: Cambridge Univ. Press.

Packer, C. 1979. Male dominance and reproductive activity in *Papio anubis*. Anim. Behav. 27: 37–45.

Parker, G.A. 1970. Sperm competition and its evolutionary effect on copula duration in the fly *Scatophaga stercoraria*. J. Insect Ph. 16: 1301–1328.

Parker, G.A. 1979. Sexual Selection and sexual conflict. In: Sexual Selection and Reproductive Competition in Insects, eds. M.S. Blum and N.A. Blum, pp. 123–166. New York; Academic Press.

Parker, G.A. 1983a. Arms races in evolution: an ESS to the opponent-independent costs game. J. Theor. Bio. 101: 619–648.

Parker, G.A. 1983b. Mate quality and mating decisions. In: Mate Choice, ed. P. Bateson, pp. 141–164. Cambridge: Cambridge Univ. Press.

Parker, G.A. 1984. Sperm competition and the evolution of animal mating strategies. In: Sperm Competition and the Evolution of Animal Mating Systems, ed. R.L. Smith, pp. 1–60. New York: Academic Press.

Parker, G.A. 1985. Models of parent-offspring conflict. V. Effects of the behavior of the two parents. Anim. Behav. 33: 519–533.

Smith, R.L. 1984. Sperm competition and the evolution of animal mating systems. New York: Academic Press.

Sutherland, W.J. 1985. Measures of sexual selection. Oxford Surv. Avol. Biol. 2: 90–101.

Trivers, R.L. 1972. Parental investment and sexual selection. In: Sexual Selection and the Descent of Man, ed. B. Campbell, pp. 136–179. Chicago: Aldine.

Walker, W.F. 1980. Sperm utilization strategies in non-social insects. Am. Natural 115: 780–799.

Sexual Selection: Testing the Alternatives
eds. J.W. Bradbury and M.B. Andersson, pp. 143–163
John Wiley & Sons Limited

Relative Roles of Intra- and Intersexual Selection

J.W. Bradbury* and N.B. Davies**

*Department of Biology, C-016
University of California, San Diego
La Jolla, CA 92093, USA
**Department of Zoology
University of Cambridge
Cambridge CB2 3EJ, England

Abstract. Whereas workers early in this century saw intrasexual conflict as an obvious and critical mechanism of sexual selection, they largely rejected intersexual choice as a significant process. Field studies in the second half of this century have clearly demonstrated that both mechanisms operate in natural systems, that they may often be in conflict, and that which plays the larger role depends very much on social and environmental contexts. To help identify some of these critical contexts, we survey several taxa and types of mating systems in which the relative roles are known to vary. In species in which males provide neither parental care nor resources to females the relative role of female choice, and the levels of sophistication in that choice, appear correlated with factors such as the duration of the breeding season, the pattern and stability of male dispersion, and some life history characteristics of the females involved. In species in which males do provide resources and/or parental care, it is apparent that female choice of resource or care, instead of male traits, tends to dominate mate assortment. Even so, females mate with other males at low, but persistent levels, and males exhibit a variety of techniques for biasing assortment in their favor. Our review suggests that nearly all systems show some mix of these two processes and the identification of critical contexts must take the possibility of such mixtures into account.

INTRODUCTION

Darwin (1871) proposed two modes of sexual selection. One consisted of competition "between individuals of the same sex, generally the males, in order to drive away or kill their rivals, the females remaining passive; whilst in the other, the struggle is likewise between the individuals of the same sex, in order to excite or charm those of the opposite sex, generally the females, which no longer remain passive, but select the most agreeable partners." In this chapter, we shall refer to these two modes as "intrasexual" and "intersexual" selection respectively. Although the process of intrasexual competition was readily acknowledged, early colleagues of Darwin such as Wallace (1889) were unwilling to accept intersexual selection, viz. female choice, as a potent evolutionary force and the resulting debate raged unabated well into the current century (cf. Huxley 1938).

There are now many examples demonstrating an important role for intersexual choice in sexual selection (see Table 1). This choice, usually by females, can be directed at specific males, at resources controlled or offered by them, or both. The apparent reason that the debate had persisted for so long is that both mechanisms of sexual selection are often present in a given species. Since intrasexual conflict among males is typically a more obvious process, its detection was invoked as sufficient explanation for the evolution of specific traits or behaviors without considering concurrent contributions via female choice. It is now clear that the two processes are not exclusive, that they can affect traits either in the same or in different directions, and that their magnitudes may vary quite independently.

This fact definitely complicates the study of sexual selection. It throws back on the investigator who attempts to examine one or the other process in isolation the burden of proving that only a single mode is involved. It also raises the question of a) what factors determine the relative magnitudes of the two processes in any given system, and b) given the different evolutionary dynamics and stabilities of the two modes (Kirkpatrick 1982; Lande 1981), how do they combine and interact when both are present at substantial levels? Are female choices congruent with the outcomes of male-male competition (Borgia 1979) or is there often conflict between the sexes as to who mates with whom? It has been difficult to treat these issues theoretically because the phenomenological patterns are not yet clear. Many field workers are biased towards one or the other and this colors interpretations of their data. Models and theory are often worked out for one taxon without considering their propriety for other taxa. Our goal in this chapter is to bracket the range of mixtures of intra- and intersexual selection in as unbiased a fashion as possible and to identify contrasts between systems that we think are especially challenging theoretically. Since we cannot review all taxa or systems here, we shall focus on several specific strategies of male-male competition and examine the degree to which female choice does or does not emerge as a potent force in each.

MALE SELF-ADVERTISEMENT

Here we consider systems in which males are relatively sedentary, defend no resources, and wait for females to approach them for mating. Classical leks and anuran choruses are obvious examples. Species in which males are less aggregated, such as bell birds or ruffed grouse, would also belong in this category. It is less obvious where one should put swarms of male insects. Many of these do not broadcast signals to females but instead aggregate at sites having some extrinsic cue "known" to both sexes (Downes 1969; Thornhill and Alcock 1983). We take up the similarities and differences between insect swarms and vertebrate leks later in this section.

In all of these systems, typical males neither control female access to resources nor follow females as they move about. Since females approach males and not vice versa, such systems ought to be prime candidates for conspicuous intersexual choice by females. Before we discuss specific examples, it is helpful to circumscribe the likely variations in female choice. The alternatives to mate selection by females are either random mating or imposition of mate allocation by males. As Sutherland (1985) has pointed out, the demonstration that mating is nonrandom must include corrections for finite "handling times." Given that choice is demonstrable, it can be exerted at several levels and using a variety of choice functions. Level becomes important when males are dispersed spatially in a heterogeneous fashion and females could first choose between male groups and then between males within a selected group. Different criteria and discrimination acuities might be used at different levels. Choice functions can be divided according to whether they favor the mean (or modal) values of the male cue distributions or one or both of the distribution tails. Whatever the optimal cue relative to modal male values, the simplest choice function is a threshold one: a female mates with any male having a cue value above (or below) some fixed value. Next is a "best of N" system in which the female mates with that male having the highest (or lowest) cue value from among the first N males encountered. Finally, one might encounter a "best of all" choice in which a female selects a male with the optimal cue from among many encountered in a finite time (Janetos 1980; Lande 1981; Seger 1985). The effort invested in the choice varies along another dimension depending on how much females correct for signal propagation effects. In "passive" choice a female approaches the most intense signal regardless of proximity; in "active" choice she corrects measures of intensity for distance between herself and the males producing the signals (Arak 1983a; Parker 1978). There can be varying levels of active choice depending on the extent and accuracy of such corrections.

Studies of self-advertisement systems typically begin with tests for nonrandomness in male mating success, identify one or more male traits which are correlated with matings, and then attempt to determine which (or how much) of male-male competition and female choice account for the correlations. Although often difficult, manipulations are then required to show that the correlations do or do not reflect causal relationships. We discuss below several systems in which there are either enough data to show the range of variation in relative intra- and intersexual selection levels or which are sufficiently intriguing to warrant special attention.

Anuran choruses

Frog and toad mating systems vary from scramble competition for females in "explosive" breeding species, where all mating and spawning may occur within a few days, to contest competition, involving male defense of a calling site or oviposition site, in "prolonged" breeders where spawning may occur over several weeks (Wells 1977). In explosive breeders, there seems to be little if any scope for female choice of mates (Halliday 1983). Males wander about in search of females, many of which are grabbed soon after emerging from hibernation. Several studies have looked at the influence of male size on mating success (size is easy to measure but age is usually a confounding and unmeasured variable). Pairing tends to be random at first, which suggests that male size has no effect on ability to find females. In some species (e.g., *Bufo typhonius*, Wells 1979) where spawning takes place soon after pairing, there is no size-related pattern in the distribution of male matings at spawning except that the smallest males do not mate. In other species where there may be several days between pairing and spawning, fights occur for the possession of females, and larger males have greater spawning success because they are better able to take over paired females (e.g., *Bufo bufo*, Davies and Halliday 1979). In some cases, there may be size assortative mating with large males paired to large females and small males to small females. Perhaps this arises because the male has a stronger grip on the female in pairs that are well-matched for size and so is better able to resist takeovers. Overall, females in explosive breeders appear not to choose mates, and any nonrandom mating appears to be the consequence of male-male competition.

In prolonged breeders, by contrast, there is good evidence for female choice. In some cases (e.g., bullfrogs *Rana catesbeiana*), males defend oviposition sites and larger males, who tend to defend the better sites for larval development, attract the most females (Howard 1978). Because male success and territory quality are correlated, it is not yet known how much of female choice is directed towards territory selection, selection of a male as a cue to territory quality, or selection of a male for some longer term "genetic" benefits (see Heisler et al. this volume).

In other prolonged breeders, the males call from specific sites which are not used for oviposition. Because female attraction is mediated by male calls, such frogs and toads provide excellent systems for experimental manipulation of cues used in choice. Gerhardt (1982) has made synthetic calls of various species of treefrogs (*Hyla spp.*) which are as attractive to females as real males. Modifications of these synthetic calls show that deviations away from average values for the species result in lower preference by females. Thus females preferred croaks of intermediate pitch to ones of higher or lower pitch. Recordings from female auditory neurons showed peak responses matching the behavioral data. Gerhardt suggested that choice exhibited by these females functions to prevent hybrid matings. Since as many as 10–20 different species may be found calling in the same locale, this seems a likely interpretation.

Different results were obtained by Ryan (1985) working on the Tungara frog (*Physalaemus pustulosus*) and Robertson (1986) on the leptodactylid *Uperoleia rugosa*. In both species, males call to attract females and oviposition then takes place away from the calling site. Observations showed that larger males were more successful in both frogs, although there was strong size assortment between the sexes in the second species. Female *Uperoleia* visit a number of males without interference before selecting a mate and, if artificially separated from the male selected, will repeatedly reselect the same male when offered a choice. In both species, playbacks suggest that females use pitch of the male call as a cue in this selection. Since pitch is inversely correlated with body size, Ryan has suggested that females use call pitch to select larger males. Robertson suggests that this may also be true of *Uperoleia* within the constraint that the female must carry her selected male to the oviposition site and a mate therefore cannot be too large. It has been argued that larger males are better able to fertilize large female clutches (Sachser 1986), but Robertson points out that this is an unlikely cause for the choice in *Uperoleia*.

Arak (1983a) made a similar study on the natterjack toad (*Bufo calamita*). Although he also found that larger males were more successful than smaller ones, there was no demonstrable female preference for lower call pitch. Neighboring males on the other hand were responsive to the pitch of a territorial male's call: the lower the croak pitch, the more likely an intruding male was to retreat from the caller's territory (see also Davies and Halliday 1978). Larger males thus commanded greater space, free from competition, around their calling sites. While females did not respond to pitch differences they did exhibit a preference for louder calls. Although larger males tended to produce louder calls, the female preference was a passive one in that no compensation was made by a female for proximity of the caller. The higher mating success of larger males thus appeared to be due to a combination of reduced interference among callers and a female preference for louder calls.

These anuran examples raise two important issues. First, the correlation between the importance of female choice relative to male-male conflict and the duration of the breeding season begs additional study. Does the limited mating period of explosive breeders shift the "control" of matings to males because a) females cannot afford to lose time and energy refusing early males encountered, or b) because males cannot afford being refused? Parker (1979, 1983) has discussed some of the factors which modulate relative persistence of the two sexes and who chooses in various contexts. Many of these arguments ought to be applied and tested in amphibians. The correlation is thus very intriguing, but as yet we have no proper economic studies of the costs and benefits of female choice in any anuran species.

The second issue is raised by the variation in male traits which are utilized by anuran females in mate choice. For whatever reason (greater risk of hybrid matings or reduced male variation in body size), some tree frogs are thought to commit all call parameters to species recognition functions. Other species show female preferences for extremes with emphasis alternatively on pitch or intensity. Discrimination can be graded, as opposed to the use of discrete thresholds, but in some forms may lack sophisticated controls for signal attentuation. Unfortunately, there are still too few studies in which female choice and male-male components of mate assortment have been teased apart quantitatively to pick up any general patterns. This is unfortunate because amphibians offer an unusual opportunity in which to study constraints on sexual selection by female choice. It is clear that the anuran ear is a limited structure: it only responds to a small set of sound parameters, at least some of which are tuned for species recognition, and elaborate discrimination within one parameter invariably contrains discrimination in others. Which traits are used when female choice is feasible, how accurately the discrimination is made (e.g., passive vs. active choice), and what concurrent costs are accepted in discrimination of other call parameters are important issues which should continue to be pursued in this group. Why do natterjack toad females not correct for sound attenuation in selecting males? Can strong female choice for a particular parameter break a species out of the more typical compromises built into the anuran ear, or do these constraints prevent such choice from ever becoming strong? Hopefully, future work on anurans will clarify such issues.

Avian and mammalian leks

Researchers on vertebrate leks are still divided concerning the relative roles of male-male competition and female choice. A proper study showing evidence of the former would include a) evidence that male conflict existed, and b) clear correlations between the outcomes of such conflict and mating success. Equivalent tests for the latter include a) evidence that females are free to sample several males, and b) identification of male cues which females do (or could) assess and which correlate with mating success. Statistical controls for correlations between the two modes or beween traits are also required. Earlier studies such as Wiley's (1973, 1974) on sage grouse and Lill's (1974a, 1974b, 1976) on manakins failed to identify any differences in male display performance or morphology which correlated sufficiently with mating success to suggest female choice of male phenotypes. On the other hand, the obvious contests between males in these species and the fact that matings were heterogeneously distributed spatially suggested that male competition for particular locations which females preferred might account for the distributions of matings observed. Lill (1974a) performed a male removal experiment which showed no significant shift in female preferences following reoccupation of the experimental sites and this is often cited as evidence in support of a "center effect." In fact, his null hypothesis was that site and not male was important, his samples were sufficiently small that a Type II error was a major risk and the trends, if anything, suggested changes in female visiting after owner replacement. Many subsequent studies have used either the existence of heterogeneous spacings of matings or high levels of male agonistic behavior as evidence that this combination of female preferences for locations and male competition for control of preferred sites dominates mate allocation on leks (cf. Avery 1984; Foster 1983; Johnsgard 1983).

On the other hand, recent work by Gibson and Bradbury (1985) on sage grouse has been unable to show any significant effect of location on mating success. This corroborates the results of Hartzler (1972) who studied one of the same leks examined by Wiley and found no evidence for locational effects. Appropriate tests for center effects must be quantitative: since most males in any sizeable cluster will be "central," it follows that by chance, the few successful males will tend to be "central." Hence apparent centrality is not sufficient evidence of an effect. Where statistical tests can be applied to mating success vs. location we have been unable to confirm significant locational effects in any lek system (Bradbury and Gibson 1983). In addition, our sage grouse studies have demonstrated neither sufficient asymmetries in the outcomes of male contests nor correlations between contests and mating success to support the idea that male-male agonistic behavior modulates access to mates (Gibson and Bradbury 1985, 1986). In contrast to Wiley, we do find significant variations in male display parameters which correlate well with mating success and which females could use in selecting mates. Careful tracks of both solitary and grouped females visiting stable leks show unhampered visits to the majority of territorial males, gradual narrowing of attention to one male, and finally solicitation by the female followed by copulation. It remains to be seen whether field manipulations of males and females can demonstrate that such choice is actually occurring. Although early work on black grouse also stressed correlations between mating success and male location (Kruijt et al. 1972), more recent studies have raised questions about the magnitude of center effects and have identified male display parameters which are apparently linked to mating success (de Vos 1979, 1983; de Vos and Bruinsma 1985).

These examples do not mean that center effects and/or male-male dominance relations may not modulate mating success in other lek forms. On the contrary, there are several cases where male-male interactions seem well established as correlates of mating success. One of the better quantitative studies is by Trail (1985a, b) on cock-of-the-rock. Females of this species normally visit only 2–4 males of the 50–60 present before mating. Males often interrupt each others' courtship displays, and females are significantly less likely to mate with a frequently interrupted male and more likely to mate with males who interrupt often. Bowerbirds of several species actively destroy the bowers of male competitors and this can effect mating success of the owners

(Borgia 1985a, b). Males of *Chiroxiphia* manakins display in pairs and trios. Within any pair or trio there appears to be dominance hierarchies that determine which male obtains the majority of the matings (Foster 1977, 1981; Snow 1963; David MacDonald, personal communication). In this case, it is important to remember that females are free to visit several such clusters of males: it is only when they have selected a cluster that their options are reduced to the dominant individual. A similar case for male dominance within displaying groups has been made for Goldie's bird of paradise (LeCroy et al. 1980). However the evidence is less persuasive here and has been challenged by Beehler (1983).

It may seem tempting to resolve the disputes by allocating species to female choice and male-male competition categories. Unfortunately, this is too simplistic. Consider again the sage gouse. Although a major emphasis on female choice seems likely, males still expend great time and energy defending territories on the lek. Territories must have some advantage to them to justify the effort. Interestingly, we have found spatial and temporal variation in the degree to which male territoriality is maintained in sage grouse. On "stable" leks males are highly site specific, and females leaving a territory or lek are not followed by males beyond their territorial boundary. On "unstable" leks males from roving groups which may shift locations hundreds of meters each hour, and females are regularly followed and courted wherever possible. Some stable leks break down at times and act like unstable ones: this may occur when a female leaves the lek without mating, but occasionally may occur when a recently mated female leaves the lek (Gibson and Bradbury 1987). This intraspecific variation in lek stability is matched by wide differences among species in the modal stability of territories and the levels of male interference (Avery 1984; Foster 1983). The examples suggest that the costs and benefits of lek territoriality are delicately balanced and that shifts in the balance can change male tactics in a large way.

Why might males establish territories on leks? One necessary condition must be the female's ability to resist rape. In sage grouse, despite persistent following of females in unstable conditions, we have never seen a female mate without her prior solicitation. Male sage grouse often ignore soliciting females; this is more likely a result of males acknowledging the coyness of females than it is a result of sperm depletion (Avery 1984; Scott 1942). Males will act identically when only a single female has been present all morning, and males who attempt a mount before females adopt a completely rigid solicitation posture invariably frighten off the female.

While the ability of females to refuse males may be a necessary condition favoring the establishment of stable male territories, it cannot be a sufficient one since stability varies so widely within species. Stable territories might function a) as refuges within which interruptions are less likely, or b) as mnemonics by which females can relocate a favored male after screening the entire assembly. A stable system of refuges requires that most males can establish their control over some fixed site regardless of invader pressure or that a set of "dear-enemy" pacts can be generated. The latter mechanism is potentially unstable to defections and requires a relatively constant composition of participating males. However, it could explain the long periods spent in unresolved "facing past" encounters between neighboring sage grouse males and the sudden breakdown into scramble competition when either females indicate a possibility of mating off the lek (e.g., sage grouse) or there is an invasion of younger nonterritorial males (as occurs at the end of the season in many lek species). It is also consistent with the higher levels of interference in species, such as buff-breasted sandpipers, which establish leks on migration resulting in high turnovers in male and female composition (Myers 1979).

Alternatively, the notion that females use male location as a cue in relocating preferred males is also compatible with much of the evidence. Certainly female manakins tend to return to specific mating sites when renesting (Lill 1974a). Sage hens typically visit a lek several days running before mating, and it may be those who show no intention of mating on the lek, clearly giving up the mnemonic, which precipitate territorial breakdown. It is possible that the correlation between breeding season and free female choice in anurans could be explained by either or both of the proposed benefits of male site specificity. Certainly time is necessary to establish "dear neighbor" relationships and prolonged breeders in anurans may exhibit repeat female visits both prior to a single mating and for subsequent matings in the same season.

Whatever the factors favoring male territoriality some of the time, the situation is apparently a labile one. If the traits favored by females and those affecting male-male competition are different, selection on specific male traits may shift back and forth within a season, not to mention within a generation. As noted, several studies have implicated components of male displays in mate choice by females (e.g., Gibson and Bradbury 1985; Pruett-Jones 1985; de Vos and Bruinsma 1985). It is not yet clear whether these same traits are important in the establishment of male territories or in mate assortment during territorial breakdown and interruption. If not, selection might alternate on different traits or shift direction on the same ones.

The correlation of behavioral traits with mating success in several avian lek species repeats the findings in anurans. However, the lack of any correlations between mating and morphological traits (e.g., body size) appears to be an important difference between the two groups. It is particularly striking that species with such pronounced sexual dimorphism and strong female choice should show no detectable relationship between male morphology and mating success. This is in some ways reassuring since female choice models suggest that there is currently insufficient variation in morphological traits of these birds to explain the uniformity in female choice observed, whereas there is sufficient behavioral variation (Bradbury et al. 1985). The de-emphasis on male size in the avian systems may or may not reflect a diminished emphasis on male-male conflict when compared to anurans.

In summary, the relative roles of female choice and male-male competition in lek species are still matters of contention. The widely accepted view that males compete for locations favored by females needs to be re-examined critically and quantitatively. In several species, it appears to be less important than originally thought; in others, there is clear evidence that male-male interactions are important determinants of mating success although again location may not play a major role. The evidence of variation in the stability of male territories, both within and between species, suggests that small changes in contexts may produce sudden and significant changes in the relative roles of the two modes of sexual selection on leks. Again, mixes seem to be the most likely rule, but the ratios may be very sensitive to extrinsic factors. Economic analysis of these shifts deserves careful study.

Insect swarms

Swarms of adult males which are visited by females solely for mating occur throughout several insect orders (Brittain 1982; Downes 1969; Thornhill and Alcock 1983). What is curious is that although such insect swarms resemble vertebrate leks in most other properties, they show little evidence of unhindered female choice of a mate (Bradbury 1985). Since some of these swarms represent explosive breeders, perhaps the reasons for limited female choice are the same as those proposed for analogous anuran choruses. However, other swarms show marked persistence and long seasons yet show little more sampling by females. If intersexual selection is constrained in these insects, it may in part explain their reduced sexual dimorphism when compared to avian and mammalian lek species.

The reasons for this apparently reduced female choice are unclear. As with the vertebrate lek literature, it has been argued that female choice and male-male competition are or ought to be congruent (Alexander 1975; Borgia 1979; Thornhill and Alcock 1983). Thus there would be no lack of choice in insect swarms: male and female arrive at the same conclusions. However, there is yet little evidence either for or against this idea and given the history of this concept in avian leks, it must be examined much more carefully in these forms as well. As an alternative, it may be that most female insects simply cannot afford to be more persistent in exercising choice than males are in forcing it (Parker 1979, 1983); most have shorter lifetimes, are less buffered against ambient conditions, and suffer greater allometric constraints on the amount of sensory processing that can be committed to mate choice than do their vertebrate counterparts. It is important to note that lek mating with relatively free female choice does appear to occur in certain insects (Kimsey 1980; Knapton 1985; Lederhouse 1982; Lloyd 1979; Parsons 1977; Spieth 1974; de Vries 1980), and some females, such as those of the true drosophilids and certain butterflies, exhibit remarkable abilities to reject unwanted matings (see reviews by Bradbury 1985; Spieth 1974; Thornhill and Alcock 1983). At the moment, we can discern no clear relationships between the levels of female choice in insects and environmental, social, or demographic conditions. However, the generally heavier emphasis on male-male conflict in insect swarms when compared to avian leks cries out for detailed study. This opportunity is only augmented by the existence of exceptional insects with leks in the vertebrate sense.

FEMALE DEFENSE AND RESOURCE DEFENSE

The male strategies of female defense and resource defense have received extensive attention in the literature. Although initial attention focused on which of these forms of male-male competition was occurring in each case, it has recently been argued that neither strategy completely excludes independent female choice. In fact, mixes are likely to be the rule. In guinea pigs, for example, the reproductive dominance of alpha males in low density populations is replaced at high densities by long-lasting pair bonds and greater evidence of female choice of a mate (Sachser 1986). The degree of mix thus appears to be a function of ambient population densities. In addition, a strong emphasis on male-male competition does not preclude choice by males for particular traits in females and this produces another type of mix. In all cases, the degree to which each process occurs in the mix is very important from the view of sexual selection. We summarize below several phenomena that constitute particularly promising systems in which the factors controlling such mixes might be studied.

Multiple mating by females

One typically assumes that males who control resources or defend females gain primary reproductive access to their mates. While this is generally true, it is now clear that some females may go out of their way to mate with more than one male. Paternity studies on monogamous birds show that 5–10% of offspring are fathered by a male not the mate of the mother (Bjorklund and Westman 1983; Gowaty and Karlin 1984; Littlefield 1981; Røskaft 1983; Sealy 1984; Werschkul 1982). In some of these examples rape by males (as occurs in ducks) is an unlikely explanation of the alternative paternity. In other taxa, such as squirrels, most females mate multiply (Schwagmeyer 1984). One explanation for multiple mating is that female choice is sufficiently discordant with male-imposed strategies that females sneak off and mate again. The notion is intriguing, but given the number of alternative hypotheses for such mixed paternities (Knowlton and Greenwell 1984; Schwagmeyer 1984), careful case studies are needed before this behavior can be invoked as evidence of sexual conflict.

Female choice of resources defended by males

This is widely quoted as the least controversial aspect of female choice. In some cases, males provide resources but no parental care. A well known example is the offering of nuptial gifts by male scorpion flies (*Panorpa*) and hangingflies (*Hylobittacus*). Females prefer to mate with males who offer more food and increase their fecundity by this preference (Thornhill 1981). In other cases, parental care seems to be a more important determinant of female choice than resource quality. For example, female mottled sculpins (*Cottus bairdi*) prefer to lay their eggs in cavities defended by large males which are better at protecting the eggs against predators. In experiments, male size influenced female choice but cavity quality did not (Downhower and Brown 1980).

In many cases, both resource quality and parental care influence female choice. For example, in polygynous birds the males defend territories on which females settle to breed. Some territories are of better quality than others as they may provide better nest sites, more food, or both, and in several species it has been shown that females prefer to settle on these good quality territories. The result is that their male owners become polygynous (e.g., red-winged blackbirds, Searcy 1979; lark buntings, Pleszczynska 1978). There is, however, a cost for females who share a male's territory, namely that they have to share the male's parental care. Orians' (1969) famous "polygyny threshold model" predicts how females should trade off territory quality and male help with rearing offspring so that their choice of mate maximizes their reproductive success. In general, the model predicts that it will pay a female to choose an already-mated male provided that the better resources on his territory more than compensate her for the loss of male help relative to monogamy.

In contrast with scramble competition where male-male competition is paramount, males cannot force females to settle on their territories, so resource defence systems seem at first sight to offer females complete freedom of choice. If females were really free to choose, then Orians' model should give a good description of the way females settle on male territories of varying quality. Testing the model critically is not easy (Vehrencamp and Bradbury 1984). Assessment of territory quality and amount of male parental care is admittedly difficult (perhaps for both investigator and female!), and especially so early in the breeding season (Wittenberger 1981). Even granting this, it is clear that the model has not often provided a very good fit to observed mating patterns. Females, for example, often settle on the territories of already-mated males even though they would apparently have enjoyed greater reproductive success from mating monogamously on another territory. There have been two reactions to these findings:

1) Females may utilize other criteria in addition to resource quality and parental care. Weatherhead and Robertson (1979) suggested that females may also be choosing for specific male traits. If the offspring of such fathers inherited the paternal ability to be attractive and thus polygynous, then this might compensate the mothers for lower reproductive success. In effect, lower reproductive success in this generation would be offset in the next generation when the female's "sexy sons" produced large numbers of grandchildren. There are theoretical problems with this model (Kirkpatrick 1985). Even so, the evidence at present does not appear to support it. In red-winged blackbirds, females show consistent preferences for particular territories irrespective of the males which defend them, which suggests that territory quality is the most important factor influencing mate choice (Searcy 1979). Where there is evidence in resource defense species that females have a preference for males with specific morphological traits, e.g., in long-tailed widow birds (Andersson 1982) or sawyer beetles (Hughes and Hughes 1985), these preferences are apparently only expressed when all other factors, such as territory quality, are equal. Hence there appears to be little cost to females. The second problem with the sexy son hypothesis is that the heritability of male polygyny which is needed to compensate a female for her reduced reproductive success is unrealistically high (e.g., for pied flycatchers, Alatalo et al. 1981). In other cases, even if heritability was one, the females would still have lower success from mating polygynously (Catchpole et al. 1985).

2) Females may not be free to choose and hence the Orian's model may not be appropriate. Although a male cannot force a female to settle on his territory, females may not always be able to assess the mating status of a male, and may unwittingly pair up with an already-mated male. Although females often suffer in polygyny when compared to monogamy, males often gain. For example, if a monogamous female raised four young (with the full help of thc male) compared to three young in polygyny (with the part-time help of a male), then the male clearly does better being bigamous (six vs. four young) even though the female does worse (three vs. four). These asymmetrical payoffs exist in many cases, and so polygny may occur not only when females choose to mate polygynously (as assumed in the Orians model) but also when males are able to gain an advantage at the expense of females (Davies and Houston 1986). In some cases, males seem to actively deceive second females into mating with them by setting up a second territory some distance from the first so the second female is unable to tell whether the male is already paired or not (e.g., Alatalo et al. 1981). Even if the male defends one large territory, the second female may be unaware of the presence of another female mate (e.g., Catchpole et al. 1985). The main point is that conflicts of interest between males and females over which mating system maximizes the reproductive success of each may limit the freedom of females to choose.

Song repertoires

The complex songs of some species of birds have been proposed as an acoustical equivalent to the peacock's tail (Catchpole 1980). In some species, such as the sedge warbler (*Acrocephalus schoenabeanus*), males with more complex repertoires pair up earlier and there is experimental evidence that more complex songs are more stimulating to females. In other cases, male repertoire size seems to influence territory defense rather than female choice. For example, in great tits (*Parus major*), experiments have shown that larger song repertoires are better at keeping out intruding males (Krebs et al. 1978). Males with larger song repertoires have greater reproductive success, not because they pair earlier or attract quality females (e.g., older females or females laying larger clutches) but because they usually defend better quality territories in terms of food and nest sites (McGregor et al. 1981). In red-winged blackbirds, males with larger song repertoires attract more females but the effect again seems to be through enhanced success in male-male competition rather than through female choice. Older males have larger song repertoires which enables them to control the better quality territories. It is the greater territory quality which attracts more females, rather than the larger song repertoire (Yasukawa et al. 1980). In short, although a number of birds exhibit a correlation between song repertoire size and either the number or quality of mates a male obtains, this correlation may be generated either because song mediates male-male competition or because females find larger repertoires more attractive. Comparative analyses of why the emphasis shifts would be of great interest.

Male vs. female choice of a mate

Males which control resources or females or which provide paternal care might be expected to exert their own choice of a mate if females differed in fecundity or maternal abilities (Thornhill and Alcock 1983). Such discrimination by males is now well established in a variety of insects: katydids (Busnel and Dumortier 1955; Gwynne 1984), weevils (Johnson and Hubbell 1984), and butterflies (Rutowski 1980). Of particular interest is the study by Gwynne (1984) on mormon crickets which shows shifts in the relative importance of female choice of males and male choice of females as a function of population densities. Gwynne suggests that these shifts, which directly affect the abilities of males to produce the large nutritive spermatophores transferred to females during mating, are largely responsible for the differences in sexual dimorphism seen in his study species.

CONCLUSIONS

We see two basic lessons in our short (and necessarily limited) review. The first

TABLE 1 Some examples of the evidence for female choice of male characteristics within a species

Species	Reference	Conclusion	Evidence
Guppy *Poecilia reticulata*	Kodric-Brown 1985 Endler 1983	Females prefer males with most carotenoid pigments and iridescent pigments	Choice experiments in tanks
	Bischoff et al. 1985	Females prefer long-tailed males	Choice experiments in tanks. Tails cut experimentally
Long-tailed Widowbird *Euplectes progne*	Andersson 1982	Males with elongated tails have greater mating success than controls and short-tailed males. Observations suggest male–male competition not responsible for the result	Experimental lengthening and shortening of tails
Satin Bowerbird *Ptilonorphynchus violaceus*	Borgia 1985	Females prefer males with highly decorated bowers (could be viewed as a "resource" or equivalent of male plumage)	a) Positive correlation between decoration of bower and mating success b) A male's success decreases when decoration removed
Sedge warbler *Acrocephalus schoenobaenus*	Catchpole 1980 Catchpole et al. 1984	Females prefer males with complex songs	a) Males with more complex songs pair earlier b) Lab experiment —female responds more to more complex song
Sage grouse *Centrocercus urophasianus*	Gibson & Bradbury 1985	Females select males on basis of their displays	Significant correlation with mating success, controlling for body size, and territory characteristics
Tungara frog *Physalaemus pustulosus*	Ryan 1985	Females prefer males with deep-pitched croaks	Playback experiment controlling for loudness
Natterjack toad *Bufo calamita*	Arak 1983b	Females show no preference for croak pitch but prefer louder calls	Playback experiments manipulating pitch and loudness independently

concerns the relationship between the patterns of sexual selection and the mating system in a given species. It is generally assumed that mating systems and patterns of sexual selection go hand-in-hand. One expects polygyny to lead to greater opportunities for sexual selection than monogamy, and correlations between the two at this level seem clear (cf. Payne 1984). However, when one looks at the relative roles of intra- and intersexual selection, the correlations with mating system are not so clear. Within specific mating systems such as leks or resource defense polygyny, the control of mate assortment can shift radically from one sex to the other depending on the contexts. These shifts may occur in evolutionary time and be reflected as species differences but they are as likely to appear within a species on much shorter time scales as in the responses of Mormon crickets and hamsters to ambient densities or the day-to-day variations in sage grouse lek stability. If mating systems are not the primary determinant of these shifts, what are the alternatives? One possibility is that the shifts are better explained as outcomes of the appropriate male-female games outlined by Hammerstein and Parker (this volume). Perhaps some combination of parental investment game, mobility game, and persistence game analyses would better reflect the conflict which clearly exists in many of the systems we have discussed above.

Whatever the appropriate causal models, the second lesson is that there is more than enough variation both between and within species to identify the determinants of these shifts and tease apart why they produce the mixes seen. The problem is that many of us working on such systems have been slow to recognize the general issue. Much could be done with the comparative method and by field experimentation. The change in focus requires more of a shift in attention that it requires new methodology. Making such contrasts within specific taxa such as chorusing anurans, stable and unstable leks in various avian species, polygamous widowbirds and their weaver relatives, or mormon crickets makes the task simpler since many possible variables are held constant within a taxon. This variation will also be of great value when theoreticians finally turn to the issues of how intra- and intersexual selection might be expected to interact when both are favored. Up to now, most theory only allowed one or the other process to be present and we had no theoretical guidelines against which this rich diversity could be compared. Hopefully this chapter will help spark more interest in the problem by both theoreticians and field workers.

REFERENCES

Alatalo, R.V.; Carlson, A.; Lundberg, A.; and Ulfstrand, S. 1981. The conflict between male and polygamy and female monogamy: the case of the pied flycatcher, *Ficedula hypoleuca*. Am. Natural. 117: 738–753.

Alexander, R.D. 1975. Natural selection and specialized chorusing behavior in acoustical insects. In Insects, Science, and Society, ed. D. Pimentel, pp. 35–77. New York: Academic Press.

Andersson, M. 1982. Female choice selects for extreme tail length in a windowbird. Nature 299: 818–820.
Arak, A. 1983a. Male-male competition and mate choice in anuran amphibians. In: Mate Choice, ed. P. Bateson, pp. 181–210. Cambridge: Cambridge Univ. Press.
Arak, A. 1983b. Sexual selection by male-male competition in natterjack toad choruses. Nature 306: 261–262.
Avery, M.I. 1984. Lekking in birds: choice, competition, and reproductive constraints. Ibis 126: 177–187.
Beehler, B. 1983. Lek behavior of the lesser bird of paradise. Auk 100: 992–995.
Bischoff, R.J.; Gould, J.L.; and Rubenstein, D.I. 1985. Tail size and female choice in the guppy, *Poecillia reticulata*. Behav. Ecol. Sociobiol. 17: 253–255.
Bjorklund, M., and Westman, B. 1983. Extra-pair copulations in the pied flycatcher (*Ficedula hypoleuca*). Behav. Ecol. Sociobiol. 13: 271–275.
Borgia, G. 1979. Sexual selection and the evolution of mating systems. In: Sexual Selection and Reproductive Competition in Insects, eds. M.S. Blum and N.A. Blum, pp. 19–80. New York: Academic Press.
Borgia, G. 1985a. Bower destruction and sexual competition in the satin bowerbird (*Ptilonorhyncus violaceus*). Behav. Ecol. Sociobiol. 18: 91–100.
Borgia, G. 1985b. Bower quality, number of decorations and mating success of male satin bowerbirds, *Ptilonorhyncus violaceus*: an experimental analysis. Anim. Behav. 33: 266–271.
Bradbury, J. 1985. Contrasts between insects and vertebrates in the evolution of male display, female choice, and lek mating. Fortschritte der Zoologie: Experimental Behavioral Ecology, eds. B. Holldobler and M. Lindauer, Bd 31, pp. 273–288. Stuttgart: G. Fischer Verlag.
Bradbury, J., and Gibson, R. 1983. Leks and mate choice. In: Mate Choice, ed. P.P.G. Bateson, pp. 109–138. Cambridge: Cambridge Univ. Press.
Bradbury, J.; Vehrencamp, S.; and Gibson, R. 1985. Leks and the unanimity of female choice. In: Evolution: Essays in Honour of John Maynard Smith, eds. P.J. Greenwood, P.H. Harvey, and M. Slatkin, pp. 301–314. Cambridge: Cambridge Univ. Press.
Brittain, J.E. 1982. Biology of mayflies. Ann. R. Entom. 27: 119–147.
Busnel, R-G., and Dumortier, B. 1985. Etude du cycle genital du male d'*Ephippiger* (Orthopteres) a des signaux acoustiques synthetiques. Compt. Rend. Soc. Biol. 149: 11.
Catchpole, C.K. 1980. Sexual selection and the evolution of complex songs among European warblers of the genus *Acrocephalus*. Behaviour 74: 149–166.
Catchpole, C.K.; Dittami, J.; and Leisler, B. 1984. Differential responses to male song repertoires in female songbirds implanted with oestrodiol. Nature 312: 563–564.
Catchpole, C.K.; Leisler, B.; and Winkler, H. 1985. The evolution of polygyny in the great reed warbler, *Acrocephalus arundinaceus*: a possible case of deception. Behav. Ecol. Sociobiol. 16: 285–291.
Darwin, C. 1871. The Descent of Man, and Selection in Relation to Sex. London: John Murray.
Davies, N.B., and Halliday, T.R. 1978. Deep croaks and fighting assessment in toads, *Bufo bufo*. Nature 274: 683–685.
Davies, N.B., and Halliday, T.R. 1979. Competitive mate searching in common toads, *Bufo bufo*. Anim. Behav. 27: 1253–1267.
Davies, N.B., and Houston, A.I. 1986. Reproductive success of dunnocks, *Prunella modularis*, in a variable mating system. II. Conflicts of interest among breeding adults. J. Anim. Ecol. 55: 139–154.

de Vos, G.J. 1979. Adaptedness of arena behaviour in black grouse (*Tetrao tetrix*) and other grouse species (Tetraonidae). Behaviour 68: 277–314.

de Vos, G.J. 1983. Social behaviour of black grouse: an observational and experimental field study. Ardea-T. Ned. 71: 1–103.

de Vos, G.J., and Bruinsma, O.H. 1985. Rookooing and mate choice in black grouse: a field study. Poster presented in 19th Intl. Ethological Congress, Toulouse, France. Aug. 24–Sept. 2, 1985.

de Vries, P.J. 1980. Observations on the apparent lek behaviour in Costa Rican rain forest *Perrkybis pyrrha* (Pieridae). J. Res. Lepidopt. 17: 142–144.

Downes, J.A. 1969. The swarming and mating flight of Diptera. Ann. R. Entom. 14: 271–298.

Downhower, J.F., and Brown, L. 1980. Mate preferences of female mottled sculpins, *Cottus bairdi*. Anim. Behav. 28: 728–734.

Endler, J.A. 1983. Natural selection on color patterns in poeciliid fishes. Env. Biol. F. 9: 173–190.

Foster, M.A., 1977. Odd couples in manakins: a study of social organization and cooperative breeding in *Chiroxiphia lineais*. Am. Natural. 111: 845–853.

Foster, M.S. 1981. Cooperative behavior and social organization of the swallow-tailed manakin (*Chiroxiphia caudata*). Behav. Ecol. Sociobiol. 9: 167–177.

Foster, M.S. 1983. Disruption, dispersion, and dominance in lek-breeding birds. Am. Natural 122: 53–72.

Gerhardt, H.C. 1982. Sound pattern recognition in some North American treefrogs (Anura: Hylidae: Implications for mate choice. Am. Zoolog. 22: 581–595.

Gibson, R., and Bradbury, J. 1985. Sexual selection in lekking sage grouse: phenotypic correlates of male mating success. Behav. Ecol. Sociobiol. 18: 117–123.

Gibson, R., and Bradbury, J. 1986. Male and female strategies on sage grouse leks. In: Ecological Aspects of Social Evolution, eds. D. Rubenstein and R. Wrangham. Princeton: Princeton Univ. Press.

Gibson, R., and Bradbury, J. 1987. Lek organization in sage grouse: variations on a territorial theme. Auk 104: 77–84.

Gowaty, P.G., and Karlin, A.A. 1984. Multiple maternity and paternity in single broods of apparently monogamous eastern bluebirds (*Sialia sialis*). Behav. Ecol. Sociobiol. 15: 91–95.

Gwynne, D.T. 1984. Sexual selection and sexual differences in mormon crickets (Orthoptera: Tettigoniidae, *Anabrus simplex*). Evolution 38: 1011–1022.

Halliday, T.R. 1983. Do frogs and toads choose their mates? Nature 306: 226–227.

Hartzler, J.E. 1972. An Analysis of Sage Grouse Lek Behavior. PhD Dissertation Thesis, Univ. of Montana, Missoula, MT.

Howard, R.D. 1978. The evolution of mating strategies in bullfrogs, *Rana catesbeiana*. Evolution 32: 850–871.

Hughes, A.L. and Hughes, M.K. 1985. Female choice of mates in a polygynous insect, the whitespotted sawyer, *Monochamus scuttellatus*. Behav. Ecol. Sociobiol. 17: 385–387.

Huxley, J.S. 1938. The present standing of the theory of sexual selection. In: Evolution: Essays on Aspects of Evolutionary Biology Presented to Professor E.S. Goodrich on His Seventieth Birthday, ed. G.R. de Beer, pp. 11–42. Oxford: Clarendon Press.

Janetos, A.C. 1980. Strategies of female mate choice: a theoretical analysis. Behav. Ecol. Sociobiol. 7: 107–112.

Johnsgard, P.A. 1983. The Grouse of the World. Lincoln: Univ. of Nebraska Press.

Johnson, L., and Hubbell, S.P. 1984. Mate choice: an experimental demonstation in a brentid weevil. Behav. Ecol. Sociobiol. 15: 183–188.

Kimsey, L.S. 1980. The behavior of male orchid bees (Apidae, Hymenoptera, insecta) and the question of leks. Anim. Behav. 28: 996–1004.
Kirkpatrick, M. 1982. Sexual selection and the evolution of female choice. Evolution 36: 1–12.
Kirkpatrick, M. 1985. Evolution of female choice and male paternal investment in polygynous species: the demise of the "sexy son." Am. Natural. 125: 788–810.
Knapton, R.W. 1985. Lek structure and territoriality in the chryxus arctic butterfly, *Oeneis chryxus* (Satyridae). Behav. Ecol. Sociobiol. 17: 389–395.
Knowlton, N., and Greenwell, S.R. 1984. Male sperm competition avoidance mechanisms: the influence of female interests. In: Sperm Competition and the Evolution of Animal Mating Systems, ed. R.L. Smith, pp. 62–85. New York: Academic Press.
Kodric-Brown, A. 1985. Female preference and sexual selection for male colouration in the guppy *Poecilia reticulata*. Behav. Ecol. Sociobiol. 17: 199–205.
Krebs, J.R.; Ashcroft, R.; and Webber, M.I. 1978. Song repertoires and territory defense in the great tit. Nature 271: 539–542.
Kruijt, J.P.; de Vos, G.J.; and Bossema, I. 1972. The arena system of black grouse. Proc. XVth Intl. Ornithol. Congress, pp. 399–423. Leiden: E.J. Brill.
Lack, D. 1968. Ecological Adaptations for Breeding in Birds. London: Methuen.
Lande, R. 1981. Model of speciation by sexual selection on polygenic traits. Proc. Natl. Acad. Sci. USA 78: 3721–3725.
LeCroy, M.; Kulupi, A.; and Peckover, W.S. 1980. Goldie's bird of paradise: display, natural history, and traditional relationships of the people to the bird. Wilson. B. 92: 289–301.
Lederhouse, R.C. 1982. Territorial defense and lek behavior of the black swallowtail butterfly, *Papilio polyxenes*. Behav. Ecol. Sociobiol. 10: 109–118.
Lill, A. 1974a. Sexual behavior of the lek-forming white-bearded manakin (*Manacus manacus trinitatis* Hartert). Z. Tierpsych. 36: 1–36.
Lill, A. 1974b. Social organization and space utilization in the lek-forming white-bearded manakin, *M. manacus trinitatis* Hartert. Z. Tierpsych. 36: 513–530.
Lill, A. 1976. Lek behaviour in the golden-headed manakin (*Pipra erythrocephala*) in Trinidad (West Indies). Fortschr. Verhaltensforsch. 18: 1–84.
Littlefield, C.D. 1981. Mate-swapping of sandhill cranes. J. Field Orn. 52: 244–245.
Lloyd, J.E. 1979. Sexual selection in luminescent beetles. In: Sexual Selection and Reproductive Competition in Insects, eds. M.S. Blum and N.A. Blum, pp. 293–342. New York: Academic Press.
McGregor, P.K.; Krebs, J.R.; and Perrins, C.M. 1981. Song repertoires and lifetime reproductive success in the great tit, *Parus major*. Am. Natural. 118: 149–159.
Myers, J.P. 1979. Leks, sex, and buff-breasted sandpipers. Am. Birds 33: 823–825.
Orians, G.H. 1969. On the evolution of mating systems in birds and mammals. Am. Natural. 103: 589–603.
Parker, G.A. 1978. Evolution of competitive mate searching. Ann. R. Entom. 23: 173–196.
Parker, G.A. 1979. Sexual selection and sexual conflict. In: Sexual Selection and Reproductive Competition in Insects, eds. M.S. Blum and N.A. Blum, pp. 123–166. New York: Academic Press.
Parker, G.A. 1983. Mate quality and mating decisions. In: Mate Choice, ed. P.P.G. Bateson, pp. 141–166. Cambridge: Cambridge Univ. Press.
Parsons, P.A. 1977. Lek behavior in *Drosophila (Hirtodrosophila) polybori* Malloch — an Australian rainforest species. Evolution 31: 223–225.
Payne, R.B. 1984. Sexual selection, lek and arena behavior, and sexual size dimorphism in birds. Ornith. Mono. 33: 1–52.

Pleszczynska, W.K. 1978. Microgeographic prediction of polygyny in the lark bunting. Science 201: 935–937.

Pruett-Jones, S.G. 1985. The Evolution of Lek Mating in Lawe's Parotia (Aves: *Parotia lawesii*). PhD Dissertation Thesis, Univ. of California, Berkeley, CA.

Robertson, J.G.M. 1986. Female choice, male strategies, and the role of vocalizations in the Australian frog, *Uperoleia rugosa*. Anim. Behav. 34: 773–784.

Røskaft, E. 1983. Male promiscuity and female adultery in the rock *Corvus frugilegus*. Ornis Scand. 14: 175–179.

Rutowski, R.L. 1980. Courtship solicitation by females of the checkered white butterfly, *Pieris protodice*. Behav. Ecol. Sociobiol. 7: 113–117.

Ryan, M.J. 1985. The Tungara Frog. Chicago: Chicago Univ. Press.

Sachser, N. 1986. Different forms of social organization at high and low population densities in guinea pigs. Behaviour 97: 253–272.

Schwagmeyer, P.L. 1984. Multiple mating and intersexual selection in thirteen-lined ground squirrels. In: Biology of Ground-Dwelling Squirrels, eds. J.O. Murie and G.R. Michener, pp. 275–293. Lincoln: Univ. of Nebraska Press.

Scott, J.W. 1942. Mating behavior of the sage grouse. Auk 59: 477–498.

Sealy, S. 1984. Extra-pair copulation in the yellow warbler (*Dendroica petechia*). Anim. Behav. 32: 295–296.

Searcy, W.A. 1979. Female choice of mates: a general model for birds and its application to red-winged blackbirds *Agelaius phoeniceus*. Am. Natural. 114: 77–100.

Seger, J. 1985. Unifying genetic models for the evolution of female choice. Evolution 39: 1185–1193.

Snow, D.W. 1963. The display of the blue-backed manakin, *Chiroxiphia pareola*, In Tobago, W.I. Zoologica 48: 167–176.

Spieth, H.T. 1974. Courtship behavior in *Drosophila*. Ann. R. Entom. 19: 385–405.

Sutherland, W.J. 1985. Chance can produce a sex difference in variance in mating success and explain Bateman's data. Anim. Behav. 33: 1349–1352.

Thornhill, R. 1981. *Panorpa* (Mecoptera: Panorpidae) scorpionflies: systems for understanding resource-defense polygyny and alternative male reproductive efforts. Ann. R. Ecol. Syst. 12: 355–386.

Thornhill, R., and Alcock, J. 1983. The Evolution of Insect Mating Systems. Cambridge, M.A.: Harvard Univ. Press.

Trail, P.W. 1985a. Courtship disruption modifies mate choice in a lek-breeding bird. Science 227: 778–779.

Trail, P.W. 1985b. Territoriality and dominance in the lek-breeding Guianan cock-of-the-rock. Nat. Geograph. Res. 1: 112–123.

Vehrencamp, S.L., and Bradbury, J.W. 1984. Mating systems and ecology. In: Behavioural Ecology: An Evolutionary Approach, 2nd ed., eds. J.R. Krebs and N.B. Davies, pp. 251–278. Oxford: Blackwell.

Wallace, A.R. 1889. Darwinism, and Exposition of the Theory of Natural Selection. London: Macmillan.

Weatherhead, P.J., and Robertson, R.J. 1979. Offspring quality and the polygyny threshold: 'the sexy son hypothesis'. Am. Natural. 113: 201–208.

Wells, K.D. 1977. The social behavior of anuran amphibians. Anim. Behav. 25: 666–693.

Wells, K.D. 1979. Reproductive behavior and male mating success in a Neotropical toad, *Bufo typhonius*. Biotropica 11: 301–307.

Werschkul, D.F. 1982. Nesting ecology of the little blue heron: promiscuous behavior. Condor 84: 381–384.

Wiley, R.J. 1973. Territoriality and non-random mating in sage grouse, *Centrocercus urophasianus*. Anim. Behav. 6: 85–169.

Wiley, R.H. 1974. Evolution of social organization and life history patterns among grouse (Aves: Tetraonidae). Q. Rev. Biol. 49: 209–227.
Wittenberger, J.F. 1981. Time: a hidden dimension in the polygyny threshold model. Am. Natural. 118: 803–822.
Yasukawa, K.; Blank, J.L.; and Patterson, C.B. 1980. Song repertoires and sexual selection in the red-winged blackbird. Behav. Ecol. Sociobiol. 7: 233–238.

Sexual Selection: Testing the Alternatives
eds. J.W. Bradbury and M.B. Andersson, pp. 165–179
John Wiley & Sons Limited

Sexual Selection in Flowering Plants

D.C. Queller

Department of Biology, P.O. Box 1892
Rice University
Houston, TX 77251, USA

Abstract. Flowering plants may be sexually selected in ways both similar to and different from familiar animals. Males (or male functions of hermaphrodites) compete for access to females, for example, by producing large amounts of pollinator attractants and rewards. However, some attractive displays may be selected not for sexual competition but because both sexes are limited by access to the other. Females (or female functions) may exercise choice. Due to their lack of mobility, however, female plants best exercise choice not by limiting matings but by mating promiscuously and choosing after pollination. Such choice can be for traits of pollen tubes, endosperms, or embryos. It differs from true mate choice because an advantage may be gained by choosing on the basis of fitness due to nonadditive (nonheritable) genetic causes. Animal studies of sexual selection might also benefit by considering male competition and female choice in a larger context along with offspring competition and maternal choice.

INTRODUCTION

Imagine for a moment that Darwin's only contribution to science had been his botanical works. What would be his place in the history of biology? A quick answer might be that he would still be ranked among the great botanists. But a better answer would have to reject the question, for Darwin did not study plants as an idle pastime in between his greater works. All of Darwin's "minor" works were efforts to elucidate features important to his theory of evolution (Ghiselin 1969). He would not have carried out most of his plant studies without his vision of organic evolution and his drive to understand it.

But if this is so, what are we to make of the fact that Darwin never really applied his theory of sexual selection to plants? His ideas on male competition and female preference were developed to account for some conspicuous male traits that he thought could not be explained by a simple struggle for existence. But while the antlers of the stag and the tail of the argus pheasant seemed to cry out for some special explanation, the plants that he knew so well apparently stood mute. This has not really changed much. Though plants are no longer entirely silent on such issues, the answers have had to be squeezed from them. The effects of sexual selection in plants are not usually obvious, and their discovery has usually required careful work generated by theoretical considerations of what we should expect to find.

As a result of this work, we have learned some things about plant evolution. That will be one theme of this essay, though I will not attempt to review all recent ideas and evidence on plant sexual selection (e.g., Charnov 1979; Stephenson and Bertin 1983; Willson 1979; Willson and Burley 1983). Nor will I treat the growing bodies of work on sex allocation (Charnov 1984) and the evolution of breeding systems (Bawa 1980; Charlesworth and Charlesworth 1979). Instead, I want to focus on a few of the special features of flowering plants and what distinguishes them from those animals that have been central to thinking on sexual selection (chiefly vertebrates and insects). Just as sexual selection theory has enlarged our view of plant evolution, so might plants inform our views on sexual selection.

MALE COMPETITION: THE FLEURS-DU-MÂLE SYNDROME

The basic cause of sexual selection, that female and male reproduction may be limited in different ways, holds for many plants as well as animals (Bateman 1948). If seed production is limited by nutrient resources, then increasing the number of pollinations beyond some minimal level may have little effect on female success. But male success still may be limited by the number of matings obtained. We should expect, then, that traits that increase pollination levels will often be maintained by selection in the context of competition among males for access to females. Here I will concentrate on pre-pollination competition, for convenience leaving consideration of all post-pollination phenomena to the section on female preference.

Flower number is perhaps the simplest trait to examine for effects on male competition, and evidence is accumulating that large numbers of flowers are selected primarily to increase male success (the fleurs-du-mâle syndrome, Queller 1983b). There are generally more male than female flowers in the species where these are separate (Lloyd and Webb 1977; Stephenson and Bertin 1983). A more cryptic form of the fleurs-du-mâle syndrome occurs in plants with hermaphroditic flowers. Research on milkweeds has shown that, though all flowers are hermaphroditic, the primary advantage of large floral displays lies in male competition (Chaplin and Walker 1982; Queller 1983b; Willson and Price 1977; Willson and Rathcke 1974). For example, when half of all flowers were removed from some plants, their male success (assayed by pollinia removal by pollinators) was significantly reduced relative to controls, but their female success (fruit production) was not (Queller 1983b). This is becuase they still obtain sufficient pollinations to raise a full crop of fruits. Similar results have now been reported for plants with the more typical, granular pollen, rather than pollinia (Stanton et al. 1986).

But there is another possible explanation. While female animals exert choice by comparing males and mating with a preferred type, plants cannot move about and compare males before mating. A plant's best way to exert choice may be to mate promiscuously and choose after pollination. Having many flowers might enhance female success in this way. There is some evidence for selective fruit abortion in plants, including milkweeds (Bookman 1984). But there has been no demonstration that such choice can generate anything like the male advantage generated by the extra flowers (see below). Moreover, comparative evidence also suggests that selection on the male function is more important. If the fleurs-du-mâle hypothesis is true and plants produce many hermaphroditic flowers to promote male competition and abort the excess fruits, then the fruit/flower ratios of these species should be lower than those of species that can raise the number of male flowers without dragging along the number of functionally female flowers (i.e., dioecious, monoecious, and andromonoecious species). This prediction is confirmed by a wide-ranging comparative study (Sutherland 1986). A similar argument comparing the seed to flower ratios of female and hermaphrodite morphs in gynodioecious species also supports the male competition hypothesis (Couvet et al. 1985).

Of course nonc of these studies excludes the possibility that there is some female advantage to having many flowers. The retention of the female function in hermaphroditic species suggests that there is some such advantage. But the evidence suggests that male competition is the more important factor. In the same way, selection on the male function may be more important in determining other aspects of the pollination system, including other attractants, amounts of pollinator rewards, and the timing and duration of flowering.

The logic of this argument depends on the female function being limited by resources, but in contrast to the animals that we usually consider in the context of sexual selection, this is not always the case in plants. Since plants cannot move about and search for mates they must rely on biotic or abiotic pollen vectors. Sometimes these vectors are not very effective and as a consequence seed set is limited by pollination rather then resources (Bierzychudek 1981). What should we expect to observe in such species where the prerequisite for sexual selection is absent? Increased reliance on self-fertilization is one possibility. Alternatively, since both sexes are now limited by access to the other, we might expect increased selection for more effective floral displays. In other words, when sexual selection no longer applies there can be increased selection for the kinds of traits that we consider to be sexually selected, traits that increase mating success. This seems paradoxical, particularly by comparison with weakly sexually selected animals such as monogamous birds, in which adornments tend to be reduced.

The paradox is not real but it does point up the risk of trusting analogies with familiar animals rather than going back to first principles. Darwin was concerned not so much with the degree of elaboration of a trait as with the difference in the trait between the sexes and particularly with the notion that the male trait had no utility beyond allowing its bearers to outcompete other males for access to females (Darwin 1871). If this is the basis for sexual selection then there are two categories of species that are not sexually selected: those in which neither sex is limited by access to the other, and those in which both are. Monogamous animals approach the first condition; both sexes may be limited by the resources they put into parental care, resulting in reduced dimorphism and more modest displays. Pollination-limited plants exemplify the opposite case in which both sexes are limited by access to the other sex, with the result that they are both selected for enhanced displays.

FEMALE PREFERENCE AND ALLIED PHENOMENA

When seed production is limited by resources rather than by receipt of sufficient pollen, then female plants, or the female functions of hermaphroditic plants, are presented with the opportunity to select among the available pollen. As with other aspects of sexual selection in plants the study of female preference is in its infancy, but there seems to be fairly general agreement on two points: that preference is unlikely to be for resources provided by the male, and that preferences are most likely to be exerted after pollination. Neither of these is an absolute certainty. Pollen could conceivably deliver some useful resource, but the amount must be very small, and in fact it seems that pollen carries so little in the way of nutrients that it must often use nutrients from the stylar tissue for pollen tube growth (Willson and Burley 1983). It is also possible that some pre-pollination preference exists, using floral traits that manipulate pollinators in ways that promote the receipt of appropriate pollen. But the amount of control that plants exert in this area may be relatively small (Stephenson and Bertin 1983). As there is little solid data on either of these points, I will concentrate on post-pollination choice which is not based on resources.

Restricting attention to this area also has the advantage of raising some particularly interesting questions. Perhaps the most controversial issue in the area of sexual selection is whether females are able to gain by choosing males with good genes. Because plant preferences are likely to be uncomplicated by resource-based choice, it is worth seeing what they may offer to the good genes debate. Focusing on post-pollination choice will have the advantage of highlighting some of the differences between plants and animals and will also serve to place female choice in a broader context of related kinds of preferences.

What do female choose?

Females do not choose the genes of their mates. They choose on the basis of phenotypic traits and in so doing can impose selection on the genes underlying those traits. Post-pollination choice in plants may be exercised for traits controlled by several different genetic entities. Some features of the pollen grain are controlled by the diploid genotype of the pollen parent, so choice based on these characteristics is genetically equivalent to choosing traits of males. Such choice is known for sporophytic self-incompatibility mechanisms (de Nettancourt 1977). All other kinds of post-pollination choice are somewhat different because the chosen traits are controlled by genotypes other than the diploid pollen parent. There may be preference for traits of the haploid pollen tube that carries the sperm nuclei through the stylar tissue to the ovule or, after fertilization, there may be preference based on traits of the embryo or the endosperm. The latter is formed by a second fertilization and serves as nurse tissue for the embryo. It is usually triploid and identical to the embryo except for an extra dose of the maternal complement of genes. Finally, there may be choice based on groups of embryos or endosperms, as when a fruit containing many seeds is either aborted or matured as a unit.

Mate choice in plants has often been defined in a way that includes all these possibilities (Stephenson and Bertin 1983; Willson and Burley 1983). I think it is better to draw some distinctions for both semantic and substantive reasons. A plant that selectively aborts embryos is not preferring certain mates, it is preferring certain embryos. True, by doing this it may determine which pollen parents successfully father offspring, but female plants do not choose embryos as a means to get a good mate. Quite the opposite is true. The advantage of choosing mates (if there is one) is to obtain fitter offspring, which can be done more directly through embryo choice. It is more accurate to regard mate choice as means of choosing offspring than vice versa.

Choice of gametophytes, endosperms, and embryos differs from true mate choice in at least three substantive ways. First, preference of, for example, a heritable endosperm trait will cause selective change in endosperms, not in the population of diploid plants that constitutes the mate pool (except perhaps indirectly through correlation). Second, these alternative forms of choice offer the possibility of choosing within the progeny of any given male, as well as between the progeny of different mates. But most importantly, choice of embryos, endosperms, and possibly gametophytes permits nonadditive genetic variance to play a role in selecting for the preference. As will be discussed below, this results in a third kind of selection for preference, distinct from both the Fisherian model and the good genes model.

Fisherian model of choice

Fisher (1958) proposed a model for the joint evolution of male traits and female preferences that has recently been put into explicit mathematical form (Kirkpatrick 1982; Lande 1981; O'Donald 1980). It is assumed that there is no direct selection on the female preference; all selection is on the male trait. There exists a line of equilibrium points, at each of which a given amount of natural selection against the trait is exactly balanced by a sufficient amount of female preference for it. Female preference evolves as an indirect response to selection on the male trait due to a genetic covariance that arises as a result of the preference (females preferring a trait will tend to have offspring with genes for both the preference and the trait). Thus, preference for a male trait increases the frequency of the trait, and this increases the frequency of the preference. Under certain conditions this mutually reinforcing effect can lead to accelerating runaway selection.

The same kind of mechanism could also operate on preferences of traits of haploid pollen tubes, triploid endosperms, or diploid embryos. Consider, for example, a plant with an excess of young embryos and a preference for some arbitrary embryo trait (see also West-Eberhard 1983). Assume that there is no direct selection on the preference which implies, among other things, that any rejected embryo is replaced by another embryo. If some females prefer embryos with this trait, their offspring will tend to possess the trait genes along with the preference genes, setting up a genetic covariance. With this genetic covariance, selection on the trait can now cause indirect selection on the preference for the trait. A line of equilibria may exist where preference for the trait balances natural selection against it, and since preference is only selected indirectly, it is at equilibrium at all the same points. This suggests that, even in the absence of direct selection for preference, there can be selection for preferences of arbitrary or deleterious traits of pollen tubes, endosperms, and embryos. I know of no evidence for such traits. If they do exist, they are likely to be physiological (e.g., release of hormones) and this might account for why they have not been detected. But I think that the absence of such traits may be real rather than apparent, a point I will return to later.

Good genes and good combinations

The good genes theory of female preference is controversial primarily because it requires heritability of fitness (Maynard Smith 1978). Females may be able to pick out fitter-than-average males, but since natural selection removes additive genetic variance for fitness, this is no guarantee that their offspring will also be fitter than average.

There is genetic variance for fitness that is not consumed by natural selection, but it is nonadditives due to dominance or epistasis. These nonadditive effects on fitness are due to specific combinations of genes that are broken up during sexual reproduction and therefore are not inherited. Choosing a mate which is fit due to nonadditive genetic causes is therefore also ineffective as a means of obtaining fit offspring.

The situation is quite different when the preference is for embryos rather than mates. While there is still little additive genetic variance to work with, the nonadditive variance component can now be used (Bertin 1986; Willson and

Burley 1983). Some embryos will have superior vigor due to heterotic and epistatic effects of their particular combinations of genes. A female that selectively matures such embryos will leave more descendants than one who matures embryos at random, as long as the gene combinations that produce embryo vigor also tend to increase fitness later in the lives of those offspring. The key point is that the sexual events that disrupted heterotic and epistatic effects on male fitness have already passed by the time embryos are chosen. The genes have already been reshuffled into the combinations that will be important in determining offspring fitness.

This may be clearer by analogy with two more familiar cases. First, suppose the embryo trait chosen is purely environmental in origin: by chance some seeds are damaged by insects in a way that makes it unlikely that they will produce mature plants. A parent would leave more surviving offspring if it aborted its damaged seeds and used the resources to mature additional undamaged seeds. The preferred characteristic, lack of damage need not be heritable because the choosy parent receives an immediate benefit by having higher offspring survival. The same would apply to choosing offspring on the basis of nonheritable genetic advantages.

The most familiar cases of preference for genetic complementarity are self-incompatibility mechanisms, which are widespread in plants (de Nettancourt 1977). Here, the plant rejects or retards pollen that is genetically too similar because the resulting inbred offspring would probably be less fit than those obtainable by outcrossing. But the opportunities for exerting this sort of preference are much wider if, instead of choosing on the basis of genetic similarity, the plant tests directly for vigor of the embryos. This would enable plants to select not just between selfed and outcrossed progeny but also among progeny that differ for reasons other than overall genetic similarity, including selection among the progeny of a single pollen parent.

The same basic rationale would apply to choice on the basis of endosperm vigor. Endosperm vigor would presumably not be quite as accurate a predictor of offspring success because it is not genetically identical to the offspring. However, since the endosperm develops before the embryo, it may offer an earlier opportunity for choice and it is similar enough to the embryo that it would not be surprising if endosperm vigor were correlated with offspring fitness.

Such a correlation is less plausible when choice is based on pollen tube vigor. The vigor of the haploid pollen tubes may not be a good predictor of the nonadditive genetic component of fitness of offspring because the combinations of genes that generate this component have not yet been put together (fertilization has not occurrred). However, the pollen tube must grow through the stylar environment, and it has been argued that this alone may provide an arena for testing the complementarity of pollen genes with those of the maternal parent (Mulcahy 1979).

Choice on the basis of nonadditive genetic variance can be favored, but is it sexual selection? Sexual selection has been defined in so many different ways that this may not be a profitable question to ask. But it is useful to ask how this selection for good combinations compares with phenomena that are generally included as sexual selection.

As noted above, it is not strictly correct to call this process sexual selection since something other than mates is being chosen. It may lead to differential reproduction of males (or male functions) but it does not necessarily do so. But it is analogous in the sense that it does lead to differential success of pollen or offspring. Perhaps more to the point is whether the preferences are consistent and whether they lead to selective change in the trait, criteria that are sometimes applied to sexual selection. The answer here is both yes and no. A "no" answer asserts that each plant prefers different genes (those which happen to combine well with its own genes) and that preference leads to little change in gene frequencies. A "yes" answer could be justified in that there is consistent preference for a phenotypic trait (e.g., embryo vigor), even if there is no consistency in the genes underlying the trait. There is also selective change in the trait within each generation; the average offspring is more vigorous than it would have been without the preference. Thus, the consistency and efficacy of the preference depends upon whether one considers the phenotypic trait or the genes underlying it. Whether this is called sexual selection or not, we will have to bear both the similarities and differences in mind when interpreting the results of mate choice experiments in plants.

Studies of preference and its advantages

Controlled pollinations provide a simple way to investigate post-pollination selectivity in plants. When a plant is hand pollinated with pollen from several parents, the seeds that are matured are often a nonrandom subset of them (e.g., Bertin 1986; Bookman 1984; Marshall and Ellstrand 1986; Mulcahy 1971), although this is not always so (Lee and Bazzaz 1982; Bawa and Webb 1984). The selectivity is sometimes general, with all plants tending to prefer certain pollen donors, and sometimes specific to particular maternal plants. It can arise through differential pollen tube growth or seed abortion. The existence of such selectivity raises two questions: is it due to choice or competition, and is there any selective advantage to the maternal plant? Note that these are quite distinct questions. The outcome of competition between pollen tubes, endosperms, or embryos could be either beneficial or detrimental to the maternal plant. Similarly, choice by the maternal plant may be selected because it is advantageous, but it may also be favored by the Fisherian mechanism in the absence of any direct advantage. Here I will discuss the evidence for a maternal advantage temporarily leaving aside the choice versus competition question.

There is little evidence from animals that females obtain direct genetic advantages through mate choice, the notable exception being Partridge's (1980) study of *Drosphila*. In contrast, there is a growing body of plant research that at least seems to suggest that choice can result in fitter offspring (Bertin 1986; Bookman 1984; Marshall and Ellstrand 1986; McKenna and Mulcahy 1983; Mulcahy 1971; Stephenson and Winsor 1986). The evidence comes from two kinds of studies. The first is similar to Partridge's experiment: plants given a wider choice of pollen produce offspring that are superior in some respect to the offspring of plants given a more restricted choice. In the other, preferred pollen donors are shown to sire fitter offspring than nonpreferred donors.

However, none of these studies meet all the requirements necessary to firmly estabish that choice results in genetically superior offspring. Studies of crop plants, with their long histories of artificial selection, might not be entirely reliable models for natural populations. Similarly, crosses between different natural populations present choices that are not usually made and therefore not subject to selection in nature. It is also desirable to exclude self-incompatibility mechanisms as the cause of nonrandom seed maturation in order to establish whether more novel forms of preference are responsible.

Some studies have not randomized treatments with respect to position on the maternal plant, which might affect the fitness of the seeds produced. This problem is easily remedied by careful experimental design, but there are more general and more difficult problems of dealing with maternal effects on offspring fitness. Clearly increased seed size is not an adequate indication that choice is advantageous to the maternal plant. If seed size is under maternal control, then increased seed size of preferred pollen parents may just be another indication of preference: certain pollen parents may be preferred both in the fraction of the seed crop they are allowed to sire and in the resources allocated to their seeds. This is probably advantageous to those pollen parents, but does not say anything about an advantage to the chooser, who must expend additional resources on larger seeds. Even if seed size is controlled by offspring genotypes, it is not necessarily the case that the best genotypes for attaining large seed sizes will also be the best for attaining high lifetime fitness. And, even if this condition is met, it is still not necessarily a parental advantage to give more resources to the fitter seeds. It can sometimes be more cost-effective to devote resources to the less fit seeds if those seeds stand to gain more from such investment (Temme 1986).

This problem extends to offspring fitness measures other than seed size. The seed's initial endowment can affect its germination success and its early growth. Therefore, if one wants to show that better offspring genotypes have been chosen, it is necessary to first eliminate any maternal effects. The effect of seed size should be statistically removed from measures of offspring fitness.

While no study has yet met all of these requirements, some have met most of the conditions and it seems probable that there may often be genetic advantages to choice. It is at least worth asking if plants may differ from animals in this respect.

Choice in plants and animals

If the Fisherian selection mechanism can operate in plants for choice of pollen, endosperm, or embryo traits, why do we not see the kinds of detrimental or arbitrary traits that we see in male animals? Perhaps the answer is simply that the chosen traits have escaped our notice because they are physiological in nature, but I think there is reason to believe that there may really be fewer of these traits in plants. The simple Fisherian mate choice model that predicts lines of stable (or unstable) equilibria rests on the assumption that there is no direct selection on female preference, that it is only selected indirectly by its nonrandom association with the male traits. When direct selection on preference is added to the model (e.g., choice based on territorial quality that will affect number or quality of the female's offspring), then the only stable equilibrium lies at the point that optimizes female fitness (Kirkpatrick 1985).

Maternal choice that affects the number or quality of offspring can be viewed as a form of directly selected preference. Though explicit models of such choice are needed, analogy with the mate choice models suggests that selection will tend to optimize maternal fitness. If the analogy is correct then there may only rarely be preferences for traits that lower offspring fitness. They should only be expected when the trait lowers competitive ability within the brood but does not lower the mean brood fitness. For embryo traits that do affect competitive ability between broods, preferences will evolve to optimize maternal success through choice of fit offspring. As argued above, such traits could be due to good genes, good gene combinations, or even environmental causes such as insect damage.

I am arguing here that plants may not choose arbitrary characters because they can make more effective choices. But if this is so, why do animals seem to rely more heavily on true mate choice rather than the more effective offspring choice? These are not entirely exclusive alternatives, but it does seem that animals that are very choosy about their mate will limit their opportunities for offspring choice. In many animals, female investment occurs only before fertilization so that there is no real opportunity to choose offspring on the basis of something like vigorous growth. But in many others, notably mammals and social insects, investment could be directed towards the fittest offspring. There is some evidence for choice of offspring on the basis of genetic endowment in both of these groups (Bernds and Barash 1979; Woyke 1967). But is such choice widespread and, if so, does it mean that we should expect to see less evidence of deleterious or arbitrary male characters in these groups? Mammals seem to possess fewer bright or bizarre male ornaments than birds, but perhaps this is only because we, like birds, are visually oriented and can see colors.

Of course, choosing among offspring probably has some costs. If offspring are to demonstrate their fitnesses, they must grow for a period of time at the parent's expense before the choice can be made. In addition, offspring that are rejected may resist such treatment, at some cost to the parent (of course, this may also be a problem in choosing mates). If these costs are significant, the better strategy may be to forego offspring choice. Mobile animals may be able to obtain most of the advantages of offspring choice just by avoiding matings with close kin (or perhaps also with genetically distant individuals (Bateson 1983; Price and Waser 1979). Perhaps plants choose offspring largely because they are less able to exclude such undesirable partners before mating or perhaps they do so because the uncertainties of pollination often lead to initiation of excess offspring anyway. This does not seem to be the whole answer since choice of offspring with heterotic or epistatic fitness advantages could be advantageous even after mating with a single male.

To conclude, plant studies may be both less useful and more useful than we might have hoped. On the one hand, if post-pollination choice allows them to choose good gene combinations, then plants may not be very helpful in resolving the good genes controversy. Showing an advantage to choice in plants is not the same as showing a genetic advantage to mate choice in animals. On the other hand, plant studies suggest a new set of questions concerning the properties and interactions of different kinds of choice. It seems likely that studies of sexual selection in animals could also benefit by considering mate choice in a wider context along with offspring choice.

CHOICE VERSUS COMPETITION

I have argued that plants may make several kinds of choices, ranging from true mate choice to offspring choice. In the same way, competition may take place at all the same levels: among males, among male gametophytes, among endosperms or among embryos. The problem of distinguishing between male competition and female choice is echoed at each of these levels. For example, embryo competition is not easy to distinguish from maternal preference for certain embryos.

Distinguishing male competition from female choice in animals is no easy task but at least animal behaviors are familiar enough that we can rely somewhat on our intuition. Fighting among males immediately suggests the presence of competition and conflict. Similarly, females may be observed to look over several males, resisting the advances of some and accepting others. Such behaviors can give us a reasonable feel for the extent of conflict and whether the outcome is due to male competition or female choice.

Plant behavior is less familiar to us and therefore our intuitions may be less reliable. There are a number of plant traits that seem to suggest competition or choice. For example, the speed and vigor of pollen tube growth certainly seem to have been selected in the context of competition for ovules. Pollen is often deposited in amounts in excess of what is required, sometimes from different pollen parents (Ellstrand 1984; Handel 1983). Pollen tube growth differs for different pollen parents and this can sometimes be correlated with their relative success at fathering seeds (Bookman 1984; Mulcahy 1971). Still, choice could also be involved, with female plants creating a stylar environment that favors the growth of some pollen tubes over others. The same applies to invasive haustorial structures, which seem to function in acquiring nutrients for the embryo and have been characterized as "aggressive" by plant embryologists. A competitive function is suggested by the fact that haustoria do not develop from maternal seed tissues, which are not expected to be in conflict with other seeds, but do develop from endosperms and embryos (Queller 1983a). Since endosperms and embryos are genetically different from other endosperms and embryos on the same plant, they are expected to be in conflict over which seed gets nutrient resources.

Other characters suggest the operation of maternal choice. The elongated style of the flowering plants forces pollen tubes to grow longer distances, increasing opportunities for choice among them, but the choice might be entirely passive, with females accepting the outcome of pollen tube competition (Mulcahy 1979). The seed integument is a maternal tissue that may serve to control seed growth (Westoby and Rice 1982). The widespread occurrence of abortion (Stephenson 1981) suggests some level of maternal control over seed maturation but it is possible that abortion only controls the number of seeds matured, with embryo or endosperm competition determining which seeds survive.

Because such traits do not seem to allow a rigorous distinction between choice and competition, some studies have attempted to use the results of artificial crossing experiments. As noted above, the presence of a female advantage does not allow such a distinction. Neither does consistency of choice. All females might make similar choices or they might differ in their preferences, particularly if they are choosing good gene combinations.

Even the most promising ideas for isolating choice have their shortcomings. Bertin (1986) showed that trumpet creepers preferentially matured fruits of certain males and the preference increased during the course of the season. This is consistent with a bet-hedging strategy of being relatively nonselective early, when full pollination is not yet certain, and increasing selectivity as fruit production approaches capacity. However, it also consistent with the females controlling only the numbers of fruits to be matured in different parts of the season, with the stronger competitors becoming increasingly dominant later in the season when competition is more intense.

Marshall and Ellstrand (1986) showed that multiply-sired fruits of wild radish weighed more than even the best of singly-sired fruits, and they argued that this probably indicates preference for multiply-sired fruit. The result is inconsistent with at least the most commonsense view of competition, since we would expect multiply-sired fruits to represent some average of their sires' competitive ability.

Even if these two methods can sometimes identify choice, neither is very powerful since they depend either on a seasonally increasing selectivity or on a preference for multiply-sired fruits. It is likely that many cases of preference do not involve either of these special conditions.

Trying to distinguish choice from competition may not be successful for many plants because the two processes may often lead to the same result. It may often be the the best strategy to "choose" the best competitor. This is especially true for plants because an advantage can be gained from post-pollination choice of good gene combinations. In true mate choice, accepting the best male competitor brings no genetic advantage unless fitness is heritable. Nevertheless, we cannot know that choice and competition lead to the same end until we try to distinguish them. It is clear that more general and more rigorous methods are needed. Such methods, if they could be developed, would also help lessen the reliance, in animal studies, on intuitive interpretations of behavior.

Acknowledgements

For comments on the manuscript, I would like to thank B. Charlesworth, C.R. Boake, P.H. Gouyon, K.G. Heller, R.D. Howard, R. Warner, G.S. Wilkinson, and J.E. Strassmann.

REFERENCES

Bateman, A.J. 1948. Intra-sexual selection in *Drosophila*. Heredity 2: 349–368.
Bateson, P. 1983. Optimal outbreeding. In: Mate Choice, ed. P. Bateson, pp. 257–277. Cambridge: Cambridge Univ. Press.
Bawa, K.S. 1980. Evolution of dioecy in flowering plants. Ann. R. Ecol. Syst. 11: 15–39.
Bawa, K.S., and Webb, C.J. 1984. Flower, fruit, and seed abortion in tropical forest trees: implications for the evolution of paternal and maternal reproductive patterns. Am. J. Botany 71: 736–751.
Bernds, W.P., and Barash, D.P. 1979. Early termination of parental investment in mammals, including humans. In: Evolutionary Biology and Human Social Behavior: An Anthropological Perspective, eds. N.A. Chagnon and W. Irons, pp. 487–506. North Sciutate, MA: Duxbury.
Bertin, R.I. 1986. Non-random fruit production in *Campsis radicans*: between-year consistency and effects of prior pollination. Am. Natural. 126: 750–759.
Bierzychudek, P. 1981. Pollinator limitation of plant reproductive effort. Am. Natural. 117: 838–840.
Bookman, S.S. 1984. Evidence for selective fruit abortion in *Asclepias*. Evolution 38: 72–86.
Chaplin, S.J., and Walker, J.L. 1982. Floral display of a forest milkweed: energetic constraints and adaptive value of the floral arrangement. Ecology 63: 1857–1870.

Charlesworth, B., and Charlesworth, D. 1979. Population genetics of partial male-sterility and the evolution of monoecy and dioecy. Heredity 41: 137–154.
Charnov, E.L. 1979. Simultaneous hermaphroditism and sexual selection. Proc. Natl. Acad. Sci. USA 76: 2480–2484.
Couvet, D.; Henry, J.-H.; and Gouyon, P.-H. 1985. Sexual selection in hermaphroditic plants: the case of gynodioecy. Am. Natural. 126: 294–299.
Darwin, C. 1871. The Descent of Man and Selection in Relation to Sex. London: Murray.
de Nettancourt, D. 1977. Incompatibility in Angiosperms. Berlin: Springer-Verlag.
Ellstrand, N.C. 1984. Multiple paternity within the fruits of the wild radish, *Raphanus sativus*. Am. Natural. 123: 819–828.
Fisher, R.A. 1958. The Genetical Theory of Natural Selection. New York: Dover.
Ghiselin, M.T. 1969. The Triumph of the Darwinian Method. Berkeley: Univ. of California Press.
Handel, S.N. 1983. Pollination ecology, plant population structure, and gene flow. In: Pollination Biology, ed. L. Real, pp. 163–211. New York: Academic Press.
Kirkpatrick, M. 1982. Sexual selection and the evolution of female choice. Evolution 36: 1–12.
Kirkpatrick, M. 1985. Evolution of female choice and male parental investment in polygynous species: the demise of the "sexy sons". Am. Natural. 125: 788–810.
Lande, R. 1981. Models of speciation by sexual selection on polygenic traits. Proc. Natl. Acad. Sci. USA 78: 3721–3725.
Lee, T.D., and Bazzaz, F.A. 1982. Regulation of fruit maturation pattern in an annual legume, *Cassia fasciculata*. Ecology 63: 1374–1388.
Lloyd D.G., and Webb, C.J. 1977. Secondary sex characters in plants. Botan. Rev. 43: 177–216.
Marshall, D.L., and Ellstrand, N.C. 1986. Sexual selection in *Raphanus sativus*: experimental data on non-random fertilization, maternal choice, and consequences of multiple paternity. Am. Natural. 127: 446–461.
Maynard Smith, J. 1978. The Evolution of Sex. Cambridge: Cambridge Univ. Press.
McKenna, M.A., and Mulcahy, D.L. 1983. Ecological aspects of gametophytic competition in *Dianthus chinensis*. In: Pollen: Biology and Implications for Plant Breeding, eds. D.L. Mulcahy and E. Ottaviano, pp. 419–424. New York: Elsevier.
Mulcahy, D.M. 1971. Correlation between gametophytic and sporophytic characteristics in *Zea mays* L. Science 171: 1155–1156.
Mulcahy, D.M. 1979. The rise of the angiosperms: a genecological factor. Science 206: 20–23.
O'Donald, P. 1980. Genetic Models of Sexual Selection. Cambridge: Cambridge Univ. Press.
Partridge, L. 1980. Mate choice increases a component of offspring fitness in fruit flies. Nature 283: 290–291.
Price, M.V., and Waser, N.M. 1979. Pollen dispersal and optimal outcrossing in *Delpinium nelsoni*. Nature 227: 294–297.
Queller, D.C. 1983a. Kin selection and conflict in seed maturation. J. Theor. Bio. 100: 153–172.
Queller, D.C. 1983b. Sexual selection in a hermaphroditic plant. Nature 305: 153–172.
Stanton, M.L.; Snow, A.A.; and Handel, S.N. 1986. Floral evolution: attractiveness to pollinators increases male fitness. Science 232: 1625–1627.
Stephenson, A.G. 1981. Flower and fruit abortion: proximate causes and ultimate functions. Ann. Rev. Ecol. Syst. 12: 253–279.
Stephenson, A.G., and Bertin, R.I. 1983. Male competition, female choice, and sexual

selection in plants. In: Pollination Biology, ed. L. Real, pp. 109–149. New York. Academic Press.
Stephenson, A.G., and Winsor, J.A. 1986. *Lotus corniculatus* regulates offspring quality through selective fruit abortion. Evolution 40: 453–458.
Sutherland, S. 1986. Patterns of fruit-set; what controls fruit-flower ratios in plants? Evolution 40: 117–128.
Temme, D.H. 1986. Seed size variability: a consequence of variable genetic quality among offspring? Evolution 40: 414–417.
West-Eberhard, M.J. 1983. Sexual selection, social competition, and speciation. Q. Rev. Biol. 58: 155–183.
Westoby, M., and Rice, B. 1982. Evolution of seed plants and inclusive fitness of seed tissues. Evolution 36: 713–724.
Willson, M.F. 1979. Sexual selection in plants. Am. Natural. 113: 777–790.
Willson, M.F., and Burley, N. 1983. Mate Choice in Plants: Tactics, Mechanisms, and Consequences. Princeton: Princeton Univ. Press.
Willson, M.F., and Price, P.W. 1977. The evolution of inflorescence size in *Asclepias* (Asclepiadaceae). Evolution 31: 495–511.
Willson, M.F., and Rathcke, B.J. 1974. Adaptive design of the floral display in *Asclepias syriaca* L. Am. Midl. Natural. 92: 47–57.
Woyke, J. 1967. Diploid drone substance — cannibalism substance. Proc. Int. Apicult. Congr. 21: 471–472.

Standing, left to right:
Pierre Gouyon, Dave Queller, Norbert Sachser, Peter Hammerstein, Jack Bradbury

Seated (center), left to right:
Christian Vogel, Barbara König, Nick Davies, Mary Jane West-Eberhard

Seated (front), left to right:
Tore Slagsvold, Fritz Trillmich, Geoff Parker

Sexual Selection: Testing the Alternatives
eds. J.W. Bradbury and M.B. Andersson, pp. 180–195
John Wiley & Sons Limited

Conflicts between and within the Sexes in Sexual Selection
Group Report

M.J. West-Eberhard, Rapporteur
J.W. Bradbury
N.B. Davies
P.-H. Gouyon
P. Hammerstein
B. König
G.A. Parker
D.C. Queller
N. Sachser
T. Slagsvold
F. Trillmich
C. Vogel

INTRODUCTION

Sexual conflict began with the primal divergences in size and function of gametes — with the origin of anisogamy. Since that moment competition between and within the sexes has produced a wondrous array of forms and behaviors in both plants and animals, seeming to manifest every conceivable permutation of male vs. female pursuit, manipulation, cooperation, and control. This report summarizes our brief and stumbling efforts to canalize our own verbal conflict and cooperation in a way that might illuminate pattern and cause in the diversity of sexual competition.

CONCEPTS AND DEFINITIONS: RECOGNIZING THE PHENOMENA

Sexual selection is differential reproduction that is the result of competition between the members of one sex for mates. The mating success of an individual can be measured in terms of the number of zygotes formed by unions of his or her gametes with those of the opposite sex, a definition which allows for the possibility of post-copulatory gamete competition. Following Darwin, we recognize that there are two major forms of sexual competition: one involving direct interaction (e.g., combat) of competitors and one involving "choice" (differential mating) following comparison of competitors, or of resources produced or controlled by competitors (e.g., nuptial gifts or territories). These two categories are often referred to as "intrasexual" and "intersexual" competition, respectively. However, all sexual selection fundamentally involves competition among the members of a single sex and thus could be called "intrasexual" competition, whether via direct interaction or comparison of competitors. "Choice" means differential responsiveness to potential mates.

Since females are often the limiting sex, sexual competition is often strongest among males. In this report, therefore, we will often refer to "male-male competition" and "female choice" even though female-female competition for mates, male choice of females, and mutual sexual selection can occur.

We found it useful in theoretical discussions to make the following distinctions regarding female choice.

Active versus passive choice (Lloyd 1979; Parker 1985).

Active choice involves some comparison or scanning activity by females which reduces the possibility that they are responding differentially to different males due to biases of perception rather than to differences in the male phenotype (the signal or resource itself). Passive choice is a differential response without bias-reducing comparison. For example, in a species where the female is choosing on the basis of song loudness, a passively choosing female would go to the male with the loudest apparent song but might thereby choose a song that seemed loud simply because the singer was nearby. A female exercising active choice, on the other hand, might move about or evaluate distance in order to compare the relative loudness of songs actually being produced by males. Passive choice is analogous to male-male "scramble" competition, with the loudest or best located (with respect to females) males capturing more females in their sonic traps. The terms "active" and "passive" to refer to this distinction may be confusing, since "passive" could be "active" in the sense of involving a decision accompanied by some movement or energy expenditure.

Sexually selected preference versus preexisting preference.

Female choice may involve discrimination evolved in the context of sexual selection (sexually selected preference) or it may involve female responses evolved in other contexts, which males exploit during courtship (preexisting preference). For example, male mimicry of an infant cry might prove an effective courtship signal by causing females to pause or become less aggressive, and a maternal response might then serve as a sort of "preadaptation" yielding differential responsiveness to males differing in their ability to imitate infant cries. Preexisting preference (or transfer of responses to the sexual context) could conceivably produce either passive or active choice, since preexisting responses (such as that to babies' cries) could be preadapted for discrimination of phenotypic (e.g., signal) differences.

Sexual reproduction can give rise to both cooperation and conflict between the sexes. Obviously, the interests of both male and female are served by mating and zygote production. However, conflict between the sexes arises from purely sexual competition when members of the competing sex manipulate the responses of potential mates in ways that are costly to them (e.g., forcing prolonged copulation in sites exposed to predators, producing deceptive signals of resource superiority, or successfully bypassing signals of true phenotypic of genetic quality). The "battle of the sexes" arising from sexual manipulation can come to include conflict over parental investment if females select mates on the basis of resources or behaviors benefiting offspring. The conflict over parental investment can then begin to affect sexual selection, with males selected to give deceptive signs of costly donations and females selected to detect deception and choose truly endowed males. We will discuss conflict between the sexes over parental investment only as it affects competition for mates.

MATING GAMES: WHAT SETS THE RULES?

One of the most striking aspects of sexual conflict, aside from its spectacular effects on the phenotypes of males (plumes, antlers, pheromones, and other accoutrements and antics of courtship and battle), is the very great diversity of arrangements under which sexual competition occurs. The mating behavior of a species can be synchronized or sporadic and can occur in aggregations or in isolated pairs. Courtship can be simple or complex and copulation rapid or prolonged. Males may seek females while females either hide or signal their whereabouts, or these sex roles may be reversed. All of these patterns may vary independently and do so even among related species. What causes this enormous diversity in the circumstances and conditions of sexual competition?

Factors that influence the evolution of mating arrangements have been studied under several major headings, which are really interrelated topics basic to the study of sexual conflict: parental investment theory, mating systems (especially in relation to ecology), sex allocation theory, and sex ratio theory (especially considerations of the operational sex ratio). In our discussions we have deliberately excluded a detailed examination of these topics, which would require whole Dahlems within a Dahlem. However, we have seen that these factors are complexly interrelated in setting the rules of sexual selection and that it may prove fruitful in the future to consider their relationships in a formal theory. We record the following thoughts in the hope that they may help point the way toward the development of improved concepts.

Environmental conditions (such as the distribution, availability, and defensibility of limiting resources) can determine the relative investment of the two sexes in their offspring and thus influence the degree of sexual selection and sexual dimorphism. The influence of environmental conditions on the mating system is discussed elsewhere (e.g., Emlen and Oring 1977; Thornhill and Alcock 1983). A mating system is sometimes taken as a good indicator of whether male-male competition or female choice will predominate in a species.

This is, however, clearly inadequate. Among lek species, for example, males may be chosen without interference in some species (e.g., in sage grouse), whereas in other species male-male interruption may predominate (e.g., in buff-breasted sandpipers). We thus need a better set of predictors for sexual selection patterns (see Bradbury and Davies, this volume).

Game theory is one avenue of approach. Maynard Smith (1977) has used game theory to investigate the relationship between parental investments and the freedom of each sex to compete for mates. He has shown that in the many cases where more than half the offspring can be raised successfully by one parent alone, the other one should leave to search for a new mate if there is a reasonable chance to find one. However, he was unable to decide which of the two alternative evolutionary stable strategies (ESSs) the system would evolve — the "stickleback" (female leaves) or the "duck" (male leaves) ESS. Such questions may sometimes be resolved by considering a sequence of related games. Whichever parent leaves the offspring, both then face a "mobility game" (see Hammerstein and Parker, this volume) whose logic suggests that males rather than females are likely to search for mates (the "roaming male" ESS). Perhaps the mobility game, coupled in some way with the parental investment game, may provide a more general answer as to why it is more likely that females have to care for offspring. More complex arrays of alternatives might be analyzed by dividing sexual options into a series of games, e.g., the parental investment game, the mobility games, and the persistence game (Hammerstein and Parker, this volume). As with the determination of mating systems, each of these games has an outcome in part dictated by ecological and phylogenetic contexts. The difference is that factors such as mobility and persistence may be more important in deciding the relative roles of female choice and male-male competition than more structural decisions such as whether to form a harem by resource defense or by female defense (see Bradbury and Davies, this volume). The first game must be the parental investment game: who gets left with the baby? Certainly the outcome of this game will constrain the latitude of one sex to exercise its sexual options. This is not sufficient, however. The mobility game is then involved: where males are more mobile most species show some emphasis on male-male competition; in contrast, female mobility appears to be a necessary, but not always sufficient, condition for wide female choice as on leks. This is suggested by the parallel evolution of some insect swarms and vertebrate leks where the parental investment game left the female with the baby and the mobility game left males localized in groups to await females, *but*, the two taxa appear to show different levels of female choice. The persistence game might then be involved as a final factor: which sex has the time the energy to court or resist courtship longest? Thus, although not always independent, these games considered to relation to ecological factors affecting their outcomes may predict patterns of sexual selection better than does the classical approach of investigating the influence of ecology on the evolution of mating systems.

MALES VERSUS FEMALES: WHO WINS?

Sexual conflict has so far been analyzed by studying game theoretic models. The game theory approach lacks any explicit treatment of the underlying genetics. However, it has the virtue of allowing a greater variety of strategies and thus more phenotypic complexity than might be possible with the standard framework of population genetics. Put simply, in constructing theory to explain and predict patterns of sexual behavior we are often forced to choose between complexity on the one hand and genetic rigor on the other.

There are many kinds of conflict between the sexes, from the primordial conflict over anisogamy to those just discussed involving investment in searching for mates, mate choice, and parental care. Who wins depends on a number of factors, according to the type of conflict concerned. In anisogamy the very existence of males in multicellular organisms implies that, in an evolutionary sense, males have "won" this primordial conflict. In this section we first present a conceptual framework for viewing conflict between the sexes regarding the costly activities and decisions associated with mating. We then discuss the mechanisms by which males and females manipulate each other in their own interests.

A conceptual framework for the study of male-female sexual conflict

Conflict over which sex will be mobile (or incur the costs of mate search), like conflict of interest over parental investment, can commonly have more than one stable solution for a given set of conditions. In such cases perhaps the most important consideration concerning which outcome is likeliest is the starting condition (i.e., evolutionary history). If we can deduce the plausible starting conditions, the equilibrium attained can usually be predicted. For example, suppose that in a parental investment game the two ESSs were male guard/female desert and female guard/male desert. We might expect to find males sometimes guarding offspring in fish, where males and females are usually equally equipped to do so, whereas female guard/male desert might be more common in mammals in which only females lactate.

A second way to deduce the probable outcome — and one that may be particularly useful if no a priori assumptions can be made about starting conditions — concerns the payoff-relevant asymmetries present in the game. If, for instance, it were much more costly for male fish to guard eggs than female fish, then even though the opportunity to guard may be similar for both sexes we would perhaps predict a greater "zone of attraction"* for the female guard/male desert ESS. If the zone of attraction for one of the ESSs is sufficiently small, then drift alone may be enough to ensure that we would

* range of initial conditions that will lead eventually to the establishment of a particular ESS.

eventually attain the alternative ESS and stay there.

Games between the sexes will always be asymmetric in the sense that one player is irreversibly labeled male and one player is labeled female. However, there will usually also be a payoff-relevant asymmetries, which will be important in establishing the outcome on the "zone of attraction" for alternative ESSs.

The solution to some sexual conflict games can be a single ESS (e.g., some of the mating decision conflicts by Hammerstein and Parker, this volume). Which sex generally wins can then be deduced from a knowledge of the payoff-relevant asymmetries. For the sexual war of attrition, these would be the cost to a male of time spent in attempting rape versus the cost to the female of equivalent time spent avoiding rape and the relative value of winning the contest for each player. For the sexual arms race the costs would be morphological. However, in both of these games, we would expect to see examples in a given population where one sex appears to "win," and other examples where the other sex appears to "win," although one outcome may be overwhelmingly predominant.

In some cases it seems likely that the formal models proposed by Hammerstein and Parker (this volume) are inadequate since they may properly be dealt with as repeated games. For example, a female may be able to eject or resorb an unwanted ejaculate after mating (see *MATE CHOICE AFTER MATING* below). Sometimes males show no inclination to persist in courting a female that is uninterested in them. Such phenomena could be attributable to the persistence game or sexual arms race. However, such solutions could imply a repeated game, in which the female has the strategic possibility of rejecting unwanted sperm. A set of models of repeated games in sexual conflict might be useful, especially in parental investment models. Even when the female has the last, controlling move regarding mate choice, the entire sequence of preceding games has significance for sexual selection as a series of circumstances in which males compete with each other to be the beneficiaries of the final decision.

The mobility game is relevant to the question of how sexual selection is achieved — whether via male-male interaction or female choice. There is a general correlation between male-male competition and mobile males on one hand and fixed males and female choice on the other. With regard to leks, the outcome of the mobility game may be *preset* by ecological conditions (e.g., those leading to very large, overlapping ranges in females: the correlation between female range size and male aggregation for mating is quite good). Unfortunately, the correlation between male aggregation and "strong" female choice is less strong.

It would be of interest in this context to know if female choice occurs in the "mating swarms" of insects, which in some respects resemble leks. "Explosiveness" of mating may be one factor affecting the likelihood of female choice, (Emlen and Oring 1977; Linsenmair, personal communication). Factors affecting relative persistence of the sexes may also play a role, and this also may be related to seasonal synchrony (explosiveness) of breeding. In short, the weight is thrown from male-male interaction to female choice given ecological generation of high female mobility and a) sufficient breeding time to accomplish choice and/or b) factors favoring female persistence in mating activity over male persistence.

Manipulation of mate choice

Whatever the (often complex) causes of limited access to mates, the common occurrence of asymmetry in this regard leads to two sets of manipulations, one performed by the sexually limited competing sex (usually the males) and the other by the limiting sex (usually the females). In this section we will restrict discussion to manipulations of the opposite sex regarding mate choice and exclude other possible kinds of male-female manipulations (e.g., of parental investment or the sex ratio). As will be clear from the examples discussed below, "manipulation," or the exercise of control over the responses of another, need not imply conflict, force, or coercion although these negative descriptions may sometimes apply.

Males employ a very large array of behavioral and morphological features which can influence or restrict choice of mates. Well-studied examples seem to fall into two categories:

1) mechanisms involving force including rape, persistent pursuit or "harrassment," intromittent genitalia, and infanticide (further discussed below); and
2) material enticements including nuptial gifts, nutritive spermatophores, participation in parental care, protection against predators, and prevention of harrassment by other males.

In addition, ethological studies and theoretical considerations suggest that a third kind of manipulation is likely, namely deception (sensu Otte 1975) or mimicry (Lloyd 1980). Examples include inflation of the body or plumage erection simulating large size, the bubbles added to balloon-fly nuptial gifts enlarging their appearance, imitation of non-sexual signals, and stimulation of behavioral or hormonal responses evolved in nonsexual contexts, but increasing the likelihood of mating or gamete acceptance (see "Post-mating mate choice" below).

On the other hand, individual females may increase the opportunities of effectiveness of choice via:

1) increased mobility (searching, comparison);
2) inciting congregation and/or contests among males (e.g., via female aggregation or female attraction to male aggregation and interaction);
3) multiple mating in species which manipulate sperm (e.g., in which there is sperm precedence or sperm destruction by females).

There is a striking lack of information on the activities and abilities of females compared to those of males as regards these activites, despite their great potential importance in sexual conflict. The following points may help to focus research in this area.

Inciting contests among males may enable comparison of strength and vigor of potential mates at little cost to the female, thus superimposing the "truth" of male contest on potentially deceptive or male coercive courtship. However, inciting male-male contests may sometimes "backfire" on females, if fighting males are selected for increased size (as they often are in fighting species). This could lead to trapping females into parental care expenditures in producing large male offspring which constitute larger investments than are desirable under natural selection so that selection on *females* might limit the exaggeration of the male trait (large size). Partridge and Endler (this volume) discuss other aspects of the costs of sexually selected traits under natural selection in different life history stages.

As already discussed, part of the cost of female choice (when males offer resources) may be the loss of access to males who are busy garnering resources at the time the female is making a choice. Or choice may require that she spend more time searching for these males. In effect, the operational sex ratio (and freedom of "choice") for such a female is lowered by her choosiness regarding the males.

Multiple mating does not, of course, necessarily lead to increased female choice, since special mechanisms (such as sperm manipulation) must be involved and multiple mating may have many different explanations (such as avoidance of harrassment, formation of protective or parental care alliances with multiple males, inability of females to resist copulation, etc.). Studies of the pied flycatcher (*Ficedula hypoleuca*) exemplify the kinds of research that can illuminate conflict between the sexes. Males of this species may be residential at the nest or may go off for further matings. Females mated monogamously have a higher fitness than do females mated with a polygynous male because of the greater parental investment of monogamous males. However, polygynous mating occurs even when there are some unmated males in the population. The polygynous males may deceive females into mating against their own interests (see Alatalo et al 1981); alternatively, the second females may accept their status due to the costs of searching for unmated males (Slagsvold 1986).

MATE CHOICE AND INTRASEXUAL COMPETITION AFTER MATING

Most familiar forms of mate choice occur before mating, but there is also potential for choice after mating. The choice could be of male gametes, of offspring, or (in plants) of endosperms. In animals, gametic choice would amount to delayed mate choice since gametic genes are not expressed. Such forms of choice can be quite cryptic. For example, Thornhill (Thornhill and Alcock 1983) studied female scorpion flies that mated randomly but used sperm differentially by laying eggs only after mating with large males who offered large nuptial gifts. Post-mating mate choice is considered at length by Eberhard (1985), who shows that females manipulate sperm following copulation in four different ways: prevention of complete intromission; restriction of sperm transfer within the female reproductive tract; sperm inactivation, removal, and storage; and by remating. There is also evidence that these processes are influenced by male phenotype and/or behavior prior to or during courtship in some species.

In plants, gametic choice is potentially different because the male gametophytic stage does express its own haploid genotype. Such choice is well-known for gametophytic self-incompatibility mechanisms but needs to be investigated for finer kinds of discrimination. Differential success of pollen genotypes is well known (Jones 1928), but distinguishing female choice from simple pollen competition will be difficult.

Selection of offspring should also be investigated. Strictly speaking this is not sexual selection, but it shares some properties with sexual selection, and it can result in favoring one father over others in broods of mixed paternity. There is one crucial difference from true mate choice in that the mothers might choose good offspring on the basis of non-additive genetic effects (Queller, this volume).

Intrasexual competition likewise does not necessarily end with the fertilization of the egg, as long ago appreciated by Haldane (1932) in discussing "certation," or pollen-tube competition, in plants. Infanticide can represent either male-male competition (e.g., in langurs, lions, etc.) or female-female competition (e.g., prairie dogs). Observations of langur monkeys (*Presbytis entellus*) (Sommer and Mohnot 1985 and references therein) illustrate how infanticide can be the result of male-male competition for mates. Some langur populations breed in harems. There are few harem leader positions within the population, resulting in severe male-male competition for harems and short tenureships for each harem-possessing male. Under these conditions a male can increase his reproductive success by killing those infants sired by his predecessor which still hold their mothers in the state of lactational amenorrhea, thereby bringing those females quickly into estrus again. Field data strongly support the hypothesis of Hrdy that male infanticide in langurs is thus an adaptive reproductive strategy evolved under sexual selection on males.

In a model considering female reproductive characteristics and the fequency distribution of male tenure lengths within the population, it can be shown that a mixed ESS may be established, stabilizing a rather fixed proportion or infanticidal or non-infanticidal "take-overs" of harems. Langur females seldom (if ever) succeed in their efforts to defend their infants against infanticide. Thus, in this case males appear to be the winners in a dramatic example of postfertilization conflict.

SPECIAL PROPERTIES OF SEXUAL CONFLICT IN PLANTS

In plants, sexual selection may act on gametes (gametophytes) or embryos, but most frequently it acts on sex allocation to male versus female function as pointed out by Queller (this volume). In quite the same way, hermaphroditic animals are under sexual selection for the allocation of resources to male versus female function. In monogamously mating hermaphrodites this may lead to minimal development of the testes and maximal egg production, while under a promiscuous mating system the majority of the resources may be allocated to the male function. Similarly, in sequential hermaphrodites, sexual selection may determine the time or body size at which sex change occurs as well as the direction of the change (i.e., from male to female or vice versa).

Although we recognize that sex allocation in hermaphrodites as well as in animals with separate sexes is strongly influenced by sexual selection, we chose not to discuss this issue further since it has been reviewed recently (Charnov 1982) and is too vast a subject to be included in our discussion. Nor did we cover the large subject of how sex allocation and sexual selection influence the evolution of plant breeding systems (Charlesworth and Charlesworth 1979).

Table 1 compares some important differences between higher animals and plants. Sedentary animals would share some features of both groups.

TABLE 1 Differences between plants and higher animals relating to the nature of mate choice in plants

Plants	Higher animals	Consequences
Fixed	mobile	pre-mating choice difficult in plants
Usually hermaphrodite Developmental canalization* weak	dioecious strong canalization	sexual differentiation rarer, weaker in plants; recognition of male and female traits more difficult (even if dioecious, higher similarity between male and female)
High fecundity Common genes expressed in n and $2n$ phase	lower fecundity fewer such genes expressed	post-mating choice of genetic quality theoretically easier (on pollen tubes or on zygotes)

* Cellular differentiation is much stronger and more definitive in animals (as revealed by regeneration from tissue culture possible in plants and still impossible in most higher animals).

Pre-mating choice in plants would be difficult to achieve. However, females (or a female function of a hermaphrodite) can choose their mates by flowering time (demonstrated only in cases of hybridization or incest avoidance), floral morphology (e.g., distyly demonstrated in cases of incest avoidance), gamete choice (e.g., gametophytic self-incompatibility demonstrated in cases of incest avoidance), and zygote choice (e.g., self-incompatibility in Cacao-demonstrated in cases of incest avoidance).

In *Thymus vulgari* (Gouyon et al., in preparation), hermaphrodites have bigger flowers than the female. If one measures seeds, pollen, and flowers one finds that in the hermaphrodite, pollen number and flower size are positively correlated, while both are negatively correlated with seed number. In females, flower size is negatively correlated with seed number. Growth hormones of the flower, however, are produced in the anthers. These are more or less completely aborted in the female. These observations could possibly be interpreted as the result of developmental constraints. However, regarding sexual conflict, it is of interest that the flower growth is controlled by a male organ.

It seems likely that males (or male functions) in plants may compete for pollinator attraction and overproduce flowers (see Queller, this volume). It also seems likely that pollen competitively enters the style. But no *effect* of this selection is yet known. A runaway process might be possible affecting the efficiency of pollen in the style (females favoring a certain type of pollen and therefore their own sons as better males), but in hermaphrodites this could be in conflict with mechanisms preventing inbreeding.

In conclusion, although possible mechanisms of sexual selection are known in plants, sexual selection has rarely been demonstrated to act. All suggestive examples involve self-incompatibility mechanisms or prevention of interspecific crosses. Is this because ideas regarding sexual selection have only recently been applied to plants? Or does it indicate a true difference in the biology of these organisms when compared to animals? Similarly, one might ask why plant-like (sessile) animals such as sponges and corals lack some of the outstanding sexual features of plants, such as agents of sperm transport ("pollinators").

SUGGESTIONS FOR FUTURE STUDY

During our discussions it became clear that certain kinds of information are simply not available that are essential for the analysis of conflict within and between the sexes. There is a shortage not so much of hypotheses but of certain kinds of critical information that could contribute to both the formulation and testing of theory. Some of these gaps in our knowledge or organisms were mentioned repeatedly in this and other working groups. We therefore propose a list of areas where new information is urgently needed and seems possible to obtain.

1) Observational and experimental evidence that mate choice occurs, of both the characters chosen and the "protocols" or procedures (behaviors) used to effect choice, including the timespan and number of comparisons made during choice. We have very few data on how females behave when choosing among

males. How many males does a female sample before she mates? Does she actively reject some males in favor of others? Frog choruses and vertebrate leks would seem excellent systems to study this in detail.

2) What are the precise effects of male displays on female behavior? In some peacocks and pheasants the ocelli on the tail and tail coverts are argued to have a hypnotic effect on females. Following Dawkins and Krebs' (1978) ideas on "communication or manipulation," it seems possible that the elaboration of plumage and displays are an end result of male-male competition for better and better manipulation of females. Some males may therefore be inducing copulation through sensory-psychological effects in much the same way that other males use physical force for the same ends. Recent reviews (Eberhard 1985; West-Eberhard, 1984) provide evidence that sexual signals manipulate particular widespread behavioral or physiological responses of females in ways likely to affect male mating success. This provides an alternative to the common view that differential female responsiveness ("female choice") originates under selection on females to choose their mates (whether in a mate-quality or species-recognition context).

3) Measurements are needed of the costs of female choice. For example, if predation pressure is high at the breeding site it may be costly for a female to sample several males before she chooses. Other possible costs are time and energy spent in sampling. It is important to document costs because they influence the outcome of certain theoretical models for the evolution of female preference. It is important here to distinguish between costs of choice and costs of mate-seeking.

4) In cases where males defend a resource, are females choosing the male or the resource? Some experiments suggest that females are influenced by the male character even when resource differences are important influences on female reproductive success (Andersson 1982). Other experiments have shown that it is the resource rather than male display which influences female choice (Slagsvold 1986).

5) Evidence of gamete and zygote competition, choice in plants and higher animals, and post-copulatory female choice (manipulation of male gametes). In plants, do larger pollen grains have an advantage? What is the function of differences in pollen structure? Do female structures have features encouraging or otherwise influencing pollen competition? Since the haploid genome actually affects the phenotypic characteristics of pollen (see Haldane, 1966), there is an opportunity in plants (not know to be present in animals) for female choice of male genotype. Is there any evidence that female plants take advantage of this opportunity?

6) Information on role of both sexes in influencing mating decisions. Many studies focus on either male behavior or female behavior but not both. There may be differences not only in the relative importance of male-male competition versus female choice both between species (e.g., explosive breeders with a

low probability of female choice versus prolonged breeders with a higher probability) but also within the same species due to demographic factors. We need comparative studies of:

a) different populations of the same species under different ecological or demographic conditions;
b) different species of the same taxonomic group living in different habitats to determine factors which influence the importance of male-male competition versus female choice (see example).

7) Comparative study of related species designed to show whether or not there is an evolutionary progression toward refined sensory discrimination of male traits by females in a sexual context. Female choice (differential sexual responsiveness) of females due to stimulation by males of response repertoires evolved in a nonsexual context is a neglected and possibly very widespread means of sexual coercion. Comparative study may allow discrimination between responses evolved under selection for female choice, and those "preadapted" for it: if, in a group of closely related species, most show a female response in a nonsexual context, whereas in only one or a few species it is stimulated by males during courtship in addition (leading to differential mating success), this would provide positive evidence for the occurrence of preexisting preference.
8) There is much scope for future development of models of sexual conflict. On the one hand there is much potential for increasing the strategic complexity of some of the existing models (e.g., sequential game analyses of parental investment and conflict over mating decisions). On the other hand simple ESS models should be examined for their robustness with regard to explicit genetic interpretation. In particular we should like to establish whether certain sexual conflict games are less robust than the classical ESS models in this respect. There would certainly be difficulties in attempting to model the central problem in Fisherian runaway processes with its assortative mating and correlation of signal and response, without explicit genetics.

CONCLUSION

Given the very great diversity of patterns of sexual selection, which we have noted over and over again in our comments, the testing of the plausibility of particular theories may depend not on the kind of single-stroke tests possible in physics and chemistry and certain experimental areas of biology but on the accumulation of information like that just listed for a large number of species. This has been pointed out by Hamilton in discussing the parasite model of female choice, and it is a general feature of many questions in evolutionary biology, including the general question of the importance of Darwinian natural and sexual selection in nature.

In the case of alternative hypotheses for the basis of female choice, the "alternative" hypotheses may not be truly competing generalizations but may prove complementary: all may be needed to explain the diversity of courtship signals observed. The relative importance of these hypotheses depends on the regularity with which their requirements and predictions are fulfilled within species in nature.

One role of sexual selection hypotheses, in addition to promoting broadly applicable generalizations, is to draw attention to features of behavior and morphology whose functions and even existence have gone unnoticed in the past. Some of these features may contribute yet unimagined "alternative hypotheses and critical tests" of sexual selection theory. Should those fail, we can always return to the way we were:

The Way We Were

A sweet and friendly porcupine,
Snuffled up; his eyes met mine,
I pickled him in bright red wine,
And labelled each and every spine.

A thousand more I then did nail,
and measured all from head to tail,
and plotted on a log/log scale,
. . . but everything to no avail.

An orgy of regression failed—
So algebra I then assailed,
With all the fun that this entailed
I modelled . . . and results prevailed.

My future now was quite secure,
And to the field I go no more,
Papers, papers by the score
Full of sums and such a bore.
–G.A. Parker

REFERENCES

Alatolo, R.V.; Carlson, A.; Lundberg, A.; and Ulfstrand, S. 1981. The conflict between male polygamy and female monogamy: the case of the pied flycatcher, *Ficedula hypoleuca*. Am. Natural. 117: 738–753.

Andersson, M. 1982. Female choice selects for extreme tail length in a widowbird. Nature 229 (5886): 818–820.

Charlesworth, B., and Charlesworth, D. 1979. Population genetics of partial male-sterility and the evolution of monoecy and dioecy. Heredity 41: 137–154.

Charnov, E.L. 1982. The Theory of Sex Allocation. Monographs in Pop. Biol. 18. Princeton: Princeton Univ. Press.

Dawkins, R., and Krebs, J.R. 1978. Animal signals: information or manipulation? In: Behavioral Ecology: An Evolutionary Approach, eds. J.R. Krebs and N.B. Davies, pp. 282–309. Sunderland, MA: Sinauer.

Eberhard, W.G. 1985. Sexual Selection and Animal Genitalia. Cambridge, MA: Harvard Univ. Press.

Emlen, S.T., and Oring, L.W. 1977. Ecology, sexual selection, and the evolution of mating systems. Science 197: 215–223.

Haldane, J.B.S. 1966. The Causes of Evolution. Ithaca, NY: Cornell Univ. Press.

Jones, D.F. 1928. Selective Fertilization. Chicago: Univ. of Chicago Press.

Lloyd, J.E. 1979. Sexual selection in luminescent beetles. In: Sexual Selection and Reproductive Competition in Insects, eds. M.S. Blum and N.A. Blum, pp. 293–342. London: Academic Press.

Lloyd, J.E. 1980. Male *Photuris* fireflies mimic sexual signals of their females' prey. Science 210: 669–671.

Maynard Smith, J. 1977. Parental investment: a prospective analysis. Anim. Behav. 25: 1–9.

Otte, D. 1975. On the role of intraspecific deception. Am. Natural. 109: 239–242.

Parker, G.A. 1985. Phenotype limited evolutionary stable strategies. In: Current Problems in Sociobiology, ed. Kings College Sociobiology group, pp. 173–201. Cambridge: Cambridge Univ. Press.

Slagsvold, T. 1986. Nest site settlement by the pied flycatcher: does the female choose her mate for the quality of his house or himself? Ornis Scand. 17: 210–220.

Sommer, V, and Mohnot, S.M. 1985. New observations on infanticides among hanuman langurs (*Presbytis entellus*) near Jodhpur (Rajasthan/India). Behav. Ecol. Sociobiol. 16: 245–248.

Thornhill, R., and Alcock, J. 1983. The Evolution of Insect Mating Systems. Cambridge, MA: Harvard Univ. Press.

West-Eberhard, M.J. 1984. Sexual selection, competitive, communication and species-specific signals in insects. In: Insect Communication, ed. T. Lewis, pp. 283–321. New York: Academic Press.

Sexual Selection: Testing the Alternatives
eds. J.W. Bradbury and M.B. Andersson, pp. 197–207
John Wiley & Sons Limited

Measuring Sexual Selection

M.J. Wade

Department of Biology
University of Chicago
Chicago, IL 60637, USA

Abstract. Some of the methods for measuring selection in natural populations are described with emphasis on the application of these techniques to measures of sexual selection. *Parameter estimation* in contrast to *hypothesis testing* is also discussed and some of the important differences are illustrated by the reanalysis of an example of sexual selection from the published literature. The variance measures and gradient measures of sexual selection are further discussed in relation to two different research questions that are often confounded: a) *How* is selection operating? and b) *Why* does selection operate the way it does?

INTRODUCTION

Darwin defined sexual selection as that component of natural selection arising from the variance among males in the numbers of mates or the variance among males in reproductive success owing to competition for mates (where competition can be either for numbers or quality). He identified two proximal causes of variance in the numbers or quality of mates of males: a) male-male competition, and b) female choice. The effects of these processes on sex differences in morphology, physiology, and behavior constitute one of the most active areas of current evolutionary research. In this paper I present background information concerning the measurement of sexual selection in natural and laboratory populations. The most useful measures are those that best permit field observations to be brought to bear incisively on formal evolutionary theories of sexual selection. Although many different kinds of measurements can help to confirm our intuition or quantify plausible conclusions, the issue is not quantification for quantification's sake. The issue is how best to relate observation and theory. In this regard, I argue that the measures of selection that I discuss here do *not* require changes in the kinds of observations that field or laboratory researchers presently make but rather changes in the ways that data are analyzed and reported. We are always confronted with the problem of choosing appropriate summary statistics for extensive data sets. I am not prescribing a research question; I am advocating a method of analysis because of its relationship to formal evolutionary theory.

MEASURING SELECTION VERSUS HYPOTHESIS TESTING

Hypothesis testing is one of the most frequently used experimental techniques in studies of adaptive evolution. When the research question can be reduced to an either/or or if/then statement, hypothesis testing is the empirical method of choice, often contrasted in a superior tone of voice with "mere descriptive study." In biological systems we are often confronted, however, with a situation in which several different processes are acting simultaneously. The research question is often *not* which *one* process is most important, but rather, what is the *relative strength* of each operating force.

Similarly, evolutionary biologists often make use of the tremendous taxonomic diversity of living systems to answer questions of adaptation using the comparative method. In making such comparisons between populations, species, genera, or other more inclusive taxonomic groupings, the research question in many cases is not in *which* group a process is operating but rather *to what degree* does this process operate in each group. Differently put, there are a small number of different forces that operate on a large number of different species and most research questions are ones of the relative strengths of common forces rather than the detection of unique forces.

Thus, hypothesis testing can be viewed as an attempt to determine whether or not a process should be included in an evolutionary description whereas parameter estimation can be viewed as quantifying its role in such a description. The most important parameters to estimate are those that play a central role in formal evolutionary theory. In this way theory exerts a powerful constraining influence on our measures of selection. In the next sections I will describe (but not derive) a series of basic measures of selection and discuss their relationship to formal evolutionary theory. I will discuss the kinds of experimental studies necessary for measuring selection in natural or laboratory populations. Finally, using an empirical example from the published literature on sexual selection, I will illustrate the interrelationship of hypothesis testing and parameter estimation in the study of adaptive evolution.

MEASURES OF SEXUAL SELECTION

The *opportunity for selection* is the variance in relative fitness associated with a particular component of the life history of an organism. In keeping with the important distinction between selection and the response to selection, the variance in relative fitness as a measure of selection (and not of response to selection) includes *all* sources of fitness variance, environmental as well as genetic. In the formal theoretical literature, relative fitness is that functional transformation by which a phenotypic distribution at one time becomes the phenotypic distribution at a later time *within the same generation*. If we are to restrict our definition of relative fitness to only the genetic component of fitness, as has been advocated by some authors (e.g., Endler 1986; Sutherland 1985), then we could not predict the phenotypic distribution at a later time within the generation from knowledge of the present phenotypic distribution and relative fitness. Some have argued that we should restrict our attention to only the genetic components of fitness because evolution across generations depends upon there being heritable variance in fitness. If we adopt an equilibrium view of nature, as is often done in constructing adaptive arguments, then the lack of heritable variance in fitness expected as a consequence of Fisher's Fundamental Theorem would prevent the study of evolution by natural selection. However, natural selection is not evolution (Fisher 1931) and it can be studied in its own right. Indeed, if we view selection as arising from the interaction of the phenotype with the environment and if we grant the basic continuity of the abiotic and biotic environment over time, then we can study today the selective forces that must have characterized at least the recent past as well as the immediate future. At the very least, no matter how dynamic a view of the environment we adopt (as opposed to an equilibrium view), by measuring selection we can characterize the spatial and temporal scale of environmental changes in a way relevant to evolutionary theory. (It is difficult, however, to maintain an equilibrium view of nature, let alone a belief in optimal adaptation, without assuming some continuity of the environment.)

The variance in relative fitness plays a central role in evolutionary theory (see, for example, Fisher's Fundamental Theorem). It is a necessary *though not sufficient* cause of evolution by natural selection. It is related to the adaptive transformation of the phenotypic distribution in the following way. The most common criterion of evolutionary change is the change in the mean of the distribution of a phenotypic trait. Even in comparative studies it is the mean differences between populations, species, or some other taxonomic groupings that are both the basis for erecting a classification and the objects in the search for adaptive patterns. A formal and general equation describing the change in the mean of the phenotypic distribution is thus of some interest. The change in the mean within a generation is always equal to the correlation between relative fitness and the value of the trait in question. No trait can covary with fitness to a greater degree than fitness covaries with itself. The variance in relative fitness (i.e., the covariance of fitness with itself) is the *opportunity for selection* and it measures the change in mean fitness within a generation or within a portion of the life history. It sets an upper limit to the amount of adaptive change that natural selection can effect. The variance in fitness sets only an upper limit to adaptive change for three reasons. First, the existence of variance in fitness does not guarantee the existence of a covariance between fitness and any phenotypic trait. Variance in fitness guarantees that selection *can* operate not that selection *does* operate. Second, only a fraction of the covariance between fitness and character is transmitted across generations. This fraction is given by the heritability of the trait whch is always less than or equal to one. Third, the existence of a covariance between fitness and character does not of itself explain *why there should be* such a covariance. It is this latter limitation that places an important conceptual limitation on our use of variance and covariance measures of selection. These measures only tell us that selection is acting and not why it is acting. Despite these caveats, the opportunity for selection is an important measure of selection with a direct interpretation in the light of formal evolutionary theory.

In regard to sexual selection, the sex *difference* in the opportunity for selection is given implicitly in the following equation,

$$I_{males} = RI_{females} + I_{mates} \tag{1}$$

I_{males} is the variance in the reproductive success of males divided by the square of the mean reproductive success; it is the variance in *relative fitness* among males. R is the sex ratio expressed as the number of female mates per male and $I_{females}$ is the variance in relative fitness among females. Lastly, I_{mates} is the variance among males in the numbers of mates divided by the square of the average number of mates per male, i.e., the component of the total variance in male relative fitness owing to variance among males in the numbers of mates. Because each term in equation (1) is a variance, each is positive and, clearly, the magnitude of I_{mates} is the cause of the *sex difference* in the opportunity for selection. It is also clear that both male-male competition for mates and female choice of males are processes that influence the magnitude of I_{mates}.

I derived equation (1) by making some simplifying assumptions about the ways in which mating and production of offspring take place. These assumptions limit the applicability of the equation to organisms whose biology conforms to the assumptions. One important assumption was that each female mates with a single male and that the male is the sire of all of her progeny. Females of many organisms in nature, however, produce multiple clutches and seek different mates between clutches (e.g., the imported willow leaf beetle, *Plagiodera versicolora* (Wade and Breden 1986; McCauley and O'Donnell 1984). The production of multiple clutches by females diminishes the sex difference in the opportunity for sexual selection. In this case, equation (1) becomes

$$I_{males} = R(I_{females} - I_{clutches}) + I_{mates} \tag{2}$$

where $I_{clutches}$ is the variance among females in the numbers of clutches produced divided by the square of the mean number of clutches per female (i.e., the fraction of the variance in relative fitness among females owing to variance between females in the numbers of clutches produced). The variance between females in the numbers of clutches plays a role very similar to that of the variance in numbers of mates of males. It clearly *diminishes the sex difference in the opportunity for selection*. Indeed, if the variance among females in the numbers of clutches were greater than the variance among males in the numbers of mates, then we might expect that those behaviors and morphologies that we associate traditionally with sexually selected males would instead be manifested in the female sex.

Fisher's insight into the necessary relationship of the *mean* reproductive success of the male sex to that of the female sex has been developed into an elaborate theory of sex ratio evolution and parental investment in the two sexes. I believe that the necessary relationship of the *variance* in reproductive success of the male sex to that of the female sex, as illustrated in equations (1) and (2), can provide insight into problems of sexual selection.

THE COVARIANCE BETWEEN FITNESS AND PHENOTYPE

Change in the mean of a phenotypic distribution is the most commonly used criterion of evolutionary change. Although changes in other moments of the phenotypic distribution have been discussed, such as changes in the variance in response to "stabilizing" or "disruptive" selection, by and large evolutionary theory has concerned itself with changes in the location of phenotypic distributions. In the study of sexually selected traits, it is the extreme exaggeration of male phenotypes that has prompted so much empirical and theoretical research. These cases attract research attention because of the *sex difference* in the mean of the phenotypic distribution. For this reason the following remarks will be concerned solely with the theoretical measurement of directional change in the average of the phenotypic distribution. (An important caveat to remember is that each moment of the phenotypic distribution can formally be considered as a separate trait often correlated, phenotypically or genetically, with other traits or moments.)

There are two different questions to be asked when measuring the effects of selection in natural populations: a) How does selection change the mean of the phenotypic distribution? and, b) Why does selection change the mean of the phenotypic distribution? Both of these questions can be answered by comparing the phenotypic distribution at one time with the phenotypic distribution at some later time within the same generation. (We deliberately want to avoid across generation comparisons because the effects of genetic transmission on the phenotypic distribution are confounded with those of selection.) However, these two questions are different in an important way that dictates a manipulative approach to the second question.

The first question can be answered by careful observations of natural populations and depends upon the location of the phenotypic distribution in relation to the fitness function. This question might be rephrased as "What is selection doing now?" By comparing the phenotypic distribution at one time

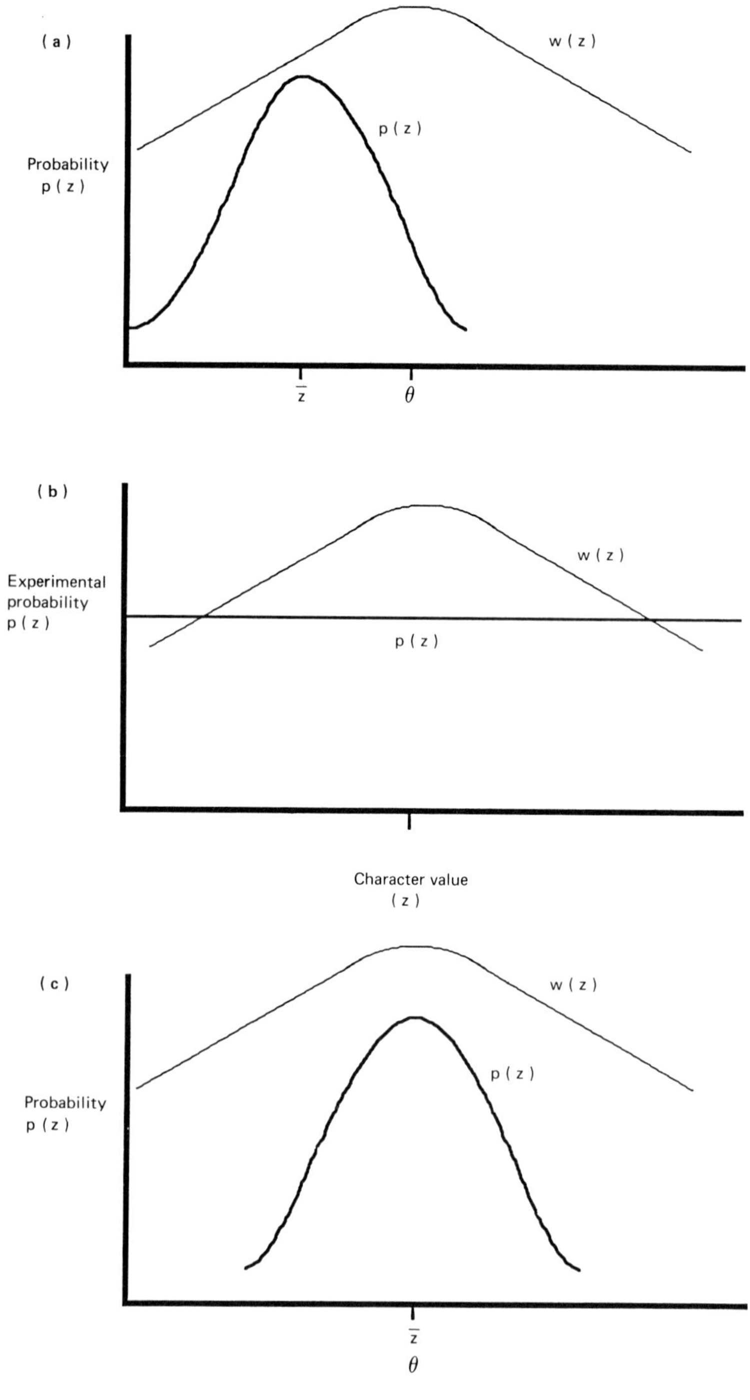
(a)
w(z)
p(z)
Probability
p(z)
z̄
θ
(b)
w(z)
Experimental
probability
p(z)
p(z)
Character value
(z)
(c)
w(z)
p(z)
Probability
p(z)
z̄
θ

with that at a later time, we can draw an inference about the operation of selection by considering the fitness function to be that linear or quadratic form that best describes the observed phenotypic transformation (see Figure 1a). The fitness function inferred in this manner depends critically upon the position of the extant phenotypic distribution in relation to the fitness function. Differently put, two phenotypic distributions located at different positions under the same *actual* fitness function will lead to two different *inferred* fitness functions (see Figure 1a and c). For example, consider a Gaussian fitness function as illustrated in Figure 1a. A Gaussian phenotypic distribution with a mean located some distance from that of the fitness function will experience strong directional selection toward the mean of the fitness function. In this case, we would infer that the fitness function contains a large linear term. However, if the mean of the phenotypic distribution were coincident with that of the (same) fitness function we would infer that there was no linear term at all. This is what I mean by saying that the effects of selection on the extant distribution of phenotypes is dependent upon the location of the phenotypic distribution *relative to the fitness function*. In each case, we still get the correct answer to the question, "How is selection acting now?", but we would be likely to formulate very different hypotheses about *why* selection should operate the way it does.

The second question concerns the nature of the fitness function *independent of the phenotypic distribution*. If we knew the answer to this question, then we could predict the effects of selection on *any* phenotypic distribution, including the extant one. To answer the question of why selection is operating we use the phenotypic distribution as a tool for investigating the selection function. Unfortunately, the extant phenotypic distribution may not be the most powerful tool if the rare phenotypic classes are coincident with the curvature of the actual fitness function. In this case the use of a *uniform* phenotypic distribution can be a more efficient experimental methodology rather than use of the naturally occurring distribution of phenotypes. It is more efficient because a priori we are ignorant of the existence or degree of curvature in the fitness function and a uniform distribution provides the most statistically powerful use of a finite number of degrees of freedom for the purposes of estimation. Furthermore, if we treat the inferred linear and quadratic terms in the fitness function as *hypotheses* for additional empirical investigation (e.g., manipulations of the phenotypic mean independent of the variance), then several other questions concerning unmeasured but possibly correlated traits could be addressed. One could use the natural distribution of phenotypes to address this question but the answer will necessarily be contingent upon the (possibly fortuitous) relative locations of the distributions of phenotype and fitness as discussed above.

A special kind of a problem arises in sexual selection because fitnesses are often considered to be highly context dependent, that is, dependent upon the particular shape of the phenotypic distribution itself. I do not consider this problem to be different in kind from that discussed above. It requires that *context* itself be varied in a systematic way in addition to, say, the morphological distribution. This amounts to repeating the above manipulative experiments in different contexts judged to be relevant. This kind of study of frequency dependent fitness functions has been developed in experimental studies of kin and group selection (see, for example, Wade 1984). The amount of experimental work required, however, increases exponentially with every aspect of context-contingent mating behavior that we grant to be important.

HYPOTHESIS TESTING AND MEASUREMENT: AN EXAMPLE

Application of the above equations for measuring selection to the laboratory data of Thornhill (1980) will serve to illustrate some important differences between hypothesis testing and parameter estimation in both methodology and interpretation. Neither the data analysis nor the results obtained by Thornhill are being questioned here. Rather, I am illustrating how to apply the measures of selection and suggesting that they can provide somewhat different insights into the results and their interpretation.

Thornhill conducted an experimental study of the adaptive function of the notal organ, an abdominal clamp of males, used in mating in scorpion flies. In order to determine whether or not the notal organ increased mating success, he used beeswax to effectively oblate the organ in a number of males. He then placed these males in competition for mates with normal males who were sham manipulated with warm water instead of beeswax. Before placing the males in competition for mates, he starved them for four days in order to eliminate a second important component of mate acquisition in these flies, nuptual offerings of a salival mass. In addition, he performed the competition studies in a laboratory environment free of prey items in order to eliminate a third important component of mate acquisition, nuptual offerings of prey. Differently put, he created a phenotypic class of males without functional notal organs, a class of males well beyond the bounds of the naturally occurring (or possible) phenotypic distribution. Furthermore, he eliminated through experimental protocol two potentially confounding behaviors related to mate acquisition. These kinds of manipulations are standard in hypothesis testing. The results of this study were very clear: no male without a notal organ was successful in copulation although attempts at copulation were equally frequent in the two groups. The result confirms the hypothesis that the notal organ plays a significant role in mating success and it serves as the basis for an elaborate hypothetical scenario concerning adaptive aspects of rape and forced copulation. Thornhill even extrapolates this scenario to humans.

Hypothesis testing and parameter measurement on the same data set do not always lead to the same evolutionary inferences. For example, one can use the techniques described above for measuring selection to investigate the opportunity for selection in this laboratory study and to measure the intensity of selection on the notal organ. Note that there are only two phenotypic classes identified in this study, males with and without notal organs, and one component of fitness, successful copulation. Furthermore, owing to the perfect correlation between phenotype and fitness in the outcomes, the opportunity for selection between classes measures the amount of selection on the phenotype. The salient numbers necessary for estimating the opportunity for selection in this study are given by the experimental design. There were 10 males of each type placed in a cage with 10 females and this design was replicated 4 times. The number of copulations achieved by normal males was exactly 2 in each of the 4 replicates and the number achieved by the oblated males was 0 in every replicate. The average number of copulations per male per replicate is (2/20) or 0.10. The total variance in copulatory success about this average is 0.09 ([18/20][0.00 – 0.10]^2 + [2/20][1.0 – 0.01]^2) and the between group component of variance in copulatory success (between males with and without notal organs) is 0.01 ([0.5][0.00 – 0.10]^2 + [0.5][0.20 – 0.10]^2). The mean and the variance in copulatory success of females are 0.20 and 0.16, respectively. We calculate the total opportunity for selection in males, I_{males}, as (0.09/0.01) or 9.0 and the fraction of this opportunity for selection that is between the two experimental groups of males is (0.01/0.01) or 1.0. The opportunity for selection in females, $I_{females}$, is given by the ratio (0.16/0.04) or 4.0. In the terms of equation (1), we have

$$
\begin{aligned}
I_{males} &= RI_{females} + I_{mates} \\
9.0 &= (2)(4.0) + (1.0)
\end{aligned}
$$

as it should if we have correctly calculated all components. In this laboratory study, 11.1% of the total opportunity for selection is focused on the presence or absence of the notal organ, i.e., selection between the two phenotypic categories. To put this into some sort of perspective, let us compare these laboratory findings with those of Payne (1979) for natural populations of redwinged blackbirds. In Payne's studies the sex ratio in the redwings was 1.93 to 2.10 which compares favorably to the sex ratio of 2 used experimentally by Thornhill (1980). In the redwings the opportunity for selection owing to variance in the numbers of mates of males was 59 to 81% of the total opportunity for selection on males or three to four times greater than that observed by Thornhill. In other words, despite creating a phenotypic distribution well beyond that of the naturally occurring population and despite eliminating additional behavioral components contributing to variance in male mating success, the opportunity for selection observed in the laboratory was only a fraction of that observed in other natural instances in which these measures have been applied. Both experimental manipulations had the effect of enhancing the opportunity for selection based on the presence of absence of the notal organ. Selection on the notal organ in natural populations *must* be much less intense.

Why is there such an apparent discrepancy between the very clear-cut results from hypothesis testing and the estimate of relatively weak selection? The answer lies in the fact that *within* the category of functionally normal males there was considerable variance in copulatory success. The presence of a notal organ enhances but does not guarantee copulatory success; 32 out of 40 normal males across the 4 replicates *failed* to achieve a successful copulation. Therefore, the notal organ explains some but not all of the observed total variance in copulatory success. When these results are considered from the point of view of measuring selection rather than hypothesis testing and when the estimate of the opportunity for selection in a heavily manipulated laboratory system is found to be only a fraction of that observed in unmanipulated natural systems, then the justification for Thornhill's subsequent taxon-wide scenarios is seriously weakened.

CONCLUSION

In this report I have attempted to cover some of the basic issues involved in measuring selection in natural and laboratory populations without delving into the algebraic technicalities of the mathematics. The confounding of the question "What is selection doing?" with the question "Why *should* selection operate the way it does?" has led to some general misunderstandings regarding the application of the measures of selection proposed by Lande, Arnold, and myself and some of these problems are especially acute in the sexual selection literature. I tried to avoid the morass of argument over the definition of sexual selection by focusing instead on the description and understanding of the formal equations depicting evolutionary change. Certain components of these equations appear to me to have a natural interpretation that is coincident with Darwin's own reasoning in identifying sexual selection as a separate force from natural selection. I believe that the equations describing the sex difference in the variance in relative fitness stand in relation to sexual selection theory in much the same way that Fisher's description of the relationship between the mean fitness of the sexes stands in relation to sex ratio theory. Regardless of the consensus definition of what constitutes sexual selection and what does not, these measures quantify phenotypic change in a manner directly relevant to formal evolutionary theory and they provide a coherent framework for posing questions and designing empirical studies to answer those questions. By reanalysis of a well-known example in the literature, I have tried to illustrate that there is some merit to asking the question "How strong is a force of selection?" In addition to the question "Does this force of selection exist?"

Acknowledgements

I would like to acknowledge the hospitality provided by the Mountain Lake Biological Research Station of The University of Virginia, especially its director B. Cole, during the writing of this paper. In addition I would like to thank S. Arnold, A. Grafen, B. Sutherland, S. Skinner, R. Thornhill, and R. Howard for their thoughtful comments on the paper, especially for helping to clarify disagreements in a convivial manner. I thank D. Townsend for the Figure. Lastly, I would like to gratefully acknowledge the superb organization and dedication of Silke Bernhard and her loyal staff for their considerable assistance with this project.

REFERENCES

Endler, J. 1986. Natural Selection in the Wild. Princeton, NJ: Princeton Univ. Press.

Fisher, R.A. 1931. The Genetical Theory of Natural Selection. Oxford: Clarendon Press.

McCauley, D.E., and O'Donnell, R. 1984. The effect of multiple mating on genetic relatedness in larval aggregations of the imported willow leaf beetle (*Plagiodera versicolora*, Coleoptera: Chrysomelidae). Behav. Ecol. Sociobiol. 15: 287–291.

Payne, R.B. 1979. Sexual selection and intersexual differences in variance in reproductive success. Am. Natural. 114: 447–452.

Sutherland, W.J. 1985. Chance can produce a sex difference in variance in mating success and account for Bateman's data. Anim. Behav. 33: 1349–1352.

Thornhill, R. 1980. Rape in *Panorpa* scorpion flies and a general rape hypothesis. Anim. Behav. 28: 52–59.

Wade, M.J. 1984. The effects of genotypic interactions on evolution in structured populations. In: Proceedings of the XV International Congress on Genetics, pp. 283–290.

Wade, M.J. and Breden, F. 1986. Life history of natural populations of the imported willow leaf beetle, *Plagiodera versicolora* (Coleoptera: Chrysomelidae). Ann. Entomol. Soc. Am. 79: 73–79.

Sexual Selection: Testing the Alternatives
eds. J.W. Bradbury and M.B. Andersson, pp. 209–219
John Wiley & Sons Limited

Random and Deterministic Components of Variance in Mating Success

W.J. Sutherland

School of Biological Sciences
University of East Anglia
Norwich NR4 7TJ, England

Abstract. Time may be a useful measure of the components of sexual selection. Null models show that it is not useful to measure variance in reproductive success as this is neither evidence for sexual selection nor a good measure of its intensity. Many of the results of studies of variance (e.g., that variance is greater among males than females, or that variance amongst the males is greater if the species is polygamous, if the females form harems or if the males form leks) can be explained by null models of random mating. A search-time model is used to explain the conditions under which sexual selection is likely to be important.

INTRODUCTION

Time provides a useful medium for expressing the constraints upon animal behavior. Thus it has been found useful to consider whether animals should continue feeding within a patch or spend time moving elsewhere. Similarly animals may have a choice between spending time eating a small prey item or searching for a larger one. A similar approach may be used when studying mate-searching behavior (Parker 1978). The aim of this review is to illustrate the use of search-time models for understanding sexual selection. Consider animals as having an amount of time T which they can devote to searching for mates and all the activities linked with mating. Each mating takes up a period of time H which is required for mate guarding, copulation, parental care, replacing the gametes, etc. (thus H can be thought of as the minimum possible time between mates). With a total of F mates we can allocate the time as follows:

Total time (T) = searching time + number of mates (F)$*$ time per mate (H)

Search-time models have two applications. They can be used to create null models of mating success for comparison with data. They may also be used to predict the sorts of conditions under which sexual selection will act. I will consider these two subjects in turn.

MEASURES OF VARIANCE IN LIFETIME REPRODUCTIVE SUCCESS

Variance in lifetime reproductive success has been measured by ethologists for two main reasons. First, it is considered evidence for sexual selection: for example, if the variance in mating success between members of one sex exceeds the variance within the other then the discrepancy is interpreted as resulting from sexual selection. Second, it is used as a measure of the strength of sexual selection. A high variance implies that phenotypic differences result in very different mating successes. It is my belief that neither of these reasons are justified and that simply measuring variance is not useful.

Over the last decade ecologists have realized the importance of creating appropriate null models for comparing with data (Harvey et al. 1983). Colwell and Winkler (1984) use the term null model for those models that aim to eliminate the effects of a particular ecological process. I hope to show that null models which exclude sexual selection may explain some data sets which have been used as evidence for sexual selection. The most famous example of the use of variance in reproductive success is Bateman's (1948) study of *Drosophila melanogaster* which has been widely used as the textbook example of nonrandom mating. Using genetic markers he showed that the variance in number of offspring left by females is significantly less than the variance in number of offspring left by males. Bateman attributed the difference in variance as evidence that some males were favored while others were not. However reanalysis of the data shows that the mating success of males does not differ significantly from a binomial distribution (Sutherland 1985a).

The reason why Bateman believed his data provided evidence for sexual selection is that he compared male variance against female variance rather than treating each separately. Sutherland (1985a) shows that if one sex contributes more towards the offspring than the other (i.e., one has a longer mating time) then incorporating these mating times into a Poisson distribution leads to lower variances in mating success for longer mating times. If each mating takes up a negligible amount of time then the variance follows a simple Poisson distribution, but with a longer mating time there is little opportunity to mate with a succession of partners and the variance is reduced. Banks and Thompson (1985) showed that this model described their data on the mating success of the damselfly. *Coenagrion puella*. A sex difference in variance in the number of mates (Payne 1979; Payne and Payne 1979) clearly cannot be used as evidence that sexual selection has occurred.

This null model can be modified to apply to other studies. Sutherland and Banks (unpublished) show that the expected variance in lifetime reproductive success depends upon the mortality. With low mortality a few individuals will be very old while others will die while young. If the mortality is high then individuals are unlikely to reach old age and the expected variance is reduced.

The variance in lifetime reproductive success among male red deer *Cervus elaphus* does not differ statistically from the variance among male kittiwakes *Rissa tridactyla* (Clutton-Brock 1983), despite the fact that the red deer is a highly polygamous species in which intermale competition is obviously important (Clutton-Brock et al. 1983). Clutton-Brock (1983) has attributed this to the differences in mortality. In kittiwakes the majority of the variance in lifetime reproductive success can be explained by individual differences in lifespan (Coulson and Thomas 1985), but in red deer differences in lifespan were relatively unimportant (Clutton-Brock et al. 1983). It seems that without estimates of mortality it is difficult to draw any conclusions from a measure of variance in lifetime reproductive success.

Banks and Sutherland (unpublished) used their model to analyze Finke's (1982) data on the damselfly *Enallagma hageni*. The fit of the null model depended critically upon the estimate of lifespan used. With the value of mortality given in the paper the data differed significantly from the null model, but with the value of mortality that was compatible with the given lifespan there was no difference between the data and null model. We have no criticisms of Finke's mortality estimates (the difference between the two estimates was undoubtedly due to rounding to one significant figure) but our reanalysis shows how sensitive any test would be to the precise value of mortality. As mortality is usually difficult to measure (e.g., Lakhani and Newton 1983) it will often be difficult to draw any definite conclusions about the importance of sexual selection from measures of the variance in lifetime reproductive success.

The main point I wish to make is that these null models will account for differences in variance between species (or between sexes) and that previously these differences had been explained in terms of sexual selection. I will describe two further, simple models which elaborate the approach. These models will consider the consequences of dispersion. The first will consider the distribution of females and whether they form harems. The second will consider the distribution of males and whether they are solitary, random, or aggregated (i.e., into leks).

Previous models of random mating have assumed females are distributed at random so that mating with one female does not effect the chances of mating with others. However a number of studies have shown that one male may mate with all members of a group while other males do not mate at all during the same season (Clutton-Brock et al. 1983; Le Boeff 1974).

The effect of group size on variance can be shown by a simple model—it may be thought of as representing male seals or deer competing for the opportunity to mate with harems of females. For this simple model I assume all individuals survive for a certain reproductive lifespan. Each year each male either obtains

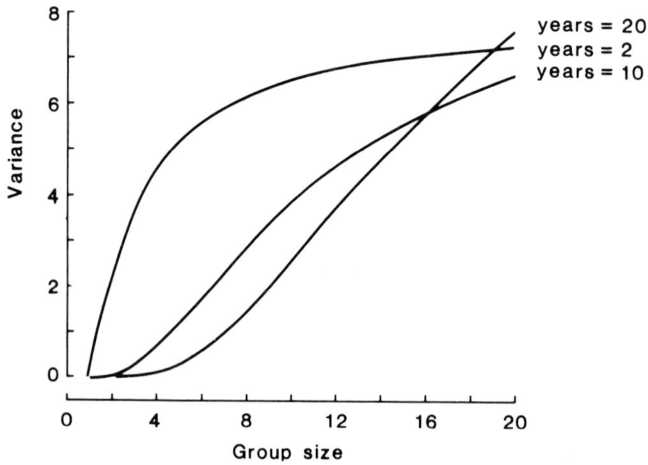

FIG. 1 The relationship between group size and variance in mating success.

a harem and mates with all members of one group or fails to mate at all. The number of years in which a male mates can then be described by a binomial distribution. With equal number of males and females the probability of a male mating is the reciprocal of group size. We can calculate the variance in mating success over the reproductive lifespan. Figure 1 shows the relationship between the variance in mating success and the number of individuals per group for different reproductive lifespans. If females are distributed singly (i.e., group size = 1) then variance will be low: if the groups are large (e.g., 20) then most males will fail to mate while others will mate repeatedly. As a consequence, variance in mating succes will increase with the size of the group. Although this model is obviously oversimplified, it shows that in species with large harems random determination of harem ownership by itself will produce considerable variance in male mating success and thus this variance cannot be used as evidence for sexual selection. The 35-fold difference in variance between male and female elephant seals recorded by Le Bouff (1985) is a case in point.

The second model assumes matings follow a binomial distribution with all individuals possessing an equal chance of mating. The consequence of dividing up the population is then explored. Thus a population of 120 can be considered as two populations of 60, or as four populations of 30, six populations of 20, or even 60 pairs of animals as found in a strictly territorial species. The variance is then simply the variance of the binomial distribution (i.e., $p[1-p]n$) where n is the number of individuals of that sex in the group and p is the probability of mating with a given member of the opposite sex (which equals $1/n$). Combining these gives variance = $1\text{-}1/n$.

This formulation shows that the larger the group the higher the within-lek variance in mating success. Species in which males aggregate on leks may be expected to show higher within-lek variance and this variance (Payne and Payne 1985) cannot be used as evidence for sexual selection. Species with solitary and territorial males may be expected to exhibit low variance. Similarly, Gwynne (1984) showed greater variance in the number of matings at high density sites than at low which is compatible with this null model.

The preceding comments might tempt one to compare the mating success of different individuals with a null model or random mating and interpret any discrepancy as evidence for sexual selection. I do not consider this useful either as it is likely to be impossible to provide a sufficient null model for comparisons with a set of field data. The reasons for this are twofold:

1) There are other sources of variance that would need to be considered. For example, Sutherland and Banks (unpublished) show that local fluctuations in the sex ratio may contribute considerably to variance and that variance is likely to be high if the population is subdivided or if movement is limited (e.g., barnacles or plants). Thus extrinsic factors, which can affect variance include mating time, mortality, local sex ratio, male dispersion, and female dispersion. It is likely that other sources of variation will be important in specific studies. For example, the study of Banks (1985) showed that the weather was of overriding importance in determining mating success in *Coenagrion puella*. Mating success was much higher on sunny days than on dull days and a full null model would have to include a measure of the weather. Any significant deviation from a null model could either be due to sexual selection or to the failure to include an essential component of variance.
2) Many of these contributions to variance will be difficult to measure accurately. We have already seen one example of how the precise value of mortality may greatly influence the fit to a null model. I suspect it will rarely be possible to measure all the variables with sufficient confidence to exclude a null model. As an example it would be easy to compare the distributions of matings at a lek on a single day against a Poisson distribution. However, even in this simple case it would be necessary to consider the time each male is present, the arrival times of females, which other males are present at each period, the probability of mating during different intervals (which may depend on the weather), etc. I thus stress that I am not advocating the use of null models and I just use null models to show the inadequacies of simple measures of variance.

Thus there are theoretical and practical problems associated with using variance which also is extremely time-consuming to measure. I would argue that once achieved it rarely does more than redescribe some aspect of the biology that could be determined more easily. Thornhill and Alcock (1983) suggest that measuring variance can be used to determine if the species is monogamous, polygamous, or polyandrous. As this measurement would involve seeing who mated with whom there are easier ways of evaluating the mating system. Showing that female *Drosophila* are less variable in mating rates than males shows that each mating results in a longer time investment for females than for males. Showing that males of lekking species are more variable than nonlekking species is another way of saying that males in leks mate with more than one female while monogamous species do not. None of these cases have shown that sexual selection is important but have simply redescribed the social system and from the social system inferred the likely importance of sexual selection.

It is not my intention to deny that sexual selection occurs or to deny that it is a major source of variance in reproductive success. Even in species which fit the null model of random mating there is evidence that certain phenotypes have a higher reproductive success than others (*Drosophila melanogaster* — Partridge and Farquhar 1983, *Coenagrion puella* — Banks and Thompson 1985).

THE STRENGTH OF SEXUAL SELECTION

Sexual selection does not seem to have operated equally on all groups. It seems to have affected some species more than others, and in many species it seems to often affect one sex more than the other. There has been considerable interest in trying to explain these patterns. Trivers' (1972) model of parental investment is probably the most famous attempt. Search time models can be used to provide a more explicit solution. Baylis (1981) uses an approach that resembles mine to account for the pattern of parental care and sexual selection in fishes. He considers the "rate limitation" which constrains the maximum rate that males and females can breed; thus a long mating time inevitably results in a slow rate. Sutherland (1985b) describes a search-time model which I shall outline here (the printing errors in that paper are also corrected here).

I shall assume that the strength of sexual selection is the degree to which different phenotypes have different fitnesses due to differences in the frequency or quality of mates. Sexual selection may often act through differences in mate quality but for simplication I will ignore this aspect and just consider the extent to which different phenotypes differ in mating rate. In common with many other studies and most field work, I will assume that mating rate is a measure of fertilization rate and will ignore differences in fertilization rate due to sperm competition. Thus this model is not a complete description of all aspects of sexual selection.

The rate at which an average searching individual finds mates is called the aptitude for encounter, λ (Parker 1979); this is comparable to the encounter rate or searching efficiency used by population biologists. Individuals may differ in their aptitude for encounter and the source of these individual differences may differ between species. Thus an individual's aptitude for encounter may depend upon color, size, fighting ability, calling rate, or pheronome production. The model considers a mutant whose aptitude for encounter is n times the average (which we set to 1) so that its aptitude for encounter is λn.

Following Holling (1959) we can calculate the number of matings F which we expect it to obtain in time T

$$\frac{F}{T} = \frac{\lambda n}{1 + \lambda n H} \tag{1}$$

where as before, H is the mating time and T is the total time devoted to activities related to mating. The rate at which the average individual, (for whom n=1) encounters potential mates can be derived by rearranging the previous equation

$$\lambda = \frac{1}{(T/F) - H} \tag{2}$$

where F is the number of matings obtained by an average individual. Substituting and rearranging gives

$$\frac{F'}{F} = \frac{1}{F\,H/T + (T - F\,H)/n\,T}\ . \tag{3}$$

Figure 2 uses this equation to show the relationship between the relative mating success of the mutant (F'/F) and the relative aptitude for encounter n. The relative aptitude for encounter expresses fighting ability or calling ability or appearance so Figure 2 illustrates the relationship between these qualities and mating success. If the relationship is steep then sexual selection is strong: the better fighters or callers mate more often. If the slope is shallow then sexual selection is weak. The slope depends upon the average fraction of time spent searching for mates (T-FH)/T. If almost all the time available is spent searching then an increase in aptidude for encounter will make a reasonable difference to the number of mates found: doubling the aptitude for encounter will almost double the number of mates. If, however, a negligible amount of time is spent searching then increasing searching ability will have no effect on total success. The effect of the fraction of time spent searching can be shown by differentiating equation (3).

$$\frac{d(F'/F)}{dn} = \frac{(T - F\,H)\,T}{(T + n\,F\,H - F\,H)^2} \tag{4}$$

Assuming most mutations are only slightly better or slightly worse than average, then n approximately equals one and equation (4) simplifies to

$$\frac{d\,(F'/F)}{dn} = \frac{T - F\,H}{T}\ . \tag{5}$$

This is equivalent to the fraction of time spent searching by an average individual. This makes intuitive sense. Male ruff *Philomachus pugnax* on a lek spend most of their time "searching." An individual whose aptitude for encounter was twice as high as average would obtain about twice as many matings

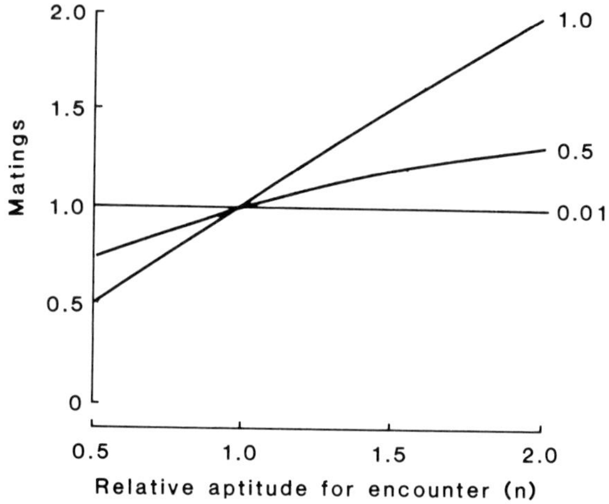

FIG. 2 The relationship between relative mating success and relative aptitude for encounter (after Sutherland 1985b). The relationship is shown for different values of the time spent searching $(T - FH)/T$. A shallow slope means that individuals differing in aptitude for encouter have similar mating successes (and thus selection is weak). A steep slope means differences result in differing mating successes (and thus selection is strong).

as average while a male common sandpiper *Tringa hypoleucos*, who helps look after the young, would not gain from a higher aptitude for encounter. It is not surprising that clearly the behavior and morphology of ruff are greatly influenced by sexual selection while that of the common sandpiper are not. In species with marked sexual dimorphism we may expect the sex with the most costly secondary sexual characteristics to be the sex which spends the higher fraction of time searching (see Figure 2). In accordance with this, in species such as the spotted sandpiper. *Actitis macularia*, American jacana *Jacana spinosa*, and phalaropes *Phalaropus* sp. the mating time for the male is longer as they take sole responsibility for the offspring and the females are larger, brighter colored, and more aggressive (Oring and Maxon 1978; Jenni and Betts 1978; Trivers 1972).

In some groups such as frogs, toads, and dragonflies it may be possible to directly measure the fraction of time spent searching. However, I do not believe that this is the main purpose of the model. I believe the advantages of this search-time model of sexual selection are that it is more simple than models using parental investment, it has a clearer algebraic justification, it provides a basis for comparative studies using operational sex ratio, amount of mate guarding, parental care, harem size, etc., and it can show the relationship between models currently in use. Sutherland (1985b) lists the relationship between the above model and five other models of sexual selection (operational sex ratio, relative parental investment, relative variance in reproductive success relative standard deviation in reproductive success, and the ratio of maximal breeding). I will elaborate here on the relationship with the operational sex ratio and show a relationship with harem size.

The operational sex ratio

This is the ratio of seeking males to seeking females. For a given absolute sex ratio S,

$$\text{the operational sex ratio} = \frac{T_m - F_m H_m}{T_f - F_f H_f} S \qquad (6)$$

From equation (5) this is equivalent to the relative strengths of sexual selection acting upon the two sexes. Although Emlen and Oring (1977), Bradbury and Vehrencamp (1977), and Thornhill and Alcock (1983) have enthusiastically supported this measure it seems to have been rather neglected in studies of sexual selection. In many cases it is considerably easier to measure than parental investment or variance in lifetime reproductive success.

Furthermore, it incorporates differences between sites, between years, and within years. The operational sex ratio may approximate to unity in some parts of the year yet be strongly biased in other seasons. We may expect the strength of sexual selection to vary in unison. Thus Arak (1983) showed that at the beginning of the season there are relatively few females and the males fight. Later in the season the ratio is more even and fighting is rare. The level of competition thus varies with the operational sex ratio. Verrell (1983) has shown how the behavior of red spotted newts *Notophthalmus viridescens* varies with the operational sex ratio.

Harem size

If one male is mating with a group of females then many other males must be unsuccessful. This may be expressed in terms of time to show that, as an approximation:

$$\text{fraction of time spent searching} \cong \frac{\text{Adult sex ratio}}{\text{Harem size}}.$$

Thus Harvey et al. (1978) have shown that the degree of sexual dimorphism in primates correlates well with the harem size.

Acknowledgements

I am most grateful to G.A. Parker for thought provoking discussions and encouragement, to R. Butlin and R. Nichols for comments on the manuscript and S. Mitchell for drawing the figures.

REFERENCES

Arak, A. 1983. Male-male competition and mate choice in anuran amphibians. In: Mate Choice, P. Bateson, 181–210 Cambridge: Cambridge Univ. Press.

Banks, M.J. 1985. Population dynamics and lifetime reproductive success of damselflies. Unpublished Ph.D. Thesis. Liverpool University.

Banks, M.J., and Thompson, D.J. 1985. Lifetime mating success in the damselfly *Coenagrion puella*. Anim. Behav. 33: 1175–1183.

Bateman, A.J. 1948. Intrasexual selection in *Drosophila* Heredity 2: 349–368.

Baylis, J.R. 1981. The evolution of parental care in fishes, with reference to Darwin's rule of male sexual selection. Env. Biol. Fish. 6: 223–251.

Bradbury, J.W., and Vehrencamp, S.L. 1977. Social organisation and foraging in emballonurid bats. III. Mating systems. Behav. Ecol. and Sociobiol. 2: 1–17.

Clutton-Brock, T.H. 1983. Selection in relation to sex. In: Evolution from Molecules to Men, ed. D.S. Bendall, pp. 457–486. Cambridge: Cambridge Univ. Press.

Clutton-Brock, T.H.; Guiness, F.E. and Albon, S.D. 1983. Red Deer. Behaviour and Ecology of Two Sexes. Edinburgh: Edinburgh Univ. Press.

Colwell, R.K., and Winkler, P.W. 1984. A null model for null models of biogeography. In: Ecological Communities. Conceptual Issues and the Evidence, ed. D.R. Strong, D. Simberloff, L.G. Abele, and A.B. Thistle, Princeton: Princeton Univ. Press.

Coulson, J.C., and Thomas, C. 1985. Differences in the breeding performance of individual kittiwake gulls, *Rissa tridactyla* (L). In: Behavioural Ecology: Ecological Consequences of Adaptive Behaviour, eds. R.M. Sibly and R.H. Smith, pp. 489–503. Oxford: Blackwells.

Emlen, S.T., and Oring, L.W. 1977. Ecology, sexual selection, and the evolution of mating systems. Science 197: 215–223.

Finke, O.M. 1982. Lifetime mating success in a natural population of the damselfly, *Enallagma hageni* (Walsh) (Odonata: Coenagrionidae). Behav. Ecol. Sociobiol. 10: 293–302.

Gwynne, D.T. 1984. Sexual selection and sexual differenes in mormon crockets (Orthoptera: Tettigoniidae, *Anarabrus simplex*) Evolution 38: 1011–1022.

Harvey, P.H.; Colwell, R.K.; Sivertown, J.W.; and May, R.M. 1983. Null Models in ecology. Ann. R. Ecol. Syst. 14: 189–211.

Harvey, P.H.; Kavanagh, M.; and Clutton-Brock, T.H. 1978. Sexual dimorphism in primate teeth. J. Zool. Lond. 186: 475–486.

Holling, C.S. 1959. Some characteristics of simple types of predation and parasitism. Can. Entomol. 91: 385–398.

Jenni, D.A., and Betts, B.J. 1978. Sex differences in nest construction, incubation and parental behaviour in the polyandrous American jacana (*Jacana spinosa*). Anim. Behav. 26: 207–218.

Lakhani, K.H., and Newton, I. 1983. Estimating age-specific bird survival rates from ring recoveries — can it be done? J. Anim. Ecol. 52: 83–91.

Le Boeff, B.J. 1974. Male-male competition and reproductive success in elephant seals. Am. Zoolog. 14: 163–176.

Le Boeff, B.J. 1985. Lifetime reproductive success in northern elephant seals. In: Abstracts of the 19th International Ethological Conference, p. 51. Toulouse, France.

Oring, L.W. and Maxon, S.S. 1978. Instances of simultaneous polyandry by spotted sandpipers, *Actitus mascularia*. Ibis. 120: 349–353.

Parker, G.A. 1978. Searching for mates. In: Behavioural Ecology: An Evolutionary Approach, eds. J.R. Krebs and N.B. Davies, pp. 214–244. Oxford: Blackwells.

Parker, G.A. 1979. Sexual selection conflict. In: Sexual Selection and Reproductive

Competition in Insects, eds. M.S. Blum and N.A. Blum, pp. 123–166 New York: Academic Press.

Partridge, L., and Farquhar, M. 1983. Lifetime mating success of male fruitflies (*Drosophila melanogaster*). Anim. Behav. 31: 871–877.

Payne, R.B. 1979. Sexual selection and intersexual differences in variance of breeding success. Am. Natural. 114: 447–562.

Payne, R.B., and Payne, K. 1979. Social organisation and mating success in local song populations of village indigobirds. *Vidua chalybeata* Z. Tierpsych. 45: 113–173.

Sutherland, W.J. 1985a. Chance can produce a sex difference in variance in mating success and account for Bateman's data. Anim. Behav. 33: 1349–1352.

Sutherland, W.J. 1985b. Measures of sexual selection. Oxford Surveys in Evol. Biol. 2: 90–101.

Thornhill, R., and Alcock, J. 1983. The Evolution of Insect Mating Systems. London: Harvard Univ. Press.

Trivers, R.L. 1972. Parental investment and sexual selection. In: Sexual Selection and the Descent of Man, ed. B. Cambell, pp 139–179. Chicago: Aldine.

Verrell, P.A. 1983. The influence of ambient sex ratio and intramale competition on the sexual behaviour of the red-spotted newt, *Notophthalmus viridescens*. Behav. Ecol. Sociobiol. 13: 307–313.

Sexual Selection: Testing the Alternatives
eds. J.W. Bradbury and M.B. Andersson, pp. 221–233
John Wiley & Sons Limited

Measuring Sexual Selection: Why Bother?

A. Grafen

Department of Zoology
Animal Behaviour Research Group
Oxford OX1 3PS, England

Abstract. Most of what we know about sexual selection comes from Darwin (1871) who never measured it. There is a difference between measuring things to do with sexual selection and measuring sexual selection itself. Many interesting problems in sexual selection are historical questions, requiring the application of historical methods for their study. The interesting intraspecific empirical question about sexual selection is what mechanisms of advantage favor extravagant characters. Fitness variance component analyses may have some weak connection with this. Quantification has no intrinsic virtues. A good field worker is nobody's poodle.

INTRODUCTION

Lord Kelvin, the second most distinguished Scottish physicist of the 19th century, was fond of asserting that until we can measure something we have no idea what we are talking about (Thomson 1889, 73). It was also Lord Kelvin who cast grave doubts on Darwin's theory of natural selection by proving that the earth was less than a hundred million years old (thus proving that even when we can measure something we may still not know what we are talking about) (Burchfield 1975). Darwin's thorough and logically sound but unquantitative methods have triumphed in the end over Lord Kelvin's calculations. The morals I draw for students of sexual selection are that quantification has no intrinsic virtues, is associated with the vice of hubris, and can set back the cause of science by giving a false air of great certainty of conclusion.

My starting point in this essay is therefore that there *may* be virtues in measuring sexual selection, but they need to be discovered, and the time and effort invested in measurement need to be justified. I begin with a consideration of the phrase "measuring sexual selection," which can be understood in two ways. The first is that it refers loosely to "measuring things that may be relevant to sexual selection," such as the length of a peacock's tail or linkage disequilibrium, and the second is that the quantity measured is "sexual selection itself."

There are dangers in claiming to measure sexual selection itself. What would we be tacitly asserting by taking some measurable quantity and naming it sexual selection? First that there exists one single quantity that captures the essence of sexual selection and second that we know what it is. There is certainly scope for argument on the second point (Wade and Arnold 1980; Sutherland 1985b), but even the existence of such a measure seems doubtful. Only very strong theoretical reasons could make us confident that our measures would be sensible in every species, with breeding systems of possibly as yet unobserved kinds.

If we decided that exactly one particular kind of numerical comparison between species, relevant to sexual selection, was always sensible and interesting, then calling that numerical measure "sexual selection" might not cause too much confusion. However, if there were no such measure, or more than one such measure, it would be best not to. It would be completely the wrong way round to choose a measure of sexual selection on other grounds and then assert that its comparison between species must be interesting. A premature christening preempts the question of the value of comparisons, among others. There is, in any case, no advantage to be gained from calling any quantity sexual selection, rather than in describing it more objectively and then arguing for a proposed connection between it and sexual selection either in a particular case or in general.

I shall, therefore, interpret my assigned title as "Measuring things to do with sexual selection: why bother?" Darwin (1871) discovered almost everything important now known about sexual selection and did so without measurement. The discovery of sexual selection and the outline of its main features are in no need of quantification: the arguments stand in our day as they did in Darwin's. We should remember that Darwin was always very careful in fitting his methods to his problems (Ghiselin 1969), that he conducted experiments in some of his enquiries, but that he eschewed decimal places and fields studies in his work on sexual selection. This is why the question of my title requires an answer.

I shall argue that the desirability of precise mathematical modelling does not imply the desirability of measuring parameters of those models in the field. Then the classes of reasons given as justification for measuring sexual selection will be considered. I shall argue that the study of sexual selection is an historical study and assess the role of measurement on that basis. I will then present the case that the interesting thing about sexual selection from an intraspecific point of view is the mechanisms of advantage, the selective forces at work on a character. Then I will discuss the relevance to these mechanisms of advantage of analyses of fitness variance components. To close, advice is offered to theoreticians and empiricists on the course of future research.

MODELLING VERSUS FIELD STUDIES

The desirability of precise models of sexual selection is beyond dispute.

Lande's (1981) partial mathematization of Fisher's (1958) runaway process confirmed that despite contrary claims (O'Donald 1982) it could work as Fisher asserted, and also showed that open-ended preferences were not necessary for the runaway process. The main technical problem which the modelling solved was the difficulty in keeping track of what happened to the genetic variances in the male trait and female preference, and how they in turn affected the selection of those two characters. Before Lande's model it was necessary to take Fisher's cryptic word for it.

Other current debates about sexual selection in which precise modelling is necessary are a) can an extravagant male character be selected *because* it is deleterious, as Zahavi (1975, 1977) claimed, and b) can host-parasite interactions maintain enough variability in a species without paternal care to ensure that females should expend time and energy in choosing a male on the basis of the fitness of the offspring the mating will produce, as Hamilton and Zuk (1982) have claimed? To attempt these problems by verbal arguments alone seems doomed to failure. Mathematics' major roles are to assist communication between different theoreticians and to help individual theoreticians marshal their thoughts.

It is tempting to believe that if a mathematical model is important in clarifying theoretical problems, then it must be interesting to measure the parameters of that model in natural populations. This modeller's hubris may be unjustified. The point of the kind of modelling of interest here is not to make precise predictions about particular species, indeed it is not to predict at all. Even if we follow Popper's view of how science works, prediction is the touchstone by which the *scientificness* (not the truth, interest, or importance) of theories is judged, but prediction is not usually the purpose of a theory. The purpose of most evolutionary models is not to foretell the future but to explain the present. Convenient assumptions are often false, and particular cases have peculiarities that would require a tailor-made model.

Even when precise predictions are not made about particular cases, the verbal conclusions from a mathematical model will often be important in interpreting observations in the field. Indeed, strong justification is required for making, in terms of equations rather than through an intermediate stage of words, the link between a model and a field study. All those explicit and implicit *caeteris paribus* assumptions that make a model worthwhile as an abstraction must be dealt with as objections to the equational link.

REASONS GIVEN FOR MEASURING THINGS TO DO WITH SEXUAL SELECTION

I have tried to understand the reasons given for measuring sexual selection by Ralls (1977), Payne (1979), Wade (1979), Arnold (1983), Lande and Arnold (1983), Clutton-Brock (1983), and Arnold and Wade (1984a,b). The reasons may be grouped into five categories, which I will discuss in turn:

1) to test predictions of sexual selection theory,
2) to measure things about selective surfaces,
3) to replace rhetorical claims,
4) to provide security of argument through precision,
5) to explain sex differences.

Testing predictions of sexual selection theory is all very well if there are such predictions. Danger lies in interpreting theoretical discussions too literally. Some authors, for example, have been under the illusion that sexual selection theory predicts a greater variance in male lifetime reproductive success (LRS) than in female LRS in species whose males (but not females) possess extravagant characters (e.g., Payne 1979). Let us examine one reasonable sounding line of argument that reaches this conclusion and spot the flaws. First, a) a female red deer's production of offspring is limited by her ability to gestate and suckle, and therefore has small scope for big gains in fitness, making selection for extravagant characters unlikely; b) as one female's production of offspring cannot be much varied by extravagant characters, different females must have similar offspring production; and c) variance in LRS is low in females. Second, a) a male red deer's production of offspring is limited by his ability to gain and defend a harem, and by the harem's size — this makes likely selection of extravagant characters such as those for male combat, and possibly attractiveness to females; b) as one male's offspring production can be varied by extravagant characters, different males must have dissimilar offspring production; and c) variance in LRS is high in males. In both cases step a) is correct, step b) is wrong, and step c) is "wronger."

Step a) is Darwin's original argument. Step b) simply doesn't follow. One flaw in the argument is the supposition that because LRS cannot be influenced by extravagant characters, it cannot be influenced at all. Factors exist which contribute to variability in LRS but which cannot be influenced by extravagant characters, and this invalidates the jump from a) to b). They include longevity, randomness (Sutherland 1985a), and individual differences in quality arising from parental and other influences while a juvenile. This means that around the core of variability, which is concerned with sexual selection, there may be much variability from other causes which would obscure comparisons.

The second flaw shows that even the core is rotten. It is wrong to conclude from the fact that LRS can be influenced by extravagant characters, that it must be variable. This step depends on the existence of variability *now* in the extravagant character. Step a) relies on the possibility of variability in LRS, not on its existence in the population now. During selection for antlers, we expect that there was variability in antler length and consequent variability in LRS. If antlers are no longer growing in evolutionary time, then the amount of heritable variability in antler length depends presumably on a balance between selection and mutation. The magnitude of this variability in LRS will depend on mutation rates, which are not obviously relevant to a measure of the scope for selection of extravagant characters.

I described step c) as "wronger" than step b), but I have just explained that step b) is completely wrong. This is because using the word variance gives the predictions a false air of scientificness and precision. Step a) is correct as a verbal argument and does not need to be made precise. Even if step b) were admitted, there is nothing to say that when it was made precise it would result in step c). Variance may not be the best measure of dissimilarity. The point of having precise predictions is lost if they are spuriously precise in the manner of step c). The invalidity of steps b) and c) shows the dangers of making perfectly good verbal arguments precise for the sake of it.

The measurement of adaptive landscapes is advanced by Lande and Arnold (1983) as a good thing. The extent of maladaption, and the frequency of multiple peaks, are facts of general interest to geneticists, particularly to those of a Wrightian bent. The topic of this volume is sexual selection and in this context, while no one would throw away measurements of an adaptive landscape, they do not seem to me to be of burning interest. Others may disagree.

A second argument of Lande and Arnold is that measurement is "the best alternative to the fabrication of adaptive scenarios . . . Our optimism that measurement can replace rhetorical claims for adaptation is founded on the growing success of field workers in their efforts to measure major components of fitness in natural populations." In my view the construction of adaptive scenarios is half of what evolutionary biology is about, and their validation is the rest. What can we make of Lande and Arnold's rhetorical claims for measurement?

Presumably the kind of replacement they have in mind is adding to an unsupported statement of the form "Character *X* is an adaptation for purpose *Y*" a set of persuasive measurements, which should include measurements of major components of fitness in natural populations. Supporting facts are good things but they are not always necessary. Furthermore, for claims about adaptations, measurements of major components of fitness in natural populations may be relevant only rarely, as hinted by the following example.

No measurement is needed to persuade us that eyes are for seeing. They may also be for the detection of the earth's magnetic field (Leask 1977). How would we test such a claim? Neurophysiological experiments on eyes, and on the optic nerves and their destination areas in the brain, would be one way (Semm et al. 1984). Testing the ability of organisms to navigate in the presence and absence of light is another as the proposed photochemical mechanism requires light in the eye. Field studies on natural variability in eyes, and its contribution to LRS, would not be high on my list of suitable methods of investigation. There are, then, at least some kinds of claim about adaptation for which field measurements of components of fitness in natural populations are irrelevant. I have argued elsewhere that these fitness measurements may be more generally rather irrelevant to the study of complex adaptations (Grafen 1988).

The fourth argument is that without measurement our arguments are insecure through imprecision. Taken as referring to theoretical arguments this is a plea for precise modelling, whose justice is discussed and affirmed above. Taken as referring to interpretation of data this rather depends on the kind of assertion desired. Certainly, no one would wish to learn only that a sex ratio "seemed to be female biassed" when counts of individuals by sex were available. On the other hand, to try to prove by measurement that the male peacock's plumage is more extravagant than the female's is unnecessary. The truth of the statement can be admitted even though the meaning of "extravagant" is still imprecise, and in many ways is better left so. The idea that seems to me to underlie the previous two arguments in favor of measuring is that "you have to be numerically precise to be right" — and I tackle this general philosophy in the next section.

We come at last to the only good reason for studying sexual selection at all, and *a fortiori*, the only good reason for measuring it (if there is one). Darwin invented sexual selection to account for puzzling characters, often associated with sex differences, that seemed to hinder males (usually) in the struggle for existence. He used it to explain much of bird coloration, antlers, human hairlessness, and much sexual dimorphism in size. Measuring is desirable to the extent that it is helpful in pursuing Darwin's aim in a logical way. Cases in which it is helpful are discussed at the end of the next section.

YOU DON'T HAVE TO BE NUMERICALLY PRECISE TO BE RIGHT

In considering the role of measurement in the study of sexual selection, it is helpful to consider which methods have been of most value so far. These are the arguments of Darwin (1871). They are of two main types: speculation as to possible mechanisms within a species, and the detection and interpretation of pattern usually between species. With this pair of techniques, Darwin invented sexual selection and discovered all its main features. I shall discuss why these techniques are so unnumerical. Then I shall outline what I consider to be a set of currently interesting problems in sexual selection and how measurement is likely to fit into the logic of their resolution.

Darwin's techniques are unnumerical partly because measurements were not available and the statistical techniques which make measurement more valuable were not available. But the major reasons is that Darwin was interested in showing the truth of historical statements, and that is because many of the questions of interest about sexual selection are historical. For example, is the generally brighter coloration of male birds explained by sexual selection? Is sexual dimorphism in ungulates the result of sexual selection? Is human hairlessness the result of sexual selection? These exemplify one central type of interesting statement about sexual selection, perhaps the most interesting type.

These questions are answered in much the same way as questions of the sort: "Does the success of a popular revolution depend on the existence of an educated and sympathetic middle class?" First, some plausible case must be made about the likely effect of the education and sympathy of the middle class on a popular revolution. Ideally, documentary evidence from particular examples would support this case. Second and most important, a pattern should be sought in the outcomes of all sufficiently studied popular revolutions. Can examples be found with an educated but unsympathetic middle class, with an uneducated but sympathetic middle class, and with neither? Do the ways in which those categories of revolution fail support the ideas about why education and sympathy are necessary?

Measurement is relevant to these questions, but obliquely and not centrally. The education of the middle classes could be measured by the number of years in school, by literacy levels, or by the number of books or magazines sold per head. Different measures might be necessary for different cases because of differences in the ways the original studies were carried out. The same measure might have to be interpreted in different ways because of special circumstances; for example, the number of books sold per head will depend on the cost and therefore available technology of production.

Darwin's arguments about sexual selection are of this type in which measurement is relevant obliquely, not centrally. The role of measurement is the same today, as I argue now in a brief discussion of my personal list of interesting topics in sexual selection:

1) Female choice or male competition?
2) Number or quality of females?
3) Does male competition lead to organs designed as weapons or displays?
4) Is female choice for genes or for paternal care?
5) If for genes, what maintains the genetic variability among males?
6) Does sexual selection work on both sexes at once?
7) Does sexual selection do for a species more of what sex already does?

These questions may have different answers for different species. Interesting answers would likely not be yes or no, but an account of in what circumstances the answer was yes, and when no, and an explanation. My concern here, however, is how measurement is relevant to answering these questions.

Answering the questions will involve recognizing a pattern or design. Take the question of whether female choice or male competition is responsible for sexually selected male body size in *Drosophila*. It is imaginable that a female mates in the same time with a male of any size when confined with him alone, but generally mates with the larger when confined with two males. This would suggest that male competition and not female choice was important. Alternatively, a comparison between different species of *Drosophila* might reveal that the extent of the influence of sexual selection on male body size correlated well with the scope for male competition but not with the scope for female choice. I stress these are hypothetical possibilities. These obvious possible ways of answering the question illustrate that measurement is obliquely relevant in that it may be necessary to measure to detect a pattern. Measurement is not central: there is no single quantity, not even one whose precise definition escapes our minds now but would become clear after much thought, whose measurement *is* the answer to the question.

Whether sexual selection acts in both sexes at once could be answered by comparative studies in the Darwinian tradition. A second reason for measurement is that the comparative method has changed since Darwin's day. A modern day exponent of this method has more measurements available, because more studies have been performed. This means that he can study finer patterns than Darwin. He is also more concerned about taxonomic artifacts, and the various methods designed to avoid them require quantification to greater or lesser degrees (Ridley 1983; Felsenstein 1985). He also lives in a community with a fetish for statistics. (By fetish I mean to suggest an irrationally strong attraction coupled with lack of understanding, as of magic.)

A third reason, which could apply to any of the questions, is to have a well-worked-out example. That old war-horse *Biston betularia* is didactically invaluable. Nothing of special interest was proved about natural selection, no new principles were discovered. To have more than a few examples of this kind would be wasteful, for it is not the job of biologists to do the same thing over and over again with different examples. But one or two well worked-out examples of how sexual selection operates through female choice, how it operates through male competition, how it can operate for good genes, how it can operate for paternal care, and so on, would be very welcome. They help in teaching, they help to clarify thought about the processes involved, and they may also turn up interesting quirks of natural history on the way. The best examples I know of in sexual selection are Endler's guppies (1980) and Andersson's widow birds (1982). Both could be improved upon as classic instances of sexual selection, but they will not easily be surpassed.

The major reason for measuring is then to assist in the recognition of patterns. Any kind of pattern will do, intraspecific, interspecific, or an anatomical pattern that we would call design. Any of these could persuade us of an important point about sexual selection, and all three techniques were used by Darwin. The logic of being persuaded by a pattern is that something intrinsically implausible demands an explanation. The nature of that explanation will ideally be closely dictated by the nature of the pattern itself.

Measuring in some other fields has a much stronger place than it does in the study of adaptations under natural and sexual selection. The body of theory in physics is very strong and is tightly related to questions of central importance in physics. We do not have such a body of theory in the study of sexual selection, and I guess never will have. This is not a matter for regret, as it arises naturally from the intrinsically historical nature of most of the interesting questions about sexual selection.

Reading discussions of exactly how sexual selection should be measured (e.g., Payne 1979; Wade 1979; Wade and Arnold 1980; Arnold 1983; Arnold and Wade 1984b; Sutherland 1985b) makes me imagine historians debating whether the significance of the Boston Tea Party should be assessed by the wet or the dry weight of the tea thrown overboard. It is not necessary to be numerically precise to be right.

WHAT A GOOD INTRASPECIFIC EMPIRICAL STUDY OF SEXUAL SELECTION SHOULD DO

The main points are that the mechanisms of advantage of the character should first be discovered, and that the question of which mechanisms count as natural and which as sexual selection should be answered afterwards, if at all.

In an imaginary crustacean species, females have a reproductive cycle and lay a batch of eggs every two weeks. They must mate in each cycle to lay fertilized eggs. Males vary in size fourfold, and the future size of young males can be determined by measuring their size at an early moult. Females are concealed in the mud at the bottom of a stream, where they spend their time feeding. Adult males search for females in the mud, helped by the detection of chemicals involuntarily released by females. Once found a female may or may not consent to mate. The consequences of male adult size are studied and it is found by making appropriate measurements that

1) young males trying to become large adults suffer higher juvenile mortality;
2) large adult males can search twice as fast because of their superior powers of locomotion;
3) large adult males are in addition twice as good at detecting females because their larger sensory apparatus can detect in lower concentrations the chemicals released by the females;
4) large adult males are twice as likely to be accepted as mates by females because females resist mating except near their moult and large males are better able to overcome this resistance;
5) females are only half as likely to remate during a cycle after mating with a large male than after mating with a small male (this is because the males insert a plug at the end of mating to prevent future matings; larger males' plugs are more efficient and there is full sperm precedence for the last male to mate);
6) large males live half as long as adults because the major predator ambushes prey, and moving twice as fast makes capture twice as likely per unit time.

Given these facts, I feel I would understand the comparative advantages of being large and small. Division into sexual and natural selection is not entirely obvious. Some traits are likely to be sexually selected, such as the sensory apparatus and the mating plugs. Whether male size is sexually selected is a scholastic question I am not sure how to answer, and I am not absolutely sure I want to. It might depend on whether the sensory apparatus and powers of locomotion and efficiency of mating plug were simply allometric with body size, or whether they improved qualitatively.

The point of this section is that *the interest of an empirical study can be extracted without having to decide whether a mechanism counts as sexual or natural selection*. Darwin's statement (1871, quoted by Arnold 1983) that ". . . in many cases it is scarcely possible to distinguish the effects of natural and sexual selection" is a sign that imprecision of definition was intended: I believe it is also desirable.

THE RELEVANCE OF ANALYSES OF COMPONENTS OF VARIANCE IN LRS

There are a number of suggested analyses of simple variance in LRS into components (Arnold and Wade 1984b; Brown 1988). (I intend to discuss only direct variance analyses and not techniques in which components of fitness are regressed on characters such as body size. The relationship between characters and fitness is of obvious biological interest). What is the relationship between such an analysis and an outline of mechanisms like that above? The analysis can be performed without any idea of the mechanisms, and the mechanisms can be studied without bothering to do the analysis or even having the data to do one. I suggest that the analysis can be considered as an accounting exercise, in which the effects of the mechanisms are coded into numbers in a complex way, rather as when a firm's assets are written down in a ledger with a valuation.

I will pursue this analogy. In investigating a firm to work out why it has been performing well recently, some questions are best answered by going over the books. What were last year's profits? Have costs been coming down, or have prices been going up? Is the volume of sales changing? But no inspection of the accounts will reveal that the reason more ice cream has been sold is the hot weather, nor that the reasons the costs have come down is that squabbles between OPEC countries have reduced the price of petrol. Now, of course, these factors are reflected *in a way* in the accounts but seeing that reflection does not really help us understand the reason an ice cream manufacturer does well in hot weather when a major energy cartel breaks down. The point here is therefore that the fitness analysis may be useful for some purposes, but very unhelpful for others.

The accounting is unhelpful because the factors of real interest are scattered in different parts of the accounts, contributing to many different headings. The weather will affect demand, the costs of heating factories, and the absentee rate among the work force. The weather could not be reconstructed from the accounts, as each of these accounting headings is also affected by many other factors.

Consider how well a firm is described by its accounts. Seen in one way, every aspect of the firm is represented there and evaluated relative to all other aspects. In another way, it is fair to say that many vital aspects of the firm are missing entirely, or if they are represented are present in a mangled and distorted way.

Understanding the reasons for a sex difference is an aspect of a species' biology not present in the fitness component analyses, although those reasons will doubtless have their influence on some of the numbers. As this is the only purpose which I consider to be a good reason for measuring things to do with sexual selection, does this mean I think that fitness variance component analyses are not worth doing?

I have given much thought to this question and can find only rather feeble arguments in favor of those analyses. The first is that they are a contribution to the natural history of the species. Of course, so would any other kind of measurement or analysis undertaken. The second is that the analysis might point to where mechanisms are likely to be found. If there is much variability between male toads in number of mates per breeding season, and rather little in number of breeding seasons, then it might be worth looking for causal connections between numbers of mates and size or beauty, while looking for causal connections with longevity are likely to be less worthwhile. These are rather weak clues, as we saw before that variance at a given stage has no necessary connection with sexual selection. So while this might be a way to use an analysis that exists, it is unlikely to be worth collecting all the necessary data just for these weak clues. Notice that having compiled all the information for the fitness variance component analysis, following up the clues involves collecting more facts. The data required for the analysis are complete in an accounting sense but by no means complete for answering questions about mechanisms of advantage.

To conclude this section, the analyses of fitness variance components suggested by various authors have little to do with the mechanisms of advantage, which are to my mind the main targets in a study of sexual selection. It is possible that I have missed the point of them. I would be persuaded of this if a close logical connection were demonstrated between a fitness variance component analysis and any important problem of sexual selection such as those I listed above. Meanwhile, my personal advice to anyone considering collecting data with the main purpose of performing such an analysis is the same as Punch's advice to those about to marry.

RECOMMENDATIONS FOR FUTURE RESEARCH

It is traditional in Dahlem conferences to give advice for future research, in which the roles of theoreticians and empiricists are pontificated upon. I see no reason not to indulge in this homely sport myself, to close this essay. I intend to redress a balance, not to be balanced myself.

To theoreticians, I would say: resist the temptation to make your models more realistic. Your function now is to provide the material which will determine what field workers will find interesting in five years' time and beyond. Particular species are in themselves boring and deserve modelling only to make general points, or for didactic purposes. Particular genetic systems are also of interest only if used to make general points.

To field workers, I would say: whatever you do, don't go out and measure a parameter in somebody else's model. Few things could be more fruitless. Your role is not to be a theoretician's technician. Instead you are at least the theoretician's equal and would be better employed doing one of three things. First, study an as yet unmodelled problem. Show the world that this problem is interesting and important, and outline what you see as its main features. Let theoreticians then busy themselves, working out in more and more detail that you had indeed got it nearly all right in the first place. Meanwhile, you can have discovered another interesting problem. Make the theoreticians come to you. After all, your function now is to provide the material which will determine what theoreticians will find interesting in five years' time and beyond.

The second useful role of the field worker is to cut the Gordian knots of the theoreticians. For example, a debate is raging overhead about whether or not sufficient variability can be maintained in heritable fitness to make it worthwhile for a female to pay costs to choose a male solely for the inherited fitness of her offspring. Ultimately, this is a question that in theory will depend on the plausibility of various parameter values and details of the models. An empirical demonstration in reasonably natural conditions that fitness is often inherited from father to offspring in the absence of paternal care would radically change the nature of the debate. No longer would the plausibility of the conditions for maintaining this heritability be the hot topic, but rather what forces are actually responsible for it, and how strong are they? The empiricist must try to stop theoreticians wasting their time on empirical questions, if it is thought they could spend their time more profitably.

The third role of the empiricist is to undermine the theoretician by casting doubt on his accepted premises. What empirical demonstration would most embarass this model? How can I make this theory seem truly pointless? And so on. Go for the jugular.

If after all three activities some field workers still have time to compile a fitness variance component analysis, then I take my hat off to their tirelessness.

REFERENCES

Andersson, M. 1982. Female choice selects for extreme tail length in a widow bird. Nature 299: 818–820.

Arnold, S.J. 1983. Sexual selection? the interface of theory and empiricism. In: Mate Choice, ed. P.P.G. Bateson, pp. 67–107. Cambridge: Cambridge Univ. Press.

Arnold, S.J., and Wade, M.J. 1984a. On the measurement of natural and sexual selection: applications. Evolution 38: 720–734.

Arnold, S.J. and Wade, M.J. 1984b. On the measurement of natural and sexual selection: theory. Evolution 38: 709–719.

Brown, D. 1988. In: Reproductive Success, ed. T.H. Clutton-Brock. Chicago: Univ. of Chicago Press.

Burchfield, J.D. 1975. Lord Kelvin and the Age of the Earth. New York: Science History Publications.

Clutton-Brock, T.H. 1983. Selection in relation to sex. In: Evolution from Molecules to Men, ed. D.S. Bendall, pp. 457–481. Cambridge: Cambridge Univ. Press.

Darwin, C. 1871. The Descent of Man and Selection in Relation to Sex. London: John Murray.

Endler, J.A. 1980. Natural selection and color patterns in *Poecilia reticulata*. Evolution 34: 76–91.

Felsenstein, J. 1985. Phylogenies and the comparative method. Am. Natural. 125: 1–5.

Fisher, R.A. 1958. The Genetical Theory of Natural Selection. New York: Dover.

Ghiselin, M.T. 1969. The Triumph of the Darwinian Method. Berkeley: Univ. of California Press.

Grafen, A. 1988. On the uses of data on lifetime reproductive success. In: Reproductive Success, ed. T.H. Clutton-Brock. Chicago: Univ. of Chicago Press.

Hamilton, W.D., and Zuk, M. 1982. Heritable true fitness and bright birds: a role for parasites? Science 218: 384–387.

Lande, R. 1981. Models of speciation by sexual selection of polygenic traits. Proc. Natl. Acad. Sci. 78: 3721–3725.

Lande, R., and Arnold, S.J. 1983. The measurement of selection on correlated characters. Evolution 37: 1210–1226.

Leask, M.J.M. 1977. A physicochemical mechanism for magnetic field detection by migratory birds and homing pigeons. Nature 267: 144–145.

O'Donald, P. 1982. Sexual selection by female choice. In: Mate Choice, ed. P.P.G. Bateson, pp. 53–66. Cambridge: Cambridge Univ. Press.

Payne, R.B. 1979. Sexual selection and intersexual differences in variance of breeding success. Am. Natural. 114: 447–452.

Ralls, K. 1977. Sexual dimorphism in mammals: avian models and unanswered questions. Am. Natural. 11: 917–938.

Ridley, M. 1983. The Explanation of Organic Diversity. Oxford: Clarendon Press.

Semm, P.; Nohr, D.; Demaine, C.; and Wiltschko, W. 1984. Neural basis of the magnetic compass: interactions of visual, magnetic and vestibular inputs in the pigeon's brain. J. Comp. Phys. A 155: 283–288.

Sutherland, W.J. 1985a. Chance can produce a sex difference in variance in mating success and account for Bateman's data. Anim. Behav. 34: 1349–1352.

Sutherland, W.J. 1985b. Measures of sexual selection. Oxford Surv. Evol. Biol. 2: 90–101.

Thomson, W. 1889. Popular Lectures and Addresses, vol 1. London: Macmillan.

Wade, M.J. 1979. Sexual selection and variance in reproductive success. Am. Natural. 114: 742–746.

Wade, M.J., and Arnold, S.J. 1980. The intensity of sexual selection in relation to male sexual behaviour, female choice and sperm precedence. Anim. Behav. 28: 446–461.

Zahavi, A. 1975. Mate selection — a selection for a handicap. J. Theor. Biol. 53: 205–214.

Zahavi, A. 1977. The cost of honesty. J. Theor. Biol. 67: 603–605.

Standing, left to right:
Paul Harvey, Eduard Linsenmair, Rick Howard, Peter Wirtz, Michael Wade

Seated (center), left to right:
Alan Grafen, Bill Sutherland, Arie van Noordwijk, Tim Clutton-Brock

Seated (front), left to right:
Uli Reyer, Achim Poethke, Jerry Wilkinson

Sexual Selection: Testing the Alternatives
eds. J.W. Bradbury and M.B. Andersson, pp. 234–245
John Wiley & Sons Limited

The Empirical Study of Sexual Selection Group Report

G.S. Wilkinson, Rapporteur
T.H. Clutton-Brock
A. Grafen
P.H. Harvey
R.D. Howard
K.E. Linsenmair
H.J. Poethke
H.-U. Reyer
W.J. Sutherland
A.J. van Noordwijk
M.J. Wade
P. Wirtz

INTRODUCTION

The process of sexual selection occurs when some aspect of an organism's phenotype covaries with mating success or fecundity by influencing intrasexual competition or mate choice. When such selection operates differentially on the sexes and heritable variation exists for the selected characters, sexual dimorphism frequently results. As with other evolutionary phenomena, sexual selection may have been important in an organism's evolutionary past but its applicability to extant populations depends on the availability of phenotypic variation. In this report we critically review methodology and specific measures which have been proposed as descriptors of sexual selection in either or both contexts. To structure the presentation the report has been divided according to the three classical methods for the study of evolution: comparative cross-species studies, observational studies within a species, and experimental manipulations in the field or laboratory. For each method we point out whether the technique is appropriate for studying how selection has operated in the past or how selection is occurring at present and provide several suggestions for promising research which has yet to be pursued.

COMPARATIVE STUDIES

Comparative studies can often provide insight into the past role of sexual selection in producing cross-taxonomic variation. Furthermore, in some situations the microevolutionary forces which act within a species can be inferred and subsequently confirmed with either observational or experimenal studies.

There are three basic components in any comparative study: a) selection of a phylogenetic group of species, b) choice of a measure suitable to the organisms which contains information about the importance of sexual selection, and c) identification of one or more ecological or social variables which the researcher may have prior reason to believe indicates the selective force. At each step several potential problems should be stressed.

Choosing an appropriate phylogenetic comparison can be difficult. Ideally, independent evolutionary events are identified so that problems of nonindependence do not bias the statistical analysis by nonrandomly inflating sample sizes. Ridley (1983) has suggested a technique, which unfortunately can only be used when the characters defining the phylogeny are dichotomous, for recognizing independent evolutionary events. Further statistical work is needed to overcome this problem. Ridley's method also necessitates knowledge of the true phylogeny. This is a problem for which systematists have no single, simple answer. Whether or not numerically intensive methods, such as Felsenstein's (1985) bootstrap technique, are required will depend on the amount of uncertainty in the proposed phylogeny.

In principle, any ecological or social character which indicates a selective agent can be used as a measure of the strength of selection. However, caution should be exercised when interpreting significant correlations because unmeasured selective forces may have produced the pattern in the data. For example, sexual dimorphism in canine length (the sexual selection measure) may correlate with harem size (the social variable implicating sexual selection) in some hypothetical group of primate species, thereby suggesting that longer canines evolved either in the context of male combat for access to females or female choice of males. It is possible, however, that the true function of canines, independent of any role in sexual selection, is to protect a male's own offspring from predators that occur commonly in the geographic range of some species more than others. Such a function might generate a spurious correlation between canine length and harem size that could only be detected by further comparative study or experimentation (see Harvey et al. 1978 for analysis of real data). Measures of sexual selection unrelated to any specific character, such as indices of the variance in mating success or the operational sex ratio, have additional problems in interpretation which will be discussed below under the section on *OBSERVATIONAL STUDIES*

Despite these difficulties, the comparative method offers a powerful tool for identifying selective agents responsible for sex differences in morphological or behavioral characters, both within and between radiations. Although such studies tend to be limited by availability of data, we can suggest a few areas in which sufficient data may be accessible; others have been suggested in the remaining group reports in this volume. All equilibrium models of sexual selection (Heisler et al. this volume) require that a sexually selected trait impose some viability cost when expressed. This assumption could be checked by comparing age-specific patterns of survival across species that vary in the expression of particular sexually selected traits. In such a comparison one must be alert to the possibility that there are two ways for species to vary in this regard: when and how much the trait is expressed during ontogeny. Also, the degree of apparent convergence in sexually selected traits, e.g., bright coloration and feather elaboration in birds, is substantially less than that observed for some selected traits. For example, we can think of no sexual selection examples which parallel the tendency for large wading birds to be white. Other examples of plumage coloration convergence due to natural selection across radiations easily come to mind. We suggest that the degree of apparent arbitrariness in the coloration of some male birds may indicate species that have experienced female choice sexual selection since one would expect some convergence in color if signal intensity (either brightness or degree of contrast) was used in intrasexual displays. Comparative study on birds and perhaps reef fishes in conjunction with observations could verify this suggestion. We admit, though, that an appropriate measure to compare coloration patterns across species is not obvious.

OBSERVATIONAL STUDIES

Unquestionably, the majority of research on sexual selection falls within this category. A wide variety of measures have been proposed to describe the intensity of sexual selection within a species. Below we point out the strengths and weaknesses of several measures and suggest how they can be used most effectively. To facilitate this discussion we recognize two categories: noncharacter-based and character-based measures. Noncharacter-based measures simply quantify the degree of variation in reproductive success without a consideration of variation in phenotypic traits. Character-based measures, on the other hand, estimate the degree of covariance between one or more phenotypic traits and fitness (selection differentials) or estimate the regression of fitness on these traits (selection gradients).

Noncharacter-based measures

The relative variance in fitness or opportunity for selection, I, (Arnold and Wade 1984b); Wade this volume) sets an upper limit on the intensity of directional selection since the covariance between a character and fitness can be no greater than the covariance (i.e., variance) between fitness and itself. However, this measure, as well as all other variance-based measures, suffers two kinds of limitations. The first limitation is that variance in fitness is a necessary but not sufficient condition for selection. For selection to occur, one or more characters must also covary with fitness. Arnold and Wade (1984a) illustrated this point using Triver's cross-sectional study of the lizard, *Anolis garmani*. In this example, the opportunity for sexual selection in females was nearly twice that in males. Nevertheless, the selection differential on male body size was five times that observed for females.

The second limitation concerns the random or "error" component in the fitness variance analysis. In order to estimate the magnitude of this environmentally induced component of fitness variance, repeated measures on individuals are needed. The "random" component of variance (see Sutherland this volume), is the within-individuals or error variance in a repeated measures study. Although some studies can estimate this error variance in the lab (e.g., Houck, Arnold, and Thisted 1985), in nature an individual has only one life history and repeated measures are not possible.

The search-time model of Sutherland (1985, this volume) is an attempt to show how the opportunity for sexual selection is dependent upon the fraction of time spent searching or competing for males. This algebraic model is similar to Triver's (1972) notion of parental investment but has the advantage that it can be related to other previously suggested measures, such as the operational sex ratio and relative variance in fitness. Sutherland's model, however, shares the same limitations as the variance-based measure discussed above. Although measuring search time may be more appealing for use in comparative studies than determining the variance in fitness, in actuality, search time as defined in Sutherland's model may be impractical to estimate in all but a few taxa.

In conclusion, it would be imprudent to initiate a study of sexual selection by measuring only variance in fitness, or some correlate of fitness such as search time, without simultaneously measuring the expression of a purported sexually selected trait. However, where variance in fitness can be estimated easily and partitioned into components which correspond to mating success and other life history stages, it might be informative to calculate some of the measures discussed above and compare them to simultaneous character-based measures of sexual selection to ascertain how much the two approaches do differ. So far, such comparisons have only confirmed the potential problems of variance measures. As mentioned above, differences in the opportunity for selection, I, between the sexes may show no relation in sign or magnitude to the differences in selection differentials. From measures of seasonal reproductive success for *Rana catesbeiana* and *R. sylvatica*, Howard (1987) calculated that $I_{males} > I_{females}$ in both species. However, no sexual dimorphism, either in body size or growth rate, occurs in *R. catesbeiana* while males are smaller and grow slower in *R. sylvatica*. In contrast to the predictions using I these patterns of dimorphism are consistent with the selection gradients estimated by regressing seasonal reproductive success on body size. The gradients are equal between male and female *R. catesbeiana* and the male gradient is less than the female gradient in *R. sylvatica*.

Character-based measures

Although a selection differential measures how much the mean value of a character has changed within a generation, it does not necessarily indicate that phenotypic variation in that character has a reproductive consequence. An observed selection differential could be solely due to selection acting on some other correlated character. To control for any phenotypic correlations a multiple regression of fitness on characters can be used to estimate the relative effect of selection on each character. The resulting partial regression coefficients are estimates of selection gradients which indicate the intensity of selection acting directly on each trait.

Such an analysis can be useful in ascertaining the mechanism by which selection is acting. For example, a selection gradient analysis could, in principle, determine that selection operates primarily on antler size rather than body size in a hypothetical ungulate. This method has also been recommended for estimating the shape of the adaptive landscape in order to predict change in phenotypic distributions over time (Lande and Arnold 1983). To infer how selection has acted in the distant past or how it will act over generations in the future requires not only that selection does not change over time but also that the genetic covariance matrix remains constant. While this latter assumption should be true as long as effects due to selection and drift are weak, there is empirical evidence (Lofsvold 1986) that it does not hold for time periods which correspond to the divergence of subspecies and species.

Selection gradient analysis requires fulfillment of at least six assumptions which are discussed briefly below. Further discussion of the problems and merits of this regression technique can be found in any regression methods text, such as Draper and Smith (1981), as well as in Manly (1985) and Endler (1986).

1) If a study neglects directly selected characters that are correlated with the included characters, the estimated gradients can be incorrect. Excluding relevant variables is, of course, a problem in any scientific study.
2) This technique makes the implicit assumption that directional selection acts on a linear combination of characters (stabilizing selection is estimated with a quadratic regression). Consequently, important variables must be included on a scale which preserves the purported relationship. If the appropriate transformation is included in the analysis, significant coefficients will emerge only on the transformed variable. For example, if sexual selection has acted directionally to increase body area and thus conspicuousness, in a reef fish then a significant gradient would appear only for the square of body length, not body length itself.
3) There may be insufficient phenotypic variation to detect the actual shape of the fitness landscape. For example, if the distribution of one character in the population is under the tail of a Gaussian fitness function, selection will appear

to be directional rather than stabilizing. The solution to this problem is either to re-estimate the gradients continually over time or artificially expand the phenotypic distribution. A rectangular distribution should maximize the power in estimating the selection surface.

4) If characters are not measured on every individual born into the population, then estimates of selection gradients can be biased. For example, if some individuals die before attaining the phenotype expressed at sexual maturity, they cannot be included in the distribution of adult phenotypes available to sexual selection. A partial solution to this problem is to divide selection into different episodes during an animal's lifetime (cf. Arnold and Wade 1984b).

5) If environmental variation causes spatial or temporal heterogeneity in selection or multiple nonadditive episodes of selection, estimates of selection gradients from pooled data sets can be misleading. This is especially relevant for sexual selection because of the inherent frequency dependence of the process. Furthermore, a nonstable age distribution necessarily creates temporal variation in selection in species with overlapping generations. Data sets therefore, must be screened carefully before pooling across sites or years. It should be noted, though, that the existence of heterogeneous selection can, in principle, be identified if contextural variables, such as indices of habitat heterogeneity or lek size, are included in the selection gradient anaylysis (Heisler and Damuth, unpublished manuscript).

6) The measured trait should not change with age otherwise effects of ontogeny and selection may be confused. This problem is particularly relevant for indeterminately growing species. Unfortunately, such species often have the advantage of revealing more morphological variation to selection. One solution to this problem is to estimate a time invariant character, such as the parameters of a function which describes the growth trajectory.

Despite these difficulties, selection gradient analysis is perhaps the most powerful method available for analysing observational data. Heisler *et al.* (this volume) discuss how this technique could be used to test assumptions and predictions of several sexual selection models. Of course, this technique requires an estimate of fitness for each individual. Lifetime reproductive success is usually recommended (Charlesworth 1980), although in age-structured populations with overlapping generations experiencing fluctuating episodes of selection it may not be the best measure of fitness. No recommendations for an alternative measure have yet been made for this complicated, but potentially frequent, case. If less than a full generation of data is collected, then the gradients will only apply to the components of fitness corresponding to the life history stage measured and should not be used to predict change across generations. In some situations, it may be possible to use surrogate measures of fitness components, such as mating speed, for seasonal or lifetime mating success if other evidence can substantiate a correlation between the surrogate measure and the fitness component. Identifying the appropriate interval over which a rate measure should be estimated can only be determined from familiarity with the reproductive schedule of the organism.

An illustration of the utility of the gradient method over simple selection differentials is provided by Price et al. (1984). Selection differentials on beak depth and beak width in the finch *Geospiza fortis* recorded in 1976, 1980, and 1982 were all either positive or zero. However, the selection gradient on beak width was consistently negative while the gradient on depth was positive over those periods. The gradient on beak depth, not the differential, is more consistent with the identified selection pressure, i.e., the ability to feed efficiently on large seeds on Isle of Daphne Major in the Galapagos.

EXPERIMENTAL STUDIES

Experimental manipulation in which the outcome can either confirm or negate a stated hypothesis is often considered to be the most convincing method for excluding alternative hypotheses and inferring causation. This method, however, shares a major problem with comparative and observational studies: most evolutionary models do not provide mutually exclusive predictions which can be tested. However, experiments do provide a means of controlling potentially correlated factors and for expanding the phenotypic distribution to detect selection. Andersson's (1982) classic experiment on widowbirds illustrates both of these advantages. By randomizing experimental birds across territories he was able to control for the confounding effect of territory quality. Then, by shortening and lengthening the tail, he expanded the tail length distribution 5 standard deviations to each side of the mean to increase the likelihood of detecting selection.

Furthermore, in the laboratory one can isolate a single process and with a simple experiment, measure its influence. For example, Wilkinson (1987) estimated the response to sexual selection on wing size in *Drosophila melanogaster* by comparing the progeny of randomly paired flies to those which had been allowed to compete for mates in a large mating chamber. By also estimating the selection differential due to a large male mating advantage, qualitative support for an equilibrium model of sexual selection was obtained in which male-male competition (as well as other fitness components) favors large body size while larval viability is enhanced by small body size.

This experiment potentially shares a limitation common to all manipulations: unless the biology of the organism is well known, artifacts of the experiment can create misleading results which prevent extrapolation to natural populations. For example, when Wilkinson (1987) estimated the selection differential due to male-male competition he arbitrarily chose a sex-ratio and density of flies to place in a mating chamber. Either the sex-ratio, the density, or confinement to a chamber could have altered mating behavior in the flies in unnatural ways. Fortunately, the sexual selection differential was independent of both sex-ratio and density. Furthermore, confinement did not apparently affect mating behavior because independent work by Partridge and her co-workers (personal communication) has demonstrated that larger males have a mating advantage when mating on rotting fruits in the wild, as well as in confinement. Such combinations of field and laboratory experimentation and observation can provide compelling arguments for how sexual selection operates.

As with estimates of selection differentials, correlations between a manipulated trait and an unmeasured trait can lead to misleading results if they are not recognized. A recent experiment on mate choice in guppies by Bischoff et al. (1985) illustrates the danger of correlated traits. In this experiment tail length was manipulated; however, tail length affects display rate which also influences mate choice. In this case, the correlation was detected and cleverly controlled by independently manipulating display rate through temperature change. From these experiments the authors concluded that mate choice by females depended twice as much on variation in tail length as on variation in display rate. Whether this experiment on artificially selected stocks of guppies is representative of how mate choice occurs in natural populations remains to be determined. A field experiment in which control over most variables is reduced but more realism is attained might add further insight and reveal the importance of other sexually selected characters. In this case, field and lab data suggest that male coloration, in addition to display rate and tail length, influences mate choice (Endler 1983; Kodric-Brown 1985; Houde personal communication).

We hope the reader recognizes that while some hypotheses can be discriminated by qualitative outcomes others require quantitative estimation of the relevant parameters. These two processes need not be exclusive. For example, some "good genes" models of sexual selection can be rejected if very little additive genetic variance for fitness exists. It should be noted, however, that the disease-oriented "good genes" model of Hamilton and Zuk (1982) makes the opposite prediction. These models can therefore be tested by estimating the magnitude of the additive genetic variance in total fitness. This subject is considered at greater length by Price et al. (this volume).

In addition to testing for additive genetic variance in fitness, we can suggest several additional areas in which experimentation is sorely needed. One problem is determining the rules for female choice. Whether females have absolute or relative preferences for a specific trait has different consequences for the evolution of the male trait than if females choose the best-of-*n* males (Seger 1985). Likewise, determining the relative intensity of male-male competition versus female choice of mates can strongly affect the evolutionary outcome. Models of male-male competition predict either a stable equilibrium point (Parker 1983, Charlesworth 1984) or continual evolution for increasing expression of the male trait (Maynard Smith and Brown 1986). In contrast, some female choice models predict a line of equilibria which can be either neutrally stable or unstable (Lande 1981, Kirkpatrick 1982). If there is a cost to preferences, this line becomes an equilibrium point with the cost minimized. As with the equilibrium line, this point also can be either stable or unstable. Thus, experiments measuring costs associated with female preferences are needed. We emphasize that one cost of mating, such as traveling to a lek to mate, is not relevant to these models; only the cost associated with choosing a particular male is critical.

PROBLEMS FOR FUTURE STUDY

A number of mechanisms other than transmission of additive genetic factors can affect the similarity between parent and offspring. These alternative forms of phenotypic transmission may have nonintuitive and unpredictable effects on both the dynamics and outcomes of a sexual selection process. In addition to maternal effects — such as the correlation between male adult body size and birth weight in red deer (birth weight is determined primarily by maternal condition), genotype environment correlations (i.e., habitat selection), and genotype environment interactions (Via and Lande 1985) — we can envision direct phenotypic transmission across and within generations in females exerting more unanimous preferences through copying or attempts by some males to mimic the behavioral traits of more successful males. In females, this process could increase the intensity of selection on males and could alter the covariances in the transmission matrix (genetic and cultural) when multiple cues are involved. Copying among males might create selection for traits which cannot be copied. The apparent prevalence of exaggerated behavioral displays which appear to be related to vigor in males of lekking species (e.g., Gibson and Bradbury 1985) suggests that these traits may be used by males (and also perhaps preferred by females) because they make copying difficult. This area needs further theoretical work before predictions can be made. Boyd and Richerson (1985) have recently suggested a framework for analyzing phenotypic transmission of quantative characters given specific learning models that might be applied profitably to the theoretical side of the problem. Also, multivariate path models which have been used to detect genotypic and phenotypic transmistion of diseases (Vogler 1985) could be used, in principle, to detect if these effects occur in animals.

REFERENCES

Andersson, M. 1982. Female choice selects for extreme tail length in a widowbird. Nature 299: 818–820.

Arnold, S.J. Wade, M.J. 1984a. On the measurement of natural and sexual selection: applications. Evolution 38: 720–734

Arnold, S.J. and Wade, M.J. 1984b. On the measurement of natural sexual selection: theory. Evolution 38: 709–719.

Bischoff, R.J.; Gould, J.L. and Rubenstein D.I. 1985. Tail size and female choice in the guppy (*Poecilia reticulata*). Behav. Ecol. Sociobiol. 17: 253–255.

Boyd, R., and P.R. Richerson. 1985. Culture and the Evolutionary Process. Chicago: Univ. of Chicago Press.

Charlesworth, B. 1980. Evolution in Age-Structured Populations. Cambridge: Cambridge Univ. Press.

Charlesworth, B. 1984. The cost of phenotypic evolution. Paleobiology 10: 319–327.

Draper, N., and Smith, H. 1981. Applied Regression Analysis, 2nd ed. New York: Wiley.

Endler, J.A. 1983. Natural and sexual selection on color patterns in poeciliid fishes. Environ. Biol. Fish. 9: 173–190.

Endler, J.A. 1986. Natural Selection in the Wild. Princeton: Princeton Univ. Press.

Felsenstein, J. 1985. Confidence limits on phylogenies: an approach using the bootstrap. Evolution 39: 783–791.

Gibson, R.M. and Bradbury, J.W. 1985. Sexual selection in lekking sage grouse: phenotypic correlates of male mating success. Behav. Ecol. Sociobiol. 18: 117–123.

Hamilton, W.D., and Zuk, M. 1982. Heritable true fitness and bright birds: a role for parasites. Science 218: 386–387.

Harvey, P.H.; Kavanagh, M.; and Clutton-Brock, T.H. 1978. Sexual dimorphism in primate teeth. J. Zool. Lon. 186: 475–486.

Houck, L.D.; Arnold, S.J.; and Thisted, R.A. 1985. A statistical study of mate choice: sexual selection in a plethodontid salamander (*Desmognathus ochrophaeus*). Evolution 39: 370–386.

Howard, R.D. 1987. Reproductive success in two species of anurans. In: Reproductive Success, ed. T.H. Cluton-Brock. Chicago: Univ. of Chicago Press.

Kirkpatrick, M. 1982. Sexual selection and the evolution of female choice. Evolution 36: 1–12.

Kodric-Brown, A. 1985. Female preference and sexual selection for male coloration in the guppy (*Poecilia reticulata*). Behav. Ecol. Sociobiol. 17: 199–205.

Lande, R. 1981. Models of speciation by sexual selection on polygenic traits. Proc. Natl. Acad. Sci. USA 78: 3721–3725.

Lande, R., and Arnold, S.J. 1983. The measurement of selection on correlated characters. Evolution 37: 1210–1226.

Lofsvold, D. 1986. Quantitative genetics of morphological differentiation in *Permomyscus*. I. Tests of the homogeneity of genetic covariance structure among species and subspecies. Evolution 40: 559–573.

Manly, B.F.S. 1985. The Statistics of Natural Selection on Animal Populations. New York: Chapman and Hall.

Maynard Smith, J. and Brown, R.W.L. 1986. Competition and body size. Theor. Pop. Biol. 30: 166–179.

Parker, G.A. 1983. Arms races in evolution — an ESS to the opponent-independent costs game. J. Theor. Bio. 101: 619–648.

Price, T.D.; Grant, P.R.; Gibbs, H.L.; and Boag, P.T. 1984. Recurrent patterns of natural selection in a population of Darwin's finches. Nature 309: 787–789.

Ridley, M. 1983. The explanation of organic diversity: the comparative method and adaptations for mating. Oxford: Oxford Univ. Press.

Seger, J. 1985. Unifying genetic models for the evolution of female choice. Evolution 39: 1185–1193.

Sutherland, W.J. 1985. Measures of sexual selection. Oxford Surveys in Evolutionary Biology 2: 90–101.

Trivers, R.L. 1972. Parental investment and sexual selection. In: Sexual Selection and the Descent of Man, ed. B. Cambell. pp. 139–179. Chicago: Aldine.

Via, S., and Lande, R. 1985. Genotype-environment interaction and the evolution of phenotypic plasticity. Evolution 39: 505–522.

Vogler, G. 1985. Multivariate path analysis of familial resemblance. Genetic Epidemiology 2: 35–53.

Wilkinson, G.S. 1987. Equilibrium analysis of sexual selection in *Drosophila melanogaster*. Evolution 41: 11–21.

Sexual Selection: Testing the Alternatives
eds. J.W. Bradbury and M.B. Andersson, pp. 247–264
John Wiley & Sons Limited

Physiological Constraints on Sexual Selection

T.R. Halliday

Animal Behaviour Research Group
The Open University
Milton Keynes MK7 6AA, England

Abstract. Data on the energetics of sexual and competitive behavior, mostly in amphibians and insects, is reviewed. For males of many species, courtship displays and fighting are energetically very expensive, but there is little direct evidence that animals reach physiological limits during a single bout of behavior. Physiological constraints do seem to be important, however, over a longer timescale, because they limit the number of times in a season or the rate at which males can engage in intense sexual activity. Physiological constraints are highly relevant to "viability-indicator" models of sexual selection, since a male's ability to overcome them is likely to be correlated with his general vigor.

For many animals, behavior patterns associated with mating appear to involve very high levels of energy expenditure. There have been, however, very few measurements made of this important component of reproductive effort and, at the start of this paper, I wish to stress the paucity of data relating to what is a neglected aspect of sexual selection.

There are three reasons why it is important to have some measure of the energetic demands imposed by sexual behavior. First, variation among males in their capacity to meet such demands may be a major determinant of variation in male reproductive success. Second, most theoretical discussions of sexual selection emphasize that there is a trade-off between benefits in terms of male mating success, from performing elaborate displays or showing high levels of competitive behavior, and a variety of costs incurred as a consequence of such activities. Some attention has been paid to measuring costs arising from predation (e.g., Burk 1982; Cade 1979; Ryan 1985) but very little to other kinds of cost. Third, there has been considerable speculation and debate about the precise nature of the possible benefits that females may derive by mating preferentially with certain males. If there is variance among males in their physiological capacity to perform those behavior patterns that attract and stimulate females, it could provide the basis for female mating preferences to act as a "physiological test" of male vigor, by which males with high quality phenotypes are more likely to attract females, as in the "viability indicator" models of Andersson (1982, 1986). If this quality is heritable, females may pass on to their offspring, of both sexes, more than the sexual attractiveness in their sons which Fisher (1930) regarded as a sufficient basis for the evolution of elaborate male characters.

The timescale on which physiological factors constrain sexual performance is likely to vary from one situation to another, so that males of different species may be constrained in different ways. A male may reach physiological limits within a single bout of sexual behavior so that the intensity and/or duration of his activity are also limited. Even when males have several opportunities to mate there may be constraints acting on a longer timescale, limiting either the frequency or the rate at which males can fight one another or court and mate with females. In other words, the physiological cost of a particular mating episode may be generally expressed as a reduced capacity to engage in subsequent sexual interactions.

Like any other component of reproductive effort, energy expended during mating and related activities cannot be expended on somatic effort. The physiological costs of reproductive activity may, therefore, be expressed, not in immediate reproductive terms but as a subsequent reduction in survival. Evidence for a simple trade-off between reproduction and survival is conflicting, however, and the relationship between them is clearly complex (Reznick 1985; Partridge and Harvey 1985). A full understanding of physiological constraints on sexual behavior may thus require consideration, not only of short-term effects but also of life history factors (see Partridge, this volume).

Most of the examples that follow concern studies of amphibians and insects. A major reason for this is that these animals tend to perform normal reproductive activities within the apparatus needed for the measurement of physiological variables more readily than do birds and mammals, for example. The methodological problems involved in making physiological measurements under natural conditions are immense and are undoubtedly a major reason for the lack of good data in this area.

MATING ENERGETICS IN ANURANS

A detailed study of the physiological costs of male sexual behavior is that by Ryan (1985) of the Tungara frog (*Physalaemus pustulosus*). Males call at night and form choruses to which females are attracted. The male advertisement call has two parts, a "whine" followed by a variable number of brief "chucks." At any one time only a proportion of the males in a chorus are calling. Males will call and build foam nests within a respirometer so that Ryan and his collaborators were able to compare their oxygen consumption during these activities to that when they were inactive (Bucher et al. 1982; Ryan et al. 1983). Oxygen consumption showed a nearly five-fold increase in calling males compared with males resting during the day.

Males called at varying rates and there was a positive correlation across males between whine rate and oxygen consumption that explained 91% of the variance in the latter. The number of chucks incorporated by males into their calls had a minimal effect on their oxygen consumption, although this is a very important factor in attracting females. The amount of oxygen expended per call decreased with call rate; this greater efficiency at higher call rates probably arises because filling the vocal sacs with air in long pauses between calls is more energetically expensive than shuttling air back and forth between the vocal sacs and the mouth during short intervals. There was no significant elevation of lactate levels in calling males, suggesting that anaerobic respiration, an important factor in the energetics of intense activity in many amphibians (Bennett and Licht 1974; Taigen et al. 1982) is not involved in calling. Males present in the chorus but not calling may be conserving energy between calling bouts.

The efficiency with which metabolic energy is coverted into sound energy in Tungara frogs is only 0.5% for a whine with one chuck and 1.2% for a whine with four chucks. These low values arise because the wavelength of the sound produced is very much greater than the length of the structures that produce the sound.

To assess the extent to which calling males are approaching possible physiological limits, Ryan and his co-workers conducted forced activity tests to determine maximum levels of energy expenditure. These suggest that calling males are operating at about 42% of their maximum aerobic capacity. Estimates of total energy expenditure by males, including both calling and nest building, over a breeding season suggest that it is less than 10% that of females, for whom egg production is very expensive.

Ryan's data suggest that, although it is energetically expensive, calling by male Tungara frogs is not an activity that imposes severe physiological constraints on their reproductive success. Of much greater significance is predation by bats which locate males through their calls. Calls containing several chucks are both more attractive to females and entail a higher predation risk than calls with few or no chucks (Ryan et al. 1982).

A recent study by Wells and Taigen (1986), of calling energetics in the gray treefrog (*Hyla versicolor*), suggests that males show a trade-off between two call parameters, duration and call-rate. These vary between males but are fairly constant within males, those producing long-duration calls tend to do so at a low rate. Longer calls are important in male-male competition; in dense choruses, and in response to playback of recorded calls, males increase their call duration and reduce their call rate. Longer calls are also more attractive to females (Gerhardt, personal communication). They appear to be more energetically expensive than shorter calls; males that produce longer calls call for fewer hours per night. Wells and Taigen suggest that the limiting physiological factor is intramuscular glycogen reserves, which become more depleted during long bouts of activity than over frequent, short bouts. They also estimated oxygen expenditure during calling and found that males tended to call at levels that approached, but were below, an estimated upper limit of 2.2 ml 0_2 g^{-1} h^{-1}.

While many studies of frogs and toads have sought evidence of variation in male mating success based on such characters as body size on various parameters of the advertisement call, several workers have found that the most significant factor is the amount of time for which males are present at a breeding site. This effect has been shown for *Bufo calamita* (Arak 1983b), *B. woodhousei* (Woodward 1982), *B. rangeri* (Cherry, personal communication), *Hyla chrysoscelis* (Godwin and Roble 1983), *H. rosenbergi* (Kluge 1981), *H. cinerea* (Gerhardt, personal communication), *Centrolenella fleischmanni* (Greer and Wells 1980; Jacobson 1985), and *C. prosoblopon* (Jacobson 1985). Wells (1977) found that *Rana clamitans* males that spent the most time in territories with good oviposition sites obtained the most matings. A correlation between time spent in mating activity and mating success could be based on variance among males in the ability to sustain sexual behavior physiologically, although Ryan's study of *Physalaemus pustulosus*, described above, offers a possible alternative explanation for certain species: sexually inactive males may be reducing their predation risks, as well as conserving energy.

Evidence suggesting a physiological constraint on male activity comes from Robertson's (1986b) study of *Uperoleia rugosa*. He observed a dynamic interchange between territorial calling males and nonterritorial silent satellites. A male loses weight while he is calling and gains weight while a satellite, until heavy enough to oust a caller from his territory.

The rate at which males lose weight varies between species and can be related to the duration of the breeding season (Table 1). These data suggest that, in species with short breeding seasons, males sustain higher daily rates of weight loss. This does not mean, however, that they incur greater physiological costs in the long term than males in prolonged-season species; since their breeding season is short species such as *Bufo bufo* have longer to recover lost

TABLE 1 Duration of the breeding season and male weight loss in various anurans

Species	Mean number of days in pond	Mean % weight loss per day	Reference
Bufo rangeri	7	1.34	Cherry (personal communication)
Bufo bufo	20	1.07	Arak (1983a)
Bufo calamita	41	0.32	Arak (1983a)
Rana clamitans	60	0.175	Wells (1978)

weight before the next season. Much of the variation between species in terms of weight loss is related to the extent to which males feed during the breeding season. In *Bufo bufo* they do not feed at all; in other species, such as *B. calamita*, they do. In bullfrogs (*Rana catesbeiana*), males grow during the long breeding season and there is no difference in growth rate between territorial and nonterritorial males (Howard 1984).

A detailed study of male reproductive energetics in two *Ranidella* species has been made by MacNally (1981). He measured the energy and ash content of somatic tissue as well as the growth of the body and individual organs and concluded that males invest heavily in current reproduction, and very little in growth. Very little male effort goes into gonads, gametogenesis, or parental behavior but instead into competition with rival males.

In species where males produce advertisement calls, a variety of call features have been found to correlate with male mating success. Fundamental frequency is sometimes inversely correlated with body size, providing a potential means by which females can mate preferentially with males of a certain size. Ryan (1985) found such a preference for large males in *Physalaemus pustulosus* and Robertson (1986a) has shown that female *Uperoleia rugosa* choose males whose weight is approximately 75% of their own. The frequency of a frog's call is a function of body size, not of the energy expended in calling, and several studies have demonstrated a tendency for females to approach preferentially calls with features, such as intensity, duration, and repetition rate which are a function of energy expenditure (Table 2). In all these examples, female preference is such as to favor those males that invest higher levels of energy in calling.

TABLE 2 Parameters of male calls for which female anurans have been found to show a preference

Species	Feature of call preferred	Reference
Bufo calamita	Louder calls	Arak (1983a)
Bufo woodhousei	Higher call rate	Sullivan (1983)
Hyla regilla	Longer duration calls	Whitney & Krebs (1975)
Hyla versicolor	Louder calls	Fellers (1979)
Physalaemus pustulosus	Calls with extra 'chucks'	Ryan (1985)

Taigen and Wells (1985) measured oxygen consumption and lactate levels in calling male *Hyla versicolor*. Anaerobic respiration made no energetic contribution to calling but aerobic metablism increased linearly with increasing call rate and duration to a level such that calling appears to be the most energetic activity that this species performs. In his study of *Rana clamitans*, Wells (1978) observed males lying apparently exhausted and helpless after a fight and he suggested that they were recovering from an oxygen debt incurred by anaerobic respiration.

It is clear from the examples quoted so far that male energetic expenditure during breeding is high in anurans, but there is only limited evidence that their sexual activity is limited by physiological factors. If it were, we might expect physiological scope and mating success to be correlated among males and two recent studies have set out to test this prediction. Sullivan and Walsberg (1985) measured the maximum aerobic scope (in the laboratory) and the call rate (in the field) of male *Bufo woodhousei* and found no correlation. Similarly, Wells and Taigen (1984) observed male activity in *Bufo americanus* on a single, high activity night and measured aerobic scope in a forced exercise laboratory test. Aerobic capacity was not correlated with any behavioral variable or with body weight. The conclusion of both studies is that mating activity is not performed at such a level in anurans that their aerobic scope is a critical constraint, mating behavior being supported by submaximal levels of energy expenditure. It would be premature, however, to conclude that male anurans are not subject to any physiological constraint. As Sullivan and Walsberg point out, laboratory tests may not accurately reflect levels of aerobic respiration in nature. Furthermore, the behavioral and physiological measurements made in these two studies were, of necessity, made over short timescales relative to the length of the breeding season. As stressed at the start of this paper, physiological constraints may act to limit a male's endurance over a prolonged period rather than his ability to perform at a peak level in the short term. By analogy, human sprinters train in a very different way to long-distance runners and acquire very different physiological adaptations.

SMOOTH NEWTS

Constraints on the reproductive success of male smooth newts (*Triturus vulgaris*) operate on three timescales:

1) *Long term* (over a lifetime). Males vary in body size and larger males have larger testes (Verrell et al. 1986). It is reasonable to assume that larger males, over their lifetime, produce more sperm than smaller males. Males may live for as long as six years, breeding for the first time when two years old, after which they continue to grow, although at a very slow rate (Verrell and Francillon 1986). Because there is large variance in male size at first breeding and slow subsequent growth, male lifetime fecundity appears to be determined primarily by prereproductive growth (Partridge, this volume, for a discussion of life history constraints on reproductive success).

2) *Medium term* (over a breeding season). During the breeding season, the testes are small, relative to their size during the rest of the year, mature sperm having been evacuated into the vasa deferentia as males enter breeding ponds (Verrell et al. 1986). At this time the testis contains immature sperm cysts which complete their development after the breeding season. This pattern of reproduction, called postnuptial gametogenesis, means that males start the

breeding season with a finite supply of sperm which cannot be supplemented by newly produced sperm during that season. Thus a male's reproductive success in a given season is limited by his sperm production in the previous summer and autumn. Because they have larger testes, larger males probably produce more sperm per season than small males.

In newts, sperm is encapsulated in spermatophores that are transferred to females during courtship, and there are four factors that tend to favor the ability in males to produce spermatophores in large numbers. First, the breeding season is long and receptive females are available, although in varying numbers, for several months (Verrell and Halliday 1985). Second, spermatophore transfer is very inefficient and many spermatophores are lost (Halliday 1974, 1977a). Third, females generally fail to pick up the first spermatophore that a male deposits but become more likely to pick up each successive one (Halliday 1974). Finally, females store sperm and can mate with more than one male, conditions under which sperm competition will occur (Halliday and Verrell 1984). Male newts thus face a problem: how to maximize immediate spermatophore production and, at the same time, conserve a finite supply of sperm for subsequent mating opportunities. The vasa deferentia of males are empty at the end of the breeding season, suggesting that they have run out of sperm, although it is possible that they have resorbed surplus sperm.

3) *Short term* (during a courtship encounter and between encounters). When a male courts a female he is subject to two constraints, his immediate supply of spermatophores (Halliday 1976) and his need to breathe from the water surface, an activity which typically terminates courtship (Halliday and Sweatman 1976). Males tested twice to sexual exhaustion (unable to put down any more spermatophores) can produce as many spermatophores in the second test as in the first if 48 hours elapses between them, but they show a significant decrease in spermatophore production if the inter-test interval is 24 hours (Verrell 1987).

Spermatophore wastage appears to be minimized by the relationship between a male's display rate and his immediate spermatophore supply. Contrary to Williams' (1975) suggestion that males should always display to females at the maximum rate possible, male newts display at a highly variable rate that is positively correlated with sperm supply (Halliday 1976; Halliday and Houston 1978). Males that carry only one spermatophore display at a low rate and are thus unlikely to court females successfully (Teyssedre and Halliday 1986). They will not, therefore, lose that spermatophore but will conserve it until such time as they have several available, at which time they will display at a high rate and will be much more likely to stimulate and transfer sperm to females.

Within a courtship encounter a male's behavior appears to be constrained by his need to breathe, inasmuch as most encounters are terminated by a breathing ascent to the water surface. The constraint is not a critical one, however. If it were, we would expect a male's tendency to breathe to be higher the more vigorously he displays, but in fact the reverse is true: the higher his display rate, the greater the delay before he breathes (Halliday and Sweatman 1976). If a male's need to breathe is increased by reducing the amount of oxygen available, his courtship performance becomes constrained and he deposits on the average fewer spermatophores before ascending to breathe (Halliday 1977b). Some males can, however, compensate for the effect of reduced oxygen by reducing the duration of certain phases of courtship and deposit as many spermatophores as they do under air. If the constraint imposed by breathing behavior is removed by increasing the oxygen available, males increase the duration of courtship but do not deposit a greater number of spermatophores. These observations suggest that, under conditions of good oxygen supply, a male's sexual performance is not constrained by the need to breathe, but that it can become so under conditions of poor oxygen supply.

Male smooth newts have an alternative mating strategy, called sexual interference, in which a male mimics female behavior at a critical point in courtship and thereby "steals" a receptive female from the male who has been courting her (Verrell 1984). For the interfering male, this behavior represents a way of obtaining an insemination at low energetic cost since, if successful, he gains access to a female who has already been made receptive by the previous male. The risk of interference raises energetic costs for courting males because their response to the presence of a rival male is to markedly increase both the intensity and duration of their courtship display.

Despite the fact that they feed during the breeding season, males lose weight dramatically. By the end of the mating period some males are 30% less heavy than when they entered the pond, though this loss tends to be recovered during the remainder of their time in the water (Verrell and Halliday 1985). These changes are paralleled by a decline followed by an increase in the size of males' fat bodies (Verrell et al. 1986).

Other urodeles

The courtship of the red-spotted newt (*Notophthalmus viridescens*) differs in a number of ways from that of the smooth newt (Verrell 1982) and males are not subject to the same constraints. Females show a higher tendency to pick up the first spermatophore than later ones, so that there is much less advantage to males from depositing several spermatophores than there is for male smooth newts. When given the opportunity to court several females, males will divide their courtship effort and their available spermatophores between them, whereas in the same situation, male smooth newts direct all their courtship to only one of the females (Verrell 1985). Their total spermatophore production does not, however, increase when they court several females, suggesting that it is limited, as in the smooth newt.

Red-spotted newts show alternative mating strategies. If a male encounters a receptive female and there are no rival males present, he performs a very brief "hula" display before initiating spermatophore transfer. If the female is unreceptive, or if a rival is present, he captures the female and there is a prolonged amplexus courtship, lasting several hours (Verrell 1982). The latency to a breathing ascent following an amplexus courtship is much shorter than that following a "hula" courtship, suggesting that amplexus is energetically more expensive (Verrell 1985).

The temporal patterning of courtship acts during amplexus suggests that a male's activity is subject to an energetic constraint, possibly related to his respiratory requirements. Male behavior during amplexus varies both in the rate at which he performs individual acts and in the overall duration of amplexus. If acts are performed at a high rate, amplexus is relatively short; long periods of amplexus are characterized by a low rate with several long pauses between acts (Verrell 1982). As in smooth newts, males will interfere in courtships, and amplectant males respond to the activities of a rival by increasing the duration of amplexus, thereby tending to increase any energetic costs involved (Verrell 1983).

Unlike newts, the lungless plethodontid salamander *Desmognathus ochrophaeus* mates on land. Bennett and Houck (1983) investigated the energetic cost of both sexual behavior and aggression between males. During both activities, there is a slight increase both in oxygen consumption and lactate production and this effect is more marked during aggression. Oxygen consumption increases by 38–56%, anaerobic respiration by 8–23% during these activities, but Bennett and Houck estimate that the energetic cost of a courtship amounts to only about 1% of the energy ingested per day.

ARTHROPODS

A number of studies provide evidence that sexual behavior can be energentically very expensive for male insects. In many species males produce loud calls to attract females and, in some, males must present large and nutrient-rich spermatophores to females in order to achieve matings. The "song" of the male bladder cicada (*Cystosoma saundersii*) has an intensity of 90.6 dB at 20 cm from its source and males generate a power output of 45 mW during calling, representing an increase of 42.55 mW over resting metabolism (MacNally and Young 1981). Less than 1% of the total power produced is converted into sound energy, with 64% being lost as heat. The energy expended by males in mating activity represents the major component of their total reproductive effort: this is about 60% of that of females, which goes primarily into egg production (MacNally and Doolan 1982). Males sing in choruses and the amount of energy that males invest in singing appears to be an important determinant of their mating success, since females respond positively (by wing-flicking) only to those males whose song is at least 9 dB louder than that of their nearest neighbor (Doolan 1981).

The production of large, energetically expensive spermatophores is likely to act as a constraint on male reproductive success, especially when selection favors those males with larger spermatophores. Such an effect has been found in heliconiine butterflies, where there is evidence that nutrients in the spermatophore are used in egg production by the female (Boggs 1981). The presence of a spermatophore in her genitalia appears to inhibit mating by the female. It therefore may be to the advantage of a male to transfer a large spermatophore since this may delay her next mating for longer and thus increase the number of her eggs that he fertilizes. In certain butterflies, the vigor of a male's display to females is correlated with the size of the spermatophore he is carrying, suggesting a spermatophore conservation mechanism comparable to that of male newts (Rutowski 1979). In some orthopterans, females eat the nutrient-rich portion of spermatophores, which are very expensive to produce; males of the mormon cricket (*Anabrus simplex*) lose up to 27% of their body weight in the process (Gwynne 1981). Females of the katydid *Conocephalus nigropleurum* show a preference for larger males that produce larger spermatophores (Gwynne 1982). As in newts, physiological constraints on male sexual performance in these animals may be expressed in the rate at which they can produce fresh spermatophores over the course of a breeding season.

In mormon crickets, the nature of the mating system differs markedly at different population densities (Gwynne 1984). At low densities, males compete with females, which accept some males and reject others. At high densities this pattern is reversed, with females competing for males and males exhibiting mate choice. In both situations, spermatophores are important to female reproduction, but Gwynne suggests that spermatophores become a more limited resource in high density populations because males have to compete for a limited supply of nutrients with which to make them. This is supported by the observation that a smaller proportion of males in high density populations have the large accessory glands in which spermatophores are produced.

In the amphipod *Gammarus lawrencianus* males grasp females in amplexus and as a result, are unable to use their gnathopods to feed (Robinson and Doyle 1985). Males that spend the majority of their time in amplexus can show as much as 45% less growth than nonamplexed males. Mating success is correlated with body size in this species and therefore, a male who shows a 45% decrease in his size increment from one molt to the next suffers a corresponding reduction in his future reproductive success. Thus males experience a continuous trade-off between current reproductive activity and future growth.

BIRDS

Like insects, many birds produce remarkably loud calls or songs in relation to their body size and there is some evidence that these vocalizations are energetically expensive. Brackenbury (1979) estimated the power output of various birds. Linnets (*Carduelis cannabina*) generate 10 mW/kg and song thrushes (*Turdus ericetorum*) 870 mW/kg. When crowing, domestic cockerels generate 60 mW/kg, a power level 1500 to 2000 times greater than that produced during normal breathing, and air flow through their lungs is twice that observed during the most active sustained exercise (Brackenbury 1980). As in anurans and insects, the efficiency of sound production is very low; only 2% of the power generated by a crowing cockerel is converted into sound energy.

Evidence that singing activity is related to a male's general vigor comes from a recent study by Gottlander (1987) of pied flycatchers (*Ficedula hypoleuca*), in which females show a preference for males with higher song rates. Song rate falls at low temperatures but can be increased experimentally in the field by providing males with extra food. These observations need not necessarily imply a physiological constraint on singing, however. Under conditions of low temperature and reduced food supply, males are likely to have to spend more time foraging and will thus have less time available for singing.

Mating success in many birds is related to their social status which is in turn related to metabolism. Roskaft et al. (1986) have examined the relationship between status and metabolic rate in great tits (*Parus major*) and pied flycatchers (*Ficedula hypoleuca*) and found a positive correlation. Dominant birds have larger hearts, relative to their body weight, than subordinates. In winter flocks of willow tits (*Parus montanus*), metabolic rate and status are also postively correlated and a change in status from low to high rank involves an increase in metabolic rate (Hogstad, personal communication). A higher metabolic rate involves greater energetic requirements, a cost that in dominant birds must be offset against the benefits of high status.

Large epigamic characters, such as elaborate plumage, must also require substantial energetic expenditure, though no estimates or measurements have been made. The very large and elaborate tail of male lyrebirds (*Menura superba*) is not completely developed until a male is seven or eight years old (Smith 1965). This may be because only older males, having completed their body growth, are able to meet the energetic demands of growing and maintaining such a large structure. Alatalo (personal communication) has measured the size of male decorations in a variety of birds and has found that decoration size in some species is more variable among small males than among large males, most of which have large decorations.

This may reflect a greater ability on the part of larger individuals to meet the physiological costs of such characters. The condition of a male's decorations will also be affected by parasites, since a large parasite load will reduce his capacity to invest energy and nutrients in his plumage. This provides the basis for the suggestion that female choice for bright male coloration may select for parasite-free males (Hamilton and Zuk 1982).

MAMMALS

In red deer (*Cervus elephus*) males show a dramatic loss of weight during the rut, with an accompanying reduction in the size of their kidney and rump fat reserves (Clutton-Brock et al. 1982). Some males lose as much as 20% of their body weight. This is largely due to the fact that the proportion of time spent feeding falls from 44% before to 5% during the rut. Males vary in the time for which they hold harems, as well as in the size of their harems, and their active involvement in rutting can stop abruptly, apparently because their condition has reached a critically low point. When two males of nearly equal size confront one another, a roaring contest and a fight are likely to ensue (Clutton-Brock and Albon 1979). The rate at which a stag roars is related to that of his opponent but also shows an upper limit that varies over the course of the rut. Males in good condition, who are to win a fight, roar at a higher rate than those whose condition has declined as a result of intense rutting activity.

Grey seals (*Halichoerus grypus*) leave the water to mate and so cannot feed at all during the mating period. Larger males have higher reproductive success and more sexually active males lose weight at a faster rate (Anderson and Fedak 1985). Because they lose weight, there appears to be a limit on the time that males spend ashore being sexually active, and larger males tend to stay ashore for longer.

There is evidence for a variety of mammals that high social status may incur various physiological and pathological costs. For example, dominant mice show elevated blood pressure and a higher incidence of aortic arteriosclerosis than subordinates (Ely 1981). Such indicators of "stress" may impose a limit on the length of time for which an animal can maintain dominant status.

IS SPERM CHEAP?

A common observation of mating dynamics at the population level is that some males mate more often than others, leading to higher variance in male than in female reproductive success. The basis of this effect is widely accepted to be anisogamy, sperm being much cheaper to produce than eggs. Males will never achieve a reproductive success equivalent to the number of their gametes, as females often will, in part because mating is very costly for males, and in part because sperm is used very inefficiently (Partridge and Halliday 1984). Where sperm competition occurs, males will tend to be selected to produce quantities of sperm greatly in excess of what is required to fertilize a female's eggs (Smith 1984).

There is accumulating evidence that males are limited in their ability to produce sperm repeatedly, suggesting that sperm may not be as energentically cheap to produce as has been widely assumed. Males typically produce sperm in discrete quantities (e.g., ejaculates in mammals and many other animals, spermatophores in many arthropods and urodeles), so that male reproductive success will be limited by the ability to produce ejaculates or their equivalent, not sperm per se (Dewsbury 1982). In newts, the time taken by males to renew their supply of available spermatophores is between one and two days (see above). It is not yet clear whether this is due to a limitation in the rate at which sperm are released from the vasa deferentia or to the secretory capacity of the glands that produce the structural parts of the spermatophore. In those insects whose spermatophores contain nutrients as well as sperm, spermatophore production rate may be limited by male ability to synthesize nutrients. In mammals, males of various species show a limited capacity to produce ejaculates, showing a sharp fall of one or more days after copulating to satiety (e.g., Dewsbury 1983). Evidence of a reduced capacity to fertilize eggs over successive matings has been found in the fish *Hyphenobrycon pulchripinnis* (Nakatsuru and Kramer 1982), the frog *Rana sylvatica* (Smith-Gill and Berven 1980), but not in the toad *Bufo americanus* (Kruse and Mounce 1982). These observations suggest that, whether or not sperm are energetically cheap to produce, there may be constraints on the rate at which males can produce them. Such a constraint may limit their reproductive success.

PHYSIOLOGICAL CONSTRAINT OR LAW OF DIMINISHING RETURNS?

Where it is found that male sexual behavior tends to reach asymptotic levels this may be due to physiological constraints, but there is an alternative explanation that should be considered: it may simply not be profitable for a male to raise his performance above a certain level. The courtship of smooth newts illustrates this effect. In the early phases of courtship, the rate at which a male newt performs one of his displays, fanning, is correlated with the number of spermatophores he has available (Halliday 1976), and because the latter value rarely exceeds three or four, fanning rate appears to have a finite limit. If a male is courting a female in the presence of a rival, however, he fans at a markedly increased rate (Verrell 1984). This suggests that during display a male does not display at a maximum rate but has the energetic scope to increase his mating effort if faced with competition from other males.

During courtship encounters, males rarely deposit more than three spermatophores before breaking off to breathe, suggesting that they have either used up their spermatophore supply or have reached some respiratory constraint. They can often, however, be experimentally induced to deposit one or more further spermatophores after they have breathed. The probabilities that a male will successfully transfer at least one spermatophore when he puts down one, two or three are 11%, 69%, and 80% respectively, suggesting that depositing an extra spermatophore increases his fitness in a decelerating fashion. It is possible to estimate how much a male's success in inseminating a female would be increased if he deposited an extra spermatophore before breathing. If he deposits four rather than three, the probability that at least one will be picked up by the female increases only by about 9%. If, however, he retains the fourth spermatophore and deposits it in courtship with another female, the probability of its being picked up is 11%. It thus appears to be more profitable for him to retain his fourth spermatophore for a second courtship, particularly because there is then a chance of his inseminating two females and, therefore, a much larger number of eggs.

This kind of explanation, invoking the law of diminishing returns, is not mutually exclusive to one based on physiological constraints. Whether or not it is advantageous to a male to increase his sexual performance above a certain level will depend on the magnitude of both costs and benefits in doing so and physiological factors are likely to be significant among the costs.

CONCLUSIONS

Detailed data on the energetic costs of sexual activity, and on the physiological capacity of animals to meet those costs, are few and often incomplete. With a few exceptions, investigators have measured a single variable, such as oxygen consumption or body weight, and have compared values obtained during mating with those found in nonsexual contexts. It is clear that for many animals sexual displays and fighting are energetically very expensive but there is very little direct evidence that animals reach or even approach physiological limits during a single bout of behavior. Where this is such evidence, it relates to fighting, not to courtship. The data do suggest, however, that physiological constraints are important over a long timescale and may well limit the rate or the total number of times in a season that males can engage in intense sexual activity. In several species, males are constrained in their ability to produce sperm repeatedly and, in those in which display rate is correlated with immediate sperm supply, this will limit their sexual performance over a season. Where male performance tends to reach an asymptote, this may imply a physiological constraint, but the alternative hypothesis that greater energy expenditure would not significantly enhance mating success should be considered.

Physiological constraints of the kind discussed in this paper are most relevant to viability-indicator models, such as those developed by Andersson (1982, 1986), Hamilton and Zuk (1982), and West-Eberhard (1979). These models are concerned primarily with morphological characters, such as elaborate plumage. Physiological constraints may influence the size or color of such characters, but their effect is perhaps more likely to be detected in variations in the displays in which they are "presented" to females. To attract a female a male must not only possess elaborate plumage, he must also be able to display it effectively. The data reviewed in this paper suggest that the most important effect of physiological constraints on male performance in many species is on the frequency or rate at which males can perform behavior at intensities that are sufficient to attract females.

Acknowledgements

I am grateful to R. Alatalo, R. Howard, N. Sachser, R. Thornhill, and P. Verrell for their comments, and I thank M. Cherry, C. Gerhardt, and O. Hogstad for allowing me to quote their unpublished work.

REFERENCES

Anderson, S.S., and Fedak, M.A. 1985. Grey seal males: energetic and behavioural links between size and sexual success. Anim. Behav. 33: 829–838.

Andersson, M. 1982. Sexual selection, natural selection and quality advertisement. Biol. J. Linn. 17: 375–393.

Andersson, M. 1986. Evolution of condition-dependent sex ornaments and mating preferences: sexual selection based on viability differences. Evolution 40: 804–816.

Arak, A. 1983a. Male-male competition and mate choice in anuran amphibians. In: Mate Choice, ed. P. Bateson, pp. 181–210. Cambridge: Cambridge Univ. Press.

Arak, A. 1983b. Sexual selection by male-male competition in natterjack toad choruses. Nature 306: 261–262.

Bennett, A.F., and Houck, L.D. 1983. The energetic cost of courtship and aggression in a plethodontid salamander. Ecology 64: 979–983.

Bennett, A.F., and Licht, P. 1974. Anaerobic metabolism during activity in amphibians. Comp. Bio. Phy. A. 48: 319–327.

Boggs, C.L. 1981. Selection pressures affecting male nutrient investment at mating in heliconiine butterflies. Evolution 35: 931–940.

Brackenbury, J. 1979. Power capabilities of the avian sound-producing system. J. Exp. Biol. 78: 163–166.

Brackenbury, J. 1980. Respiration and production of sound by birds. Biol. Rev. 55: 363–378.

Bucher, T.L.; Ryan, M.J.; and Bartholomew, G.A. 1982. Oxygen consumption during resting, calling and nest building in the frog *Physalaemus pustulosus*. Phy. Zool. 55: 10–22.

Burk, T. 1982. Evolutionary significance of predation on sexually signalling males. FLA Entomol. 65: 90–104.

Cade, W. 1979. The evolution of alternative male reproductive strategies in field crickets. In: Sexual Selection and Reproductive Competition in Insects, eds. M.S. Blum and N.A. Blum, pp. 343–379. New York: Academic Press.

Clutton-Brock, T.H., and Albon, S.D. 1979. The roaring of red deer and the evolution of honest advertisement. Behaviour 69: 145–170.

Clutton-Brock, T.H.; Guiness, F.E.; and Albon, S.D. 1982. Red Deer. Behaviour and Ecology of Two Sexes. Chicago: Univ. of Chicago Press.

Dewsbury, D.A. 1982. Ejaculate cost and mate choice. Am. Natural. 119: 601–610.

Dewsbury, D.A. 1983. Recovery from sexual satiety in deer mice (*Peromyscus maniculatus bairdi*). J. Comp. Psych. 97: 34–42.

Doolan, J.M. 1981. Male Spacing and the influence of male courtship behaviour in the bladder cicada, *Cystosoma saundersii* Westwood. Behav. Ecol. Sociobiol. 9: 269–276.

Ely, D.L. 1981. Hypertension, social rank, and aortic arteriosclerosis in CBA/J mice. Physiol. Behav. 26: 655–661.
Fellers, G.M. 1979. Mate selection in the gray treefrog. Copeia 286–290.
Fisher, R.A. 1930. The Genetical Theory of Natural Selection. Oxford: Clarendon Press.
Godwin, G.J., and Roble, S.M. 1983. Mating success in male treefrogs, *Hyla chrysoscelis* (Anura: Hylidae). Herpetolog. 39: 141–146.
Gottlander, K. 1987. Variation in the song rate of the male Pied flycatcher *Ficedula hypoleuca* – causes and consequences. Anim. Behav., in press.
Greer, B.J., and Wells, K.D. 1980. Territorial and reproductive behavior of the tropical American frog *Centrolenella fleischmanni*. Herpetolog. 36: 318–326.
Gwynne, D.T. 1981. Sexual difference theory: mormon crickets show role reversal in mate choice. Science 213: 779–780.
Gwynne, D.T. 1982. Mate selection by female katydids (Orthoptera: Tettigoniidae, *Conocephalus nigropleurum*. Anim. Behav. 30: 734–738.
Gwynne, D.T. 1984. Sexual selection and sexual differences in mormon crickets (Orthoptera: Tettigoniidae, *Anabrus simplex*). Evolution 38: 1011–1022.
Halliday, T.R. 1974. Sexual behaviour of the smooth newt, *Triturus vulgaris* (Urodela: Salamandridae). J. Herpetol. 8: 277–292.
Halliday, T.R. 1976. The libidinous newt. An analysis of variations in the sexual behavior of the male smooth newt, *Triturus vulgaris*. Anim. Behav. 24: 398–414.
Halliday, T.R. 1977a. The courtship of European newts: an evolutionary perspective. In: Reproductive biology of Amphibians, eds. D.H. Taylor and S.I. Guttman, pp. 185–232. New York: Plenum.
Halliday, T.R. 1977b. The effect of experimental manipulation of breathing behaviour on the sexual behaviour of the smooth newt, *Triturus vulgaris*. Anim. Behav. 25: 39–45.
Halliday, T.R. 1978. Sexual selection and mate choice. In: Behavioral Ecology, An Evolutionary Approach, 2nd ed., eds. J.R. Krebs and N.B. Davies, pp. 180–213. Oxford. Blackwell Scientific Publications.
Halliday, T.R., and Houston, A.I. 1978. The newt as an honest salesman. Anim. Behav. 26: 1273–1274.
Halliday, T.R., and Sweatman H.P.A. 1976. To breathe or not to breathe: the newt's problem. Anim. Behav. 24: 551–561.
Halliday, T.R., and Verrell, P.A. 1984. Sperm competition in amphibians. In: Sperm Competition and the Evolution of Animal Mating Systems, ed. R. L. Smith, pp. 487–508. New York: Academic Press.
Hamilton, W.D., and Zuk, M. 1982. Heritable true fitness and bright birds: a role for parasites? Science 218: 384–388.
Howard, R.D. 1984. Alternative mating behaviors of young bullfrogs. Am. Zoolog. 24: 397–406.
Jacobson, S.K. 1985. Reproductive behavior and male mating success in two species of glass frogs (Centrolenidae). Herpetolog. 41; 396–404.
Kluge, A.G. 1981. The life history, social organisation, and parental behavior of *Hyla rosenbergi* Boulenger, a nest-building gladiator frog. Misc. Publ. Museum Zoology, Univ. Michigan 160: 1–70.
Kruse, K.C., and Mounce, M. 1982. The effects of multiple matings on fertilization capability in male American toads (*Bufo americanus*). J. Herpetol. 16: 410–412.
MacNally, R.C. 1981. On the reproductive energetics of chorusing males: energy depletion profiles, restoration and growth in two sympatric species of *Ranidella* (Anura). Decologia 51: 181–188.

MacNally, R.C., and Doolan, J.M. 1982. Comparative reproductive energetics of the sexes in the cicada *Cystosoma saundersii*. Oikos 39: 179–186.
MacNally, R., and Young, D. 1981. Song energetics of the bladder cicada, *Cystosoma saundersii*. J. Exp. Biol. 90: 185–196.
Nakatsuru, K., and Kramer, D.L. 1982. Is sperm cheap? Limited male fertility and female choice in the lemon tetra (Pisces, Characidae). Science 216: 753–755.
Partridge, L., and Halliday, T. 1984. Mating patterns and mate choice. In: Behavioural Ecology, An Evolutionary Approach, 2nd. ed., eds. J.R. Krebs and N.B. Davies, pp. 222–250. Oxford. Blackwell Scientific Publications.
Partridge, L., and Harvey, P.H. 1985. Costs of reproduction. Nature 316: 20.
Reznick, D. 1985. Costs of reproduction: an evaluation of the empirical evidence. Oikos 44: 257–267.
Robertson, J.G.M. 1986a. Female choice, male strategies and the role of vocalizations in the Australian frog *Uperoleia rugosa*. Anim. Behav. 34: 773–784.
Robertson, J.G.M. 1986b. Male territoriality, fighting and assessment of fighting ability in the Australian frog *Uperoleia rugosa*. Anim. Behav. 34: 763–772.
Robinson, B.W., and Doyle, R.W. 1985. Trade-off between male reproduction (amplexus) and growth in the amphipod *Gammarus lawrecnianus*. Biol. Bull. 168: 482–488.
Røskaft, E.; Jarvi, T.; Bakken, M.; Bech, C.; and Reinertsen, R.E. 1986. The relationship between social status and resting metabolic rate in great tits (*Parus major*) and pied flycatchers (*Ficedula hypoleuca*). Anim. Behav. 34: 838–842.
Rutowski, R.L. 1979. The butterfly as an honest salesman. Anim. Behav. 27: 1269–1270.
Ryan, M.J. 1985. The Tungara Frog. A study in Sexual Selection and Communication. Chicago: Univ. of Chicago Press.
Ryan, M.J.; Bartholomew, G.A.; and Rand, A.S. 1983. Energetics of reproduction in a neotropical frog, *Physalaemua pustulosus*. Ecology 64: 1456–1462.
Ryan, M.J.; Tuttle, M.D.; and Rand, A.S. 1982. Bat predation and sexual advertisement in a neotropical anuran. Am. Natural. 119: 136–139.
Smith, L.H. 1965. Changes in the tail feathers of the adolescent lyrebird. Science 147: 510–513.
Smith, R.L., ed. 1984. Sperm Competition and the Evolution of Animal Mating Systems. New York: Academic Press.
Smith-Gill, S.J., and Berven, K.A. 1980. *In vitro* fertilization and assessment of male reproductive potential using mammalian gonadotropin-releasing hormone to induce spermiation in *Rana sylvatica*. Copeia: 723–723.
Sullivan, B.K. 1983. Sexual selection in Woodhouse's toad (*Bufo woodhousei*). II. Female choice. Anim. Behav. 31: 1011–1017.
Sullivan, B.K., and Walsberg, G.E. 1985. Call rate and aerobic capacity in Woodhouse's toad (*Bufo woodhousei*). Herpetolog. 41: 404–407.
Taigen, T.L.; Emerson, S.B.; and Pough, F.H. 1982. Ecological correlates of anuran exercise physiology. Oecologia 52: 49–56.
Taigen, T.L., and Wells, K.D. 1985. Energetics of vocalization by an anuran amphibian (*Hyla versicolor*). J. Comp. Phy. B. 155: 163–170.
Teyssedre, C., and Halliday, T. 1986. Cumulative effect of male's displays in the sexual behaviour of the smooth newt *Triturus vulgaris* (Urodela, Salamandridae). Ethology 71: 89–102.
Verrell, P.A. 1982. The sexual behaviour of the red-spotted newt, *Notophthalmus viridescens* (Amphibia: Urodela: Salamandridae). Anim. Behav. 30: 1224–1236.
Verrell, P.A. 1983. The influence of ambient sex ratio and intermale competition on the

sexual behaviour of the red-spotted newt, *Notophthalmus viridescens*. Behav. Ecol. Sociobiol. 13: 307–313.

Verrell, P.A. 1984. Sexual interference and sexual defense in the smooth newt, *Triturus vulgaris* (Amphibia: Urodela: Salamandridae). Z. Tierpsych. 66: 242–254.

Verrell, P.A. 1985. Is there an energetic cost to sex? Activity, courtship mode and breathing in the red-spotted newt, *Notophthalmus viridescens* (Rafinesque). Monit. Zoo. Ital. 19: 121–127.

Verrell, P.A. 1987. Limited male mating capacity in the smooth newt, *Triturus vulgaris vulgaris* (Amphibia). J. Comp. Psych. 100: 291–295.

Verrell, P.A., and Francillon, H. 1987. Body size, age and reproduction in the smooth newt, *Triturus vulgaris*. J. Zool. Lond. (A) 210: 89–100.

Verrell, P., and Halliday, T. 1985. Reproductive dynamics of a population of smooth newts, *Triturus vulgaris*, in southern England. Herpetolog. 41: 386–395.

Verrell, P.A.; Halliday, T.R.; and Griffiths, M.L. 1987. The annual cycle of the smooth newt. (*Triturus vulgaris*) in England. J. Zool. Lond. (A) 210: 101–119.

Wells, K.D. 1977. Territoriality and male mating success in the green frog (*Rana clamitans*). Ecology 58: 750–762.

Wells, K.D. 1978. Territoriality in the green frog (*Rana clamitans*): vocalizations and agonistic behaviour. Anim. Behav. 26: 1051–1063.

Wells, K.D., and Taigen, T.L. 1984. Reproductive behavior and aerobic capacities of male American toads (*Bufo americanus*): is behavior constrained by physiology? Herpetolog. 40: 292– 298.

Wells, K.D., and Taigen, T.L. 1986. The effect of social interactions on calling energetics in the gray treefrog (*Hyla versicolor*). Behav. Ecol. Sociobiol. 19: 9–18.

West-Eberhard, M.J. 1979. Sexual selection, social competition and evolution. Proc. Amer. Phil. Soc. 123: 222–234.

Whitney, C.L., and Krebs, J.R. 1975. Mate selection in Pacific tree frogs. Nature 255: 325–326.

Williams, G.C. 1975. Sex and Evolution. Princeton: Princeton Univ. Press.

Woodward, B. 1982. Male persistence and mating success in Woodhouse's toad (*Bufo woodhousei*). Ecology 63: 583–585.

Sexual Selection: Testing the Alternatives
eds. J.W. Bradbury and M.B. Andersson, pp. 265–277
John Wiley & Sons Limited

Life History Constraints on Sexual Selection

L. Partridge* and J.A. Endler**

*Department of Zoology,
Edinburgh University
Edinburgh EH9 3JT, Scotland
**Department of Biological Sciences
University of California,
Santa Barbara, CA 93106, USA

Abstract. We consider how life history variables may constrain the extent of sexual selection and the form of sexually selected traits. The demography of the sexes, the duration of the breeding season, and the degree of synchrony in breeding between the sexes are all likely to affect the intensity of sexual selection through their effects on the operational sex ratio. Predation may limit the scope for sexual selection through female choice and will also have a major effect on the form of sexually selected traits in males. The evolutionary advance in sexually selected traits may often be brought to a halt by costs of the traits in terms of an increase in juvenile and adult mortality or a reduction in adult fertility or growth. Depending upon the shape of the relationship between current reproductive success and future reproductive success, semelparity or iteroparity may evolve, and this in turn will affect the intensity of sexual selection. There is a serious need for empirical studies attempting to examine the operation and interaction of these constraints on sexual selection, because at present hard data are virtually non-existent.

INTRODUCTION

How do life history variables such as patterns of mortality and schedule of reproduction constrain the extent and nature of evolution by sexual selection? By constraints we mean not only factors which prevent or limit a response to sexual selection, but also infleunces on the nature of the traits affected. Many studies of single species have examined how variation in individual mating success is brought about. Male rivalry and female choice have both at times been implicated, and a variety of phenotypic characters have been shown to correlate with mating success in different species (Partridge and Halliday 1984). Sometimes comparative analysis of related species has revealed correlations between mating system and degree of sexual dimorphism in, for instance, body size (e.g., Selander 1982. Alexander et al. 1979), with the strong implication that sexual selection on males is responsible for the differences. In other cases simple correlations are much less obvious. It is far from clear, for instance, why particular anuran species are highly dimorphic in body size or why some fish species have extremely ornamented males when relatives with apparently similar mating systems do not. Our aim is to explore ways in which demographic characteristics might account for some of the observed variation. How do demographic patterns affect the opportunity for sexual selection and mate choice? How do patterns of growth and mortality affect the nature of sexually selected characters? How do costs of sexually selected characters constrain their evolution? We will attempt to identify some areas of interest and the kinds of data required to study them.

There are undoubtedly some purely ecological constraints on sexual selection, and many of these could be identified by using the comparative method (Clutton-Brock and Harvey 1984). For example, feeding opportunities affect patterns of dispersion, and hence the scope for monopoly of more than one member of the other sex (Vehrencamp and Bradbury 1984). However, in the present state of knowledge, broad ecological correlations with opportunity for sexual selection and the identity of the characters affected are unlikely to be generally forthcoming, and much more detailed understanding of life histories and demography of these sexes will be required before informative comparisons can be made. Cause and effect may be very difficult to separate in comparative studies of this kind because variation in mating success and sexual selection can be both a cause and a consequence of life history variation. Therefore we will confine our discussion to instances where it seems reasonable to assume that the life history pattern has evolved primarily in direct response to specific ecological factors.

Much of our material is speculative. This results partly from the rapidity with which the theory of sexual selection is developing and also from a paucity of relevant data. For instance, the sexual selection process and life history constraints on female choice may operate very differently if females are choosing male traits currently associated with fitness rather than choosing some arbitrary criterion which may have been initially established by genetic drift (Kirkpatrick 1985, 1986, this volume). Our arguments also have implications for the biological realism of the underlying assumptions of the models. For instance, selection for rapid commencement of breeding may make female choice more costly (Atatalo et al. 1981; Parker 1983), which would in turn make it much less likely to be subject to genetic drift. We also largely lack data suitable for use in comparative studies. It is unusual to find detailed studies on either life history patterns or causes of variation in mating success, and there are virtually no species in which both have been adequately studied.

Life history patterns are likely to affect sexual selection in at least three ways. First, they may infleunce the opportunity for sexual selection by affecting operational sex ratio, length of breeding season, population density, and opportunity for mate choice. Second, the identity of traits affected by sexual selection is likely to be determined partially by sources of mortality and patterns of development. Third, the life history pattern and the magnitude and timing of the costs incurred by sexually selected characters will be responsible for bringing their evolution to a halt and determining their final form.

LIFE HISTORIES AND THE OPPORTUNITY FOR SEXUAL SELECTION

Sexual selection proceeds most rapidly if there is a strong relationship between the trait value (or character state) and mating success. The extent of genetic variance for the trait will affect the rapidity of the response to phenotypic sexual selection but not the equilibrium value of the character, unless there is a perfect genetic correlation with other traits subject to sexual or natural selection. For the purposes of this paper, we will be primarily concerned with equilibrium or near-equilibrium conditions, so we will ignore the obvious constraints arising from low heritabilities. Because there is virtually nothing known at present about genetic covariances between traits, we will also ignore the covariance problem. Ultimately, genetic variances and covariances will have to be considered.

It is currently popular to estimate variance in lifetime mating success and then assume that this is a good estimate of the selection differential under sexual selection. However, this variance measures only the opportunity for sexual selection, and there is no good reason to assume that the actual rate of sexual selection should always equal or differ by the same amount from the maximum potential sexual selection (Grafen 1986; Endler 1986b). While it is true that the intensity of sexual selection cannot be inferred from variance in lifetime mating success, it is also true that without this variance there would be no opportunity for sexual selection. Ecological and life history characteristics that constrain variance in mating success are therefore of some interest to us, even if the actual value of the variance is not.

Life history patterns could affect the opportunity for sexual selection by altering the operational sex ratio (OSR), and hence the extent to which mating success can vary. This can result from demographic differences between the sexes and from variation in the length of the breeding season. Whether or not the OSR will affect the intensity of sexual selection in a predictable way will depend upon the ways in which individual mating success is determined. For instance, a female-biased OSR could either increase variance in male mating success, if one or a few males remain very successful in monopolizing mates, or decrease it if there a realized upper limit on the number of females a male can mate.

There are few cases where demographic differences between the sexes have been thoroughly investigated, and this is one of the areas where sexual selection can be both cause and effect of life history variation. For instance, juvenile mortality is fequently male-biased, resulting in female-biased sex ratios at breeding, and this mortality difference is probably a consequence of sexual selection on males (see section on *LIFE HISTORIES AND THE COSTS OF SEXUALLY SELECTED TRAITS*). In many species, males commence breeding at a later age than females, and this can further bias the OSR if it results in higher male mortality before breeding. Sex differences in age at first reproduction can be a consequence of sexual selection, especially where male sexually selected traits such as large body size or fighting ability require time to develop. Demographic differences between the sexes may be at least partly unrelated to the effects of sexual selection on males in cases where females mature later than males, for example in some fish and amphibians (Warner and Harlan 1982). Here, selection for increased body size occurs in females because of an associated increase in fecundity, and selection is less intense, or there is only a correlated response, in males. Because the commencement of breeding by both sexes is often accompanied by a marked drop in growth rate, there is hence selection for delayed maturation in females relative to males. When females mature later than males, higher female mortality before maturation can lead to a male-biased OSR and the potential for sexual selection will then be high. The latter occurs in the common toad *Bufo bufo*: males breed in their first year and females in their second and the mortality rate for both sexes is about 0.5 per year. The two factors together result in an OSR of about 3 males to 1 female at the breeding ponds, and hence a higher potenital for sexual selection than would be implied by an equal sex ratio and explosive breeding (Arak, personal communication). This type of phenomenon is only likely to be detected where longitudinal studies on individuals marked at birth can be made.

Several comparative studies have suggested that the length of the breeding season may also affect the OSR and hence the opportunity for sexual selection. The length of the breeding season is usually assumed to be determined by purely ecological factors such as the suitability of breeding sites or the availability of food for the young. The basic idea was first produced by Emlen and Oring (1977) who suggested that explosive breeders with their very short breeding seasons have a low opportunity for obtaining more than one mating. As the breeding season lengthens, this opportunity increases because at least some receptive females are available for mating over a much longer period than in explosive breeders. Depending upon the spatial distribution of females, sexual selection may then favor the evolution of various morphological and behavioral characteristics in males (Arak 1983).

In frogs and toads the varying opportunity for sexual selection associated with length of breeding period may partially account for differing degrees of size dimorphism (Woolbright 1983; Sullivan 1984; Woolbright 1985). The ratio of male to female size is greatest in species with an intermediate value of breeding period expressed as a proportion of the time available for growth. The ratio declines in species with shorter breeding periods, in accordance with the idea that this leads to less intense sexual selection. The reason for the decline in species with very extended breeding periods is less clear and may be associated with energetic constraints on the male such that very large males cannot replace losses incurred during the breeding season in the ensuing growth period (Woolbright 1985). In addition, a longer time to find mates may mean that there is enough time for all adults to obtain some matings, and energetic costs to males of attendance at breeding sites for long periods may mean that not all males are present at all times (Arak 1983). Variation in the degree of sexual dimorphism could reflect differencs in selection on females as well as on males, so that it may rarely be a reliable guide to the intensity of sexual selection. Thus in anurans with very extended breeding periods, female clutch size may be larger and hence natural selection on female body size more intense. There is an obvious need for empirical work examining the effects of length of breeding season on both the intensity of sexual selection and fertility selection on females.

The length of the breeding season may not always be a good predictor or estimator of the opportunity for sexual selection because the degree of synchrony of arrival of males and females at the breeding site may vary among species with the same breeding season length. For a given season length, if both sexes arrive simultaneously, then the OSR will be close to unity (if there is no sex difference in mortality or age at maturity). On the other hand, if males arrive first, or over a shorter time interval than females, then the OSR will be more male-biased in proportion to the degree of asynchrony between the sexes. Bird and cichlid fish leks are the most extreme examples, and the breeding congregations of many frogs are more usual examples of asynchrony. Most are associated with strong sexual dimorphism and probably with strong sexual selection. As breeding season length declines, the scope for asynchrony in arrival times declines, so it is not surprising that size of dimorphism is reduced in species with short breeding seasons. As breeding season length increases, however, the potential for asymchrony increases but it can take any of an increasing range of values.

The degree of breeding asynchrony and breeding season length may affect the mode of sexual selection. For instance, male-male interaction may be favored over female choice in species with a high degree of synchrony between the sexes and a short breeding season, whereas female choice may be favored in species with a long breeding season and asynchrony between the sexes.

Mortality from predation may affect the operation of sexual selection in several ways. If predation risk makes it dangerous to have an involved courtship behavioral sequence, then there may be less opportunity for female choice of males and hence weaker sexual selection. On the other hand, the same conditions may favor sexual selection through inter-male interaction, especially if the traits thereby selected can serve in predator defense (Arak 1982). Unfortunately too little is known about predation rates in natural populations to test if this prediction is correct. Predation can also affect the rate of sexual selection indirectly through population density. For example, in guppies (*Poecilia reticulata*) and other fishes, population densities are lower in areas with more predators and more dangerous predators (Endler 1983). In areas of higher population densities females can exercise finer discrimination in mate choice among a larger number of males than in areas of low density, so sexual selection by female choice may be stronger at higher densities (Endler 1986b).

Predation may also shape sexual selection involving male-male interactions. For example, mate guarding and long copulation time would be selected against in species or populations subject to high predation but may be favored where there is little predation. Long copulation time may also be necessary for transfer of nutrients during mating, and this may therefore also be selected against in areas of high predation.

It is clear that detailed studies of the demography of individuals, the timing and synchrony of the breeding season, and risks associated with courtship and mating are required to evalutate any predictions about their effects on the rate of sexual selection. Studies of the effects of manipulation of the OSR and population density on male and female mating behavior would be especially valuable in this respect, as would studies of the risks associated with various stages of sexual behavior in nature.

LIFE HISTORIES AND THE FORM OR DIRECTION OF SEXUAL SELECTION

Even though there may be lines of equilibrium in the dynamics of sexual selection, which can allow random factors to determine its direction (Lande 1981; Kirkpatrick 1982), ecological factors may push the outcome of sexual selection in certain directions or constrain it against evolving in others. For example, in species where juveniles require rapid growth and very high food intake, breeding can only take place at certain times of year. If, as is frequently the case, this time corresponds to times of maximum predation risk, then predation may set an upper limit to the evolutionary development of conspicuous male traits and behavior if predation risk is related to conspicuousness of display. Predation can set an upper limit to the development of secondary male traits even in species which breed continuously, such as Poeciliid fishes (Endler 1983). On the other hand, if a species can breed much earlier (or later) than when most predators are active, conspicuousness associated with courtship, which may be associated with mating success, can develop to a much greater degree. The same is true for time of day. If courtship activities can take place at hours when predation is minimal, then courtship behavior and display morphology can be much more conspicuous than if used during maximum predation risk. For example, guppies court maximally during early morning and late afternoon (intermediate to low light intensity), and this is when they display their conspicuous colors most vigorously, but maximal predation takes place at midday (other fish) and at night (fish, snakes, and crustacea). There is clearly a complex trade-off between time and energy spent of food finding, mate finding, and predator avoidance but this has not been investigated in detail in any species.

The life history consequences of predation and mating times (hour and season) have profound implications for the kinds of secondary sexual traits which are likely to evolve in any species. Particularly if females are at higher risk of predation during courtship and mating, mating preferences for male phenotypes that minimize this risk will be favored be selection. If courtship timing evolves to minimize predation (and still maximise fecundity), and this means courting at night or in conditions of poor visibility (turbid streams or dark forests), then such species are likely to develop sexually selected traits and behavior which work best in sensory modes other than vision (olfaction, chemoreception, tactile devices, electroreception, etc). Striking visual displays may evolve when there are few predators, when there is already effective antipredator defense, or in species which court in visually poor conditions where mates can be seen at short distances but predators cannot see them from longer distances. The actual sensory modes evolved, and the signalling devices using those sensory modes, will depend upon the sensory conditions during courtship and the phylogenetic constraints to evolving the modes.

Many species have evolved private channels of communication during courtship. This has usually resulted in a race between the sensory development of the predator and display traits in the prey, similar to that between models and mimics (Endler 1986a). This race for visibility to mates but not to predators, while predators are improving their sensory capabilities, is likely to constrain and direct the way evolution by sexual selection proceeds. It could go in many directions, but the initial direction is probably set by the sensory capabilities of prey and predators and the sensory conditions during courtship. Once again the sensory conditions during courtship may be set by the life history constraints on the timing of breeding. It is difficult to disentangle cause and effect in all these factors. For example, is a new courtship timing a result of feeding and predation constraints or of some individuals being more successful (more conspicuous to mates) at these new times? The interrelationships among these factors are virtually unknown and well worth detailed investigation

Pleiotropic effects, developmental interactions, or other effects yielding correlated response to selection may limit or influence the direction of sexual selection. For example, in many marine and freshwater fishes there is a positive correlation between age at maturity, body height, and body size. If sexual selection favored increased body height (larger more obvious visual or lateral line displays), then this would result in delayed maturity, which might be selected against by ecological factors limiting the progress of sexual selection. On the other hand if there were no sexual selection for body height, direct selection for age at maturity would result in a correlated response in body height, yielding false evidence for sexual selection. In addition, the correlated response in body could initiate a Fisher runaway process for body height, in the direction biassed by the correlated response. Such correlational selection effects need to be investigated.

LIFE HISTORIES AND THE COSTS OF SEXUALLY SELECTED TRAITS

A feature held in common by many models of sexual selections is the achievement of evolutionary equilibrium by a balance between sexual and natural selection (in the narrow sense of components of fitness other than those associated with mating). The balance may be stable or unstable, and the position of the equilibrium may be a point (one value) or a line (many values with movement between them affected by random factors) (Lande 1981; Kirkpatrick 1982; Seger 1985). A change in sexually selected trait away from equilibrium may increase mating success at the expense of other aspects of fitness of vice versa. Thus, at equilibrium, there are likely to be substantial costs to producing sexually selected traits. There have been few attempts to examine these costs. In life history terms, juvenile and adult mortality and adult fertility and growth rates could all be affected. How are different life history patterns likely to shape these costs and constrain the action of sexual selection?

There may be physiological or development costs associated with the growth of secondary sexual traits such as size, bright coloration, or weaponry. These may affect juvenile survivorship. For example, in many species large body size is associated with a mating advantage, but this can only be achieved by extra growth early in life because of a higher growth rate, an extended growth period, or both. Any means of achieving larger size may incur a higher mortality, and indeed across species of mammals and birds the degree of male bias in juvenile mortality is proportional to the extent of adult size dimorphism (Clutton-Brock et al. 1985). In red deer (*Cervus elaphus*) this bias increases with population density (Clutton-Brock et al. 1982). Males have higher growth and energy requirements and are therefore more susceptible to starvation during periods of food shortage. An extended growth period may also result in increased juvenile mortality. In *Dropsophila melanogaster* large males have higher mating success (Partridge and Farquhar 1983; Partridge et al. 1986). Artificial selection for large body size in this species results in an increased larval period (Robertson 1960), no change in growth rate (Robertson 1963), and increased larval mortality under some conditions (Partridge and Fowler, unpublished observations). Here the breeding site deteriorates as larval development proceeds, so that larvae which require more time to develop cannot pupate before the site becomes uninhabitable. Under natural conditions the cost of an extended growth period may well be greatly magnified. In these cases, the increase in the adult male trait due to sexual selection will eventually be balanced by the correlated increase in juvenile mortality.

Costs of sexually selected characters may be confined to the juvenile stage in some species, but this is unlikely to be the general pattern. It is extremely unlikely that any trait which increases mating success incurs no other costs in adults. In adults, costs could take the form of increased risk from predators, competitiors and parasites, or a reduction in growth in species with indeterminate growth. For example, the bright breeding colors of several fish species are known to increase mating success, but also increase predation, so there is an association in the field between low predation intensity and bright coloration (Endler 1983). In one of these species (guppies), the coloration is expressed only in adult males, so effects on the juvenile stage have been eliminated. High mating success may be associated with growth costs to the adult, as in the amphipod *Gammarus lawrencianus*, which has a precopulatory amplexus during which males do not feed (Robinson and Doyle 1985; discussed by Halliday, this volume). The optimal duration of amplexus depends on the sex ratio, the mortality rate, and the male and female size distributions. The growth cost may well reduce the optimal duration of amplexus, especially for smaller males, further affecting the OSR and the strength of sexual selection.

In addition to the direct energetic costs of sexual displays, or of mating itself, there can be indirect costs. For example, in species (such as shrews) which need to forage for a large fraction of their lives, any time used in courtship will lead to greatly increased mortality or increase in the probability of starvation. It would be interesting to know if sexually selected traits are less common or less well developed in energetically marginal species compared to species whose time/energy budgets are not so tight.

Current reproductive effort (Gadgil and Bossert 1970) can have the effect of reducing future reproductive potential through effects on either mortality or future fertility. Under some circumstances it may pay to allocate less than the maximum possible to reproduction if the current progeny lost by the reduction below the maximum is more than compensated for by the expected increase in future offspring thereby achieved (Charlesworth 1980). Depending upon the relationship between the level of current reproductive effort and expected future reproductive success, different species have evolved life history patterns varying from purely semelparous to purely iteroparous (Charlesworth 1980). Costs of sexually selected traits undoubtedly contribute to this trade-off.

There have been very few attempts to examine the cost curves for any aspect of reproductive effort and we know of none for sexually selected characters. This may be in part because of the difficulty of acquiring the right kind of data. Naturally occurring variation in sexually selected traits, mating success, survival, and growth rates are useless in this context because other uncontrolled variables, such as differences in condition between individuals, tend to affect all aspects of fitness in the same direction simultaneously (Partridge and Harvey 1985; Reznick 1985). To examine this problem, experimental manipulation of sexually selected traits is required, but even this has major difficulties. Consider, for instance, the bright coloration of male guppies. This may make nutritional demands on the male, especially for the carotenoid-based colors (red, orange, yellow), and attracts predators as well as mates. Painting spots on guppies might produce a satisfactory account of the relationship between spot number and size, mating success and predation rate, but not of the loss of time or body condition normally incurred in pigment production. For this a genetic manipulation under natural conditions would probably be necessary, but the problem is to produce morphs differing only in respect of the trait of interest (e.g., Rose 1984). A combination of genetic and phenotypic perturbations is probably the best approach.

As well as being involved in causing life history variation, the evolution of sexually selected traits is probably influenced by broad life history patterns. Since sexually selected traits are undoubtedly costly to produce, they must be regarded as a form of reproductive effort. A similar argument applies to female choice; if choice is costly because it lowers food intake, puts the female at risk, or in any other way reduces future reproductive potential, then it must also be regarded as a form of reproductive effort. Life history suggests that reproductive effort is expected to be lowest in highly iteroparous species and highest in semelparous forms (Charlesworth 1980).

Despite our lack of hard data, there are some indications of differences in sexually selected traits between semelparous and iteroparous species. For instance, Pacific salmon spawn only once, with heroic reproductive effort: male's mouths become shaped into a hook that is excellent for grappling and fighting but useless for feeding, and digestive system atrophies, and all energy is devoted to acquiring mates and spawning. A rather similar situation is found in the gladiator frog *Hyla rosenbergi* (Kluge 1981). Amphibians are not noted for their pugnacity, but in this species males indulge in escalated and dangerous fighting, slashing at one another with their spines, particularly at the eyes and ear drums of opponents. The injuries received in fights suggest that death as a direct or indirect result of fights may be common. Interestingly, this gladiator frog seems to be semelparous. Williams (1966) describes an interesting difference between annual and perennial fishes in tropical America. The annual species grow and reproduce in the wet season in temporary ponds and survive the dry season as drought-resistant eggs. Other species live in more permanent water and do not have desiccation-resistant eggs. The males of semelparous species, i.e., the killifish Genera *Aphyosemion, Aplocheilus, Notobranchius, Ptrolebiras*, and *Rollofia*, are extremely conspiciously colored, have long fancy fins (Van Ramshort and Van Den Nieuwenhuizen 1978), and are very aggressive towards other males, while the sexual dimorphism and aggression of the

"Oh, that's right! You *did* have a hat. . . .
"

The Far Side®

LAST IMPRESSIONS

—— 2002 ——

April

Sunday 14

iteroparous species is much reduced or nonexistent. The dangers of assuming that the degree of sexual dimorphism is a reliable guide to the intensity of sexual selection have been pointed out, but the data do suggest an association of semelparity with brightly colored and aggressive males, which requires further investigation. Lastly, some of the most dramatic cases of male weaponry and lethal combat in insects are found in wingless male fig wasps, which are semelparous with mating opportunities confined to the females emerging in the same fig. Winged males of the same species which disperse from their natal fig and can mate sequentially with females from several figs do not have this weaponry and do not fight (Hamilton 1979).

In one instance some quantitative data exist which are consistent with the idea of negative relationship between the intensity of sexual selection and degree of iteroparity. Among dasyurid marsupials, males become relatively larger than females across species as iteroparity declines, with the extreme semelparous species such as *Antechinus stuartii* the most extreme (Lee and Cockburn 1985). Other comparative data of this kind are highly desirable.

There is a difficulty in disentangling evolutionary cause and effect in the apparent semelparous-iteroparous pattern of sexual selection, especially for the cyprinodontid fishes. If reproductive conditions deteriorate rapidly enough to favor semelparity, as in the annual killifish, then this puts a premium on rapid growth and general pugnaciousness to maintain a territory or other means of ensuring successful competition with conspecifics for food and correct physiological conditions. This would predispose the species to strong sexual selection independent of the costs of growth and reproduction arguments just given. On the other hand, if the relative costs of growth and reproduction initially favored high reproductive effort and consequent strong sexual selection, then this may secondarily aid survival in deteriorating habitats, since sexually selected males may have a *higher* chance of surviving since they can defend larger food resources. A very large amount of work must be done on costs of reproduction and sexually selected traits, both empirically and theoretically.

CONCLUSIONS

It is clear that we know almost nothing about the factors which constrain and direct the process of sexual selection. It is also clear that it will be difficult to disentangle the causes of these constraints because they interact in complex but interesting ways and because detailed data on breeding biology and ecology of species are very difficult and time-consuming to collect. Unless we make a serious effort to find out these constraints, our understanding of sexual selection, and natural selection in general, will be very incomplete if not misleading.

REFERENCES

Alatalo, R.V.; Carlson, A.; Lundberg, A.; and Ulfstrand, S. 1981. The conflict between male polygamy and female monogamy: the case of the pied flycatcher *Ficedula hypoleuca*. Am Natural. 117: 738–753.

Alexander, R.D.; Hoogland, J.L.; Howard, R.D.; Noonan, K.M.; and Sherman, P.W. 1979. Sexual dimorphisms and breeding systems in pinnipeds, ungulates, primates and humans. In: Evolutionary Biology and Human Social Behavior, eds. N.A. Chagnon and W. Irons. North Scitate, MA: Duxbury Press.

Arak, A. 1982. Male-male competition and mate choice in frogs and toads. Unpublished Ph. D. Thesis. Cambridge: Univ. of Cambridge.

Arak, A. 1983. Male-male competition and mate choice in anuran amphibians. In: Mate Choice, ed. P. Bateson, pp. 181–210. Cambridge: Cambridge Univ. Press.

Charlesworth, B. 1980. Evolution in Age-Structured Populations. Cambridge: Cambridge Univ. Press.

Clutton-Brock, T.H.; Albon, S.D.; and Guiness, F.E. 1985. Parental investment and sex differences in juvenile mortality in birds and mammals. Nature 313: 131–133.

Clutton-Brock, T.H.; Guiness, F.E.; and Albon, S.D. 1982. Red Deer: Behaviour and Ecology of Two Sexes. Edinburgh: Edinburgh Univ. Press.

Clutton-Brock, T.H., and Harvey, P.H. 1984. Comparative approach to investigating adaptation. In: Behavioral Ecology: An Evolutionary Approach, 2nd eds., eds. J.R. Krebs and N.B. Davies, pp. 7–29. Sunderland, MA: Sinasuer Associates.

Emlen, S.T., and Oring, L.W. 1977. Ecology, sexual selection, and the evolution of mating systems. Science 197: 215–223.

Endler, J.A. 1983. Natural and sexual selection on color patterns in Poeciliid fishes. Env. Biol. Fish. 9:173–190.

Endler, J.A. 1986a. Defense against predation. In: Predator-Prey Relationships, Perspectives and Approaches from the Study of Lower Vertebrates, eds. M.E. Feder and G.V. Lauder, pp. 109–134.

Endler, J.A. 1986b. Natural Selection in the Wild. Princeton: Princeton Univ. Press.

Gadgil, M., and Bossert, W.H. 1970. Life historical consequences of natural selection. Am. Natural. 104. 1–24.

Grafen, A. 1986. On the uses of data on lifetime reproductive success. In: Lifetime Reproductive Success, ed. T.H. Clutton-Brock. Chicago: Chicago Univ. Press.

Hamilton, W.D. 1979. Wingless and fighting males in fig wasps and other insects. In: Sexual Selection and Reproductive Competition in Insects, eds. M.S. Blum and N.A. Blum, pp. 167–220. New York: Academic Press.

Kirkpatrick, M. 1982. Sexual selection and the evolution of female choice. Evolution 36: 1–12.

Kirkparick, M. 1985. Evolution of female choice and male parental investment in polygynous species: the demise of the "sexy son". Am. Natural. 125: 788–810.

Kirkpatrick, M. 1986. The handicap mechanism of sexual selection does not work. Am. Natural. 127: 222–240.

Kluge, A.G. 1981. The life history, social organisation, and parental behavior of *Hyla rosenbergi* Boulenger, a nest-building gladiator frog. Misc. Publ. Museum Zoology, Univ. of Michigan 160: 1–170.

Lande, R. 1981. Rapid speciation by sexual selection of polygenic characters. Proc. Natl. Acad. Sci. USA 78: 3721–3725.

Lee, A.K., and Cockburn, A. 1985. Evolutionary Ecology of Marsupials. Cambridge: Cambridge Univ. Press.

Parker, G.A. 1983. Mate quality and mating decisions. In: Mate Choice, ed. P. Bateson, pp. 141–166. Cambridge: Cambridge Univ. press.

Partridge, L., and Farquhar, M. 1983. Lifetime mating success of male fruitflies is related to their size. Anim. Behav. 31: 871–877.

Partridge, L., and Halliday, T. 1984. Mating patterns and mate choice: In: Behavioral Ecology: an Evolutionary Approach, 2nd ed., eds. J.R. Krebs and N.B. Davies, pp. 22–250. Oxford: Blackwell.

Partridge, L., and Harvey, P.H. 1985. Evolutionary biology: costs of reproduction. Nature 316: 20.

Partridge, L.; Hoffmann, A.; and Jones, J.S. 1987. Male size and mating success in *Drosophila melanogaster* and *D. pseudoobscura* under field conditions. Anim. Behav. 35: 468–476.

Reznick, D. 1985. Costs of reproduction: an evaluation of the emprical evidence. Oikos 219: 257–267.

Robertson, F.W. 1960. The ecological genetics of growth in *Drosophila*. 1. Body size and development time on different diets. Genet. Res. Camb. 1: 288–304.

Robertson, F.W. 1963. The ecological gentics of growth in *Drosophila*. 6. The genetic correlation between the duration of the larval period and body size in relation to larval diet. Genet. Res. Camb. 4: 74–92.

Robinson, B.W., and Doyle, R.W. 1985. Trade-off between male reproduction (amplexus) and growth in the amphipod *Gammarus lawrencianus*. Biol. Bull. 168: 482–488.

Rose, M.R. 1984. Genetic covariation in *Drosophila* life history: untangling the data. Am. Natural. 123: 565–569.

Seger, J. 1985. Unifying models for the evolution of female choice. Evolution 39: 1185–1193.

Selander, R.K. 1982. Sexual selection and dimorphims in birds. In: Sexual selection and the descent of man 1871–1971, ed. B. Campbell. Chicago: Aldine.

Sullivan, B.K. 1984. Size dimorphism in anurans: a comment. Am. Natural. 123: 721–724.

Van Ramshorst, J.D., and Van Den Nieuwenhuisen, A. 1978. Aquarium Encyclopedia of Tropical Freshwater Fish. Laussane: HP Books/Elsevier.

Vehrencamp, S.L., and Bradbury, J.W. 1984. Mating systems and ecology. In: Behavioural Ecology: An Evolutionary Approach, 2nd ed., eds. J.R. Krebs and N.B. Davies, pp. 251–278. Oxford: Blackwell.

Warner, R.R., and Harlan, R.K. 1982. Sperm competition and sperm storage as determinants of sexual dimorphism in the dwarf surfperch, *Micrometrus minimus*. Evolution 36: 44–55.

Williams, G.C. 1966. Adaptation and Natural Selection. Princeton: Princeton Univ. Press.

Woolbright, L.L. 1983. Sexual selection and size dimorphism in anuran amphiabia. Am. Natural. 121: 110–119.

Woolbright, L.L. 1985. Anuran size dimorphism: reply to Sullivan. Am. Natural. 125: 741–743.

Standing, left to right:
Manfred Milinski, John Endler, Klaus Peschke, Brian Charlesworth, Jakob Parzefall

Seated (center), left to right:
Bob Warner, Trevor Price, Linda Partridge, Bill Hamilton

Seated (front), left to right:
Rauno Alatalo, Tim Halliday, Klaus Heller

Sexual Selection: Testing the Alternatives
eds. J.W. Bradbury and M.B. Andersson, pp. 278–294
John Wiley & Sons Limited

Constraints on the Effects of Sexual Selection
Group Report

T.D. Price, Rapporteur
R.V. Alatalo
B. Charlesworth
J.A. Endler
T.R. Halliday
W.D. Hamilton
K.-G. Heller
M. Milinski
L. Partridge
J. Parzefall
K. Peschke
R. Warner

Abstract. We examine constraints on the intensity and directions of selection on sexually selected traits and also consider how the genetic basis of such traits might limit rates of evolution. We focus on topics illustrating general points that are controversial or in need of further study: a) costs of sexually selected traits, b) constraints imposed by the breeding system, in particular monogamy, c) costs to female choice, d) constraints on the male signal and female receptor system, e) directionality imposed by selection to avoid hybridization, f) parasite models of sexual selection, and g) genetic constraints on the evolution of sexual dimorphism and sexually selected characters in general. Finally, we present some ideas for future research.

INTRODUCTION

In this chapter we examine what constrains the form of a sexually selected character and what limits its exaggeration. To do this we consider selective constraints (i.e., intensities and directions of sexual and natural selection on the character), and genetic constraints (which limit the magnitude and direction of response to any selection).

Constraints on the effects of sexual selection constitute an enormous topic. Specific aspects have been considered in the background papers by Halliday, Lande, and Partridge and Endler (this volume), and the topic has been covered more generally by Kirkpatrick (this volume). Here we focus on aspects which appear to us to be worthy of further research, either because they have not been recognized as important or because they are currently controversial. We begin with a general framework for discussion, for convenience describing the male sex as the sexually selected sex throughout the paper.

Constraints on sexual selection

If a sexually selected character is at an evolutionary equilibrium, the direct forces of natural and sexual selection on the character (and/or subcomponents of these two broad categories) must balance, such that net directional selection on the character is zero. Therefore, in order to understand the limits to exaggeration of a sexually selected character, under the assumption of equilibrium, we need only to investigate the reasons for limits to intensities of natural and sexual selection. Under nonequilibrium conditions we are concerned with constraints on the rate and direction of evolution of the character. We then need to assess the intensity of direct selection and heritability of the character, and also the intensities of selection on other characters and the magnitude of genetic correlations between characters. Often an assumption of equilibrium may be a reasonable approximation, and this provides a justification for the common practice of emphasizing selective constraints.

Scope of the paper

We distinguish two kinds of selective constraints. First, there are limits to the exaggeration of male characters due to the inherent costs of these characters, and we consider some examples. Second, there are limits to the intensity of sexual selection by female choice which arise as a property of the females and the mating system. The bulk of our report is concerned with this second aspect. We first ask to what extent the mating system can limit the intensity of sexual selection, given the presence of preferences. Next we ask how the preferences themselves may be subject to direct selection. Direct selection on preferences affects the intensity of sexual selection and also often imposes biases on the direction in which selection is likely to operate. The kinds of biases in directionality which are likely to arise are the focus of many of the current debates in sexual selection. We consider three specific kinds of directionality: a) biases due to female perception; b) the impact of selection against hybridization between species; and c) the extent to which preferences become biased towards males showing high viability (the "viability indicator," or "good-genes" models (see Heisler et al., this volume). We review the last question in the context of one sub-model which is currently receiving a great deal of attention: that in which heritable variance in fitness is generated by host/parasite coevolution. Finally we discuss genetic constraints on the evolution of sexually selected characters and sex dimorphism and genetic constraints applicable to the good genes models.

SELECTIVE CONSTRAINTS ON SEXUAL SELECTION

Costs to the male

Male costs may arise through predation (Endler 1983, 1986; Ryan 1985), parasitism (Cade 1981), physiological limits (Halliday, this volume), social interactions (West-Eberhard 1983), or may arise out of male-male competition. Some examples of costs associated with male mating traits include increased probability of juvenile mortality (e.g., large body size may increase the likelihood of starvation — Clutton-Brock et al. 1985), increased probability of mortality on the breeding grounds (e.g., frog calling may attract predators — Ryan 1985), and reduced mating success (e.g., newt mating behavior which has associated physiological costs — see Halliday, this volume). In general, male costs may be reflected in viability selection, or in reduced future fertility or mating success (Partridge and Endler, this volume).

Theoretically some forms of viability selection (such as a Guassian fitness function) may not prevent the ultimate elaboration of a sexually selected character but only slow the rate of evolution along the line of equilibria (Lande 1981). In practice viability selection may be sufficiently strong to check the development of the male character altogether and/or result in favored males becoming rare, thereby placing direct selection on female preferences through search costs (see below).

Sexual selection on a character will often result in opposing natural selection not only on that character, but also on the whole phenotype. Consequently characters which are not the direct target of sexual selection may evolve and reduce any cost of sexual selection, which in turn may allow further elaboration of the sexually selected trait. This interplay between natural and sexual selection has been little appreciated and may account for some elaborate sexually selected traits showing little apparent cost. A good example where male cost has been greatly reduced comes from a study of fiddler crabs (*Uca panacea*). In males one claw is entirely used for display and one for feeding, whereas both claws are used for feeding by the females. However, the rate of food intake by males and females is similar because the single feeding claw in males is larger and also scoops food more rapidly. The buccal cavity is not sexually dimorphic; it may be limiting food intake (Caravello and Cameron 1987). More studies such as these will prove useful for understanding male costs and how they are overcome.

Variance in mating success: monogamy

In polygamous mating systems, female preferences for certain male phenotypes can lead to a high intensity of sexual selection because a few males may inseminate many females. However, with monogamy there is no variance in the number of mates a male receives. Obviously this eliminates a major component of variance in fitness capable of contributing to sexual selection and consequently generates a paradox since many monogamous species, particularly birds, appear to be intensively sexually selected. Although extreme examples of sexually selected characters are linked with polygamous mating systems, some monogamous birds (e.g., the Paradise flycatcher, *Terpsiphone paradisii*) show more dimorphism and more extravagant characters than are found in many polygamous species.

The following means by which sexual selection may occur in apparently monogamous systems have been suggested. First, there may be variation in mate quality, such as in the Darwin-Fisher model, in which females differ in time of pairing up; early breeding females have higher (nonheritable) reproductive success than later breeding females, and thus confer a sexual selection advantage on males (Darwin 1871; Fisher 1958). Second, adult sex ratios are rarely 50:50. When they are not, a female bias may lead to low levels of polygamy and a male bias to the presence of unmated males. Either way there will be variance in the number of mates a male obtains. Third, the presence of a floating population of unmated males and females (e.g., Darwin 1871) may lead to some opportunity for choice. Finally, females may engage in extra-pair bond copulation with favored males (McKinney et al. 1984). All four processes are likely to occur in nature, but the relative importance of each mechanism needs to be assessed. Heavily male-biased adult sex ratios in monogamous birds have been observed to lead to high variance in mating success, comparable with that of some polygamous systems (Payne 1984).

Currently it appears that monogamy need not be a significant constraint on sexual selection. However, sexual selection may not be the cause of the elaboration of some sexually dimorphic traits, which may arise out of the differing reproductive roles or ecological niches of the two sexes (for further discussion see Price 1984; Slatkin 1984). Also, elaborate monomorphic traits, such as the bright plumages of some mongamous parrots could arise through social selection (West-Eberhard 1983), interactions with predators, etc.

Direct selection on female preferences

By direct selection on preferences we mean that the particular choice of mate affects female fecundity and/or viability (see Heisler et al., this volume). Two general ways in which this can happen have been considered. First, different males may provide different benefits in terms of resources or parental care (Kirkpatrick 1985) and second, searching for different types of mates may incur a cost (Kirkpatrick, this volume). In all examples mating preferences are expected to evolve such that the average male phenotype among mating events in the population is that which maximizes female fecundity. The average male phenotype may deviate substantially from the viability optimum, or lie at the optimum, depending on exactly how the fitness costs and/or benefits are incurred. The Lande/Kirkpatrick line of equilibria which arises in the absence of direct selection on preferences generally collapses to a single equilibrium point. This point may, however, become unstable and lead to a runaway process. Models of search costs are in their infancy and more work is needed, particularly taking into account different types of female preference functions.

Kirkpatrick (this volume) discusses search costs in more detail.

It is important to distinguish between a) the costs of choosing per se, as opposed to making no choice, and b) differences in costs of choice associated with different preferences. Choice itself is not expected to evolve if possible costs (energy expenditure, risk of predation, etc.) are not countered by benefits, but it may still arise out of constraints on female perception, as discussed below. Given that choice is being exercised by all females and involves a cost, it is not necessarily the case that females who exhibit different probabilities of mating with males differing in some arbitrary character (e.g., bright versus dull males) have different fitnesses. The existence of the line of equilibria, rather than an equilibrium point, is sensitive only to differential fitnesses among females and not to the costs of choosing which are shared by all females.

The theoretical prediction that females should evolve preferences that maximize immediate reproductive success has been frequently confirmed in nature. Most mating systems probably incorporate some immediate cost and/or benefit to females. If situations are discovered where preferences are not directed towards males providing the greatest immediate benefits, they would be well worth further study because of the implication that there are nonequilibrium conditions and/or other factors not incorporated in these models.

Directionality arising from biases in female preferences

Many characteristics of the female and the environment will affect the intensity of sexual selection on a character. These include the number of cues a female can simultaneously assess (Burley 1981), and background matching and biases in female perception (Endler 1983, 1986; Ryan 1985)

There will always be biases in female perception arising from the way in which the sensory system is constructed. Many will reflect past, conceivably ancient, phylogenetic history. An apparent example is the super-normal (open-ended) stimulus. That this is an important constraint is at least suggested by observations of parent birds feeding young cuckoos many times their size, a clearly maladaptive behavior. The general importance of such constraints has hardly been assessed.

Some males may be more conspicuous to females than others because of inherent perception biases and/or the background against which the male signal is given. This may often give a sexual selection advantage to such males and may additionally lead to selection on the females favoring enhanced perception as a result of search costs. There is an extensive literature on the way in which male: female communication is expected to evolve in particular environments (see reviews by Wiley [1983] and Krebs and Dawkins [1984]). Signals used by males may be expected to maximize the signal to noise ratio for the female receptor. Hence the sensory modes used in the signal and detection may often be those which maximize the signal to noise ratio. In a classic study Crook (1964) showed how the form of the male signal is associated with habitat among the different species of weaver birds. For example, species of widowbirds (a genus of weavers) which live in the savannah use a conspicuous visual display enhanced by a long tail; this seems to maximize contrast with the background. In this context Andersson's (1982) results with the long-tailed widowbird (*Euplectes progne*) are expected simply because longer tails will enhance conspicuousness, while short tails decrease it. Correlations between habitat and signal appear common among animal groups; they been particularly well documented with respect to bird song (references in Dawkins and Krebs 1984). More studies are, however, needed.

A great diversity of male forms could arise among different species because the noise levels in different environments are different, even in the absence of any evolution of female preferences. The extent to which female preferences have evolved could be examined by testing female preference when a species is placed in another's environment. It has been suggested that the Fisherian sexual selection model be used as a kind of null hypothesis. Sensory biases, however, may cause the elaboration of male characters in the absence of the Fisherian process. The consequences for sexual selection arising out of the efficient communication between the sexes may more reasonably be the null hypothesis and primitive model on which to build other components of sexual selection (Endler 1986, and in preparation).

Presence of other species

The form of sexual selection within a species could be constrained by the need for reproductive isolation among closely related species. This differs from direct selection on females in that the selection pressure is usually hybrid inviability or infertility. The modern synthesis placed particular emphasis on this and led to the view that selection for premating isolation mechanisms between sympatric species (i.e., reproductive character displacement or reinforcement) was a general explanation for interspecific differences in mating preferences and the associated male secondary sexual characters (Mayr 1963; Dobkhansky 1970). More recent treatments have emphasized the importance of sexual selection in isolated populations as an explanation of species differences (Carson 1978; Lande 1981; West-Eberhard 1983; Dominey 1984). Other possibilities involve pre-mating isolation arising in allopatric populations without any sexual selection. For instance females may have a general perception (e.g., through imprinting) which results in the discrimination among large morphological differences established by natural selection, but not the discrimination of smaller within-population differencs. As another example, female preferences can evolve as a correlated response to selection on the male character without any sexual selection, as in the model of Andersson (1986).

Theoretical studies suggest that conditions for reproductive character displacement may be rarely met (see Barton and Hewitt 1981; Sved 1981; Spencer et al. 1986). This is because novel (divergent) male phenotypes will often be sexually selected against, unless preferences for those phenotypes are already established. In natural situations it is possible to reject reproductive character displacement as a mechanism when premating isolation has arisen between populations that have remained totally allopatric. Detecting evidence in favor of reproductive character displacement is far more difficult.

One potentially powerful line of evidence is where, in partially sympatric species, mating preferences and associated male characters differ more in the zone of overlap than outside it. This pattern would suggest that intersexual selection has acted differently in areas where there is danger of hybridization. However, such patterns may arise because of factors other than character displacement (see Grant [1975] for a discussion in the context of ecological character displacement). In addition, two factors may prevent the common occurrence of this pattern even when there has been reproductive character displacement. First, gene flow from areas of sympatry to allopatry could cause divergence in the allopatric areas. Second, reproductive character displacement could be cryptic if it is the preference rather than the character which has become accentuated in the zone of overlap (for an example where mate discrimination has been accentuated see Ratcliffe and Grant 1983). Thus, although there are a few cases where evidence for reproductive character displacement is strong (notably Levin 1985), it is hard to assess the general occurrence of the process from studies of pattern.

A second line of evidence bearing on the importance of reproductive character displacement would be to compare the extent of pre-mating isolation between allopatric and sympatric species. If selection for pre-mating isolation is important, then sympatric species should show greater pre-mating isolation. The problem with this approach is that sympatry may only be established among allopatric forms which already show an extensive degree of premating isolation. Otherwise, they would form a hybrid zone, merge into a single continuous species, or one species would disappear from the zone of overlap (Barton and Hewitt 1981). A greater degree of premating isolation between sympatric rather than allopatric species would therefore be predicted in the absence of any selection against hybridization (see Paterson 1982). A quantitative assessment of the importance of these processes will thus be difficult, and depend on intensive case studies.

Parasite models

Can a form of directionality arise in which males with high viability are consistently favored as mates? The theoretical issues are dealt with by Heisler et al. (this volume). We will not repeat the arguments for and against such a process here but instead will concentrate on one particular version of these so-called good genes models. In this model heritable fitness arises out of the coevolution of parasites with their hosts (Hamilton and Zuk 1982). Parasite resistance genes of the host cyclically coevolve with the parasites. Females are assumed to assess general resistance rather than a specific genotype. A specific model of the process showed that male characters which exhibit infection status can in some cases evolve once female preference exceeds a certain threshold value (Kirkpatrick 1986).

Empirical evidence for the model is weak, and studies of the role of parasites are highly desirable, especially as their importance has been generally neglected in evolutionary studies. In order to focus on specific tests we first need to clarify how this model differs from, and complements, the Fisherian models. For illustrative purposes consider two female preferences, both established at high frequency in a population, one for a character whose degree of expression measures parasite load, and one for a character which does not. Suppose also that parasite resistance is heritable. All other things being equal, the preference for the character showing parasite resistance, and hence that character also, should increase in frequency more rapidly (however, this does not exclude the possibility of future maladaptive evolution). A further bias of the spectrum of sexually selected characters towards being parasite indicators may arise by group and species selection: if there are groups and species in which heritable true fitness is raised by the process, they should survive better.

The great variation among males of closely related species is consistent with the Fisherian mechanism and/or the hypothesis that differentiation of male characters proceeds under constant female preference in different habitats, as discussed above. These hypotheses are difficult to test (Heisler et al., this volume). In contrast it should be rather easier to set up tests which will reject the hypothesis that parasites have been important in determining which sexually selected characters are observed.

First, some characters may be good indicators of the presence of specific parasites. Wattles on cock chickens may be used to indicate any diseases that alter blood pigments, e.g., malaria. Investigations of the association between the form of a sexually selected character and the particular parasites of the species of taxon are needed.

Second, we might look for correlations of parasite load and mate selection within a population. It may generally be expected that less healthy males obtain fewer matings, but the discovery of positive correlations between parasite load and mating success would refute the model. Indeed, parasites transmitted during mating could be selected to increase (i.e., manipulate) the attractiveness of their host. If parasite infection is found to negatively correlate with mating success, and parasite resistance is found to be heritable, then it is likely that the preference (if heritable) will be evolutionarily affected.

Although the comparative method should be used with care (Harvey and Mace 1982), it is potentially effective for this problem. In essence one looks at the parasitism of species sexually selected to different degrees. Replicating mate choice experiments on a variety of species which have apparently been subject to differing degrees of sexual selection might yield revealing results. Comparative studies have already implicated a role for parasites in sexual selection. For example, Hamilton and Zuk (1982) found a correlation between bright plumage and parasite load among North American passerine bird species, and in a subsequent study (unpublished) they found the same for neotropical passerines.

To summarize this section, we view parasite/host interactions as possibly playing an important role in constraining the form of taxonomic distribution of sexually selected characters. Testable subhypotheses are plentiful and empirical investigations into the role of parasites are likely to be fruitful not only for this, but for a variety of evolutionary questions. In the context of sexual selection it is important to keep in mind the complexity of the host-parasite interaction and many potential constraints.

The origin of preferences

Given the presence of a preference and a suitable male character, the character is likely to evolve under sexual selection whether or not parasites are present. It is possible that the appearance of new preferences is a rate-limiting step. Preferences and the male character do not appear likely to increase deterministically from low frequency in the parasite situation (Kirkparick 1986). Possibilities for the appearance of the preference include genetic drift, correlated response to selection (both natural and sexual), or perception-biased preferences (Kirkpatrick, this volume). Fisher championed the idea of a correlated response to natural selection. Charlesworth (this volume) shows how heritable variance in fitness can lead to a selection pressure to increase preferences when they are directed towards the most fit individuals. Given potential differences between origination and subsequent evolution, tests of mechanism of origination are likely to be highly indirect.

GENETIC CONSTRAINTS ON SEXUAL SELECTION

Previous sections dealt with some of the natural and sexual selection pressures on sexually selected traits and also noted some possible constraints on female perception. We next consider how the genetic basis of the sexually selected traits may constrain the response to selection. Unless genetic variance or covariance completely disappears genetic constraints do not prevent the eventual attainment of an equilibrium determined by the forces of selection. They do have important effects in influencing directions and rates of evolution. We consider each in turn.

Evolutionary trajectories

Effects on evolutionary trajectories are particularly important when there are multiple equilibria; the genetic parameters may make some equilibrium points more likely to be reached than others. For example, given the line of equilibria arising from the classic Fisherian model, the genetic parameters determine if and where a population returns to the line following a perturbation. Slatkin (1984) investigated the evolution of sexual dimorphism when there is a bimodal fitness surface (with two fitness peaks) and a monomorphic population centered at the selective minimum. The alternative stable equilibria are dimorphism, with each sex at one of the two selective optima, or monomorphism, with both sexes at one optimum. When there is a high genetic correlation between the sexes for the character under selection, monomorphism is far more likely to result than dimorphism.

The evolution of sexual dimorphism

Slatkin's model is an example where genetic correlations affect the likelihood of dimorphism evolving. Most often genetic correlations have been discussed in the context of constraints on rates of evolution of dimorphism. As reviewed by Lande (this volume), high positive genetic correlations between the sexes can greatly slow the rate of evolution of sex dimorphism and the male character. Some sexually selected characters may be far from an optimum as determined by the forces of selection. We amplify Lande's remarks with some comments on the genetic mechanisms by which characters become sexually dimorphic.

The most common method is by sex-limited expression of genetic effects, many of which may be autosomal (Lande, this volume). This may be a slow process, and an alternative is to select for mutations at *Y*-linked loci in species which have active *Y* chromosomes. In species with male heterogamety this will limit the relevant characters to the male sex (Fisher 1931). This path is obviously not available in groups with female heterogamety, or with nonchromosomal sex determination. The lack of genetic activity on the *Y* chromosome in many groups further restricts this possibility. However, many sexually selected color genes in fish show close linkage to the sex region of the *Y* chromosome (Haskins et al. 1970; Kallman 1975; Yamamoto 1975; Kirkpichnikov 1981).

When a character is under major gene control, as often appears to be the case for color and pattern effects (Winge and Ditlevson 1947; Clarke and Sheppard 1955; Sheppard et al. 1985), sex limitation requires that the genes concerned be either sex-limited in their expression from the start or that modifiers at other loci inducing sex limitation become fixed following the establishment of the initial genes in the population (Turner 1981). The ability of the sex-determining mechanism to permit sex limitation of gene expression must place some constraints on sexual selection but little useful can be said about this yet.

Theoretical analysis shows that selection on recessive or partially recessive favorable genes proceeds more rapidly if they are sex-linked (Rice 1984; Charlesworth et al. 1987). This applies in particular to male-limited characters (when there is male heterogamety), and so one might expect such genes to show a disproportionate concentration on the *X* chromosome in inter-species crosses. Genes with intermediate dominance will not show this effect, however. Evidence from interspecies crosses suggests that genes affecting hybrid fitness often seem to be strongly influenced by the *X* chromosome, whereas morphological and behavioral characters usually do not show such an effect. This probably reflects differences in the nature of the genetic control of the respective traits (Charlesworth et al. 1987). There is thus little reason to expect sex-linkage versus autosomal inheritance to place important constraints on the response to sexual selection.

General constraints

Genetic constraints may slow the evolution of a sexually selected character, apart from any effect on sex dimorphism. For example, if variation in the male character is due to a single locus, fixation under sexual selection will prevent further change in the male character and female preference until a new variant arises.

Absence of heritability for sexually selected characters, particularly behavioral characters, may be more common than is currently appreciated. An example of an apparent constraint comes from the *Rana* frogs of central Europe. *Rana esculenta* is a hybrid between *R. ridibunda* and *R. lessonae* (Berger 1971, 1977). The hybrid, *R. esculenta*, is usually sympatric with *lessonae* only so it is not generated by matings between *ridibunda* × *lessonae*. Instead the hybrid populations are maintained by *lessonae* males pairing with *esculenta* females. Remarkably the offspring are all purely *esculenta*. This is because *esculenta* produces only *ridibunda* gametes and the *lessonae* genome is removed before gametogenesis (Graf and Müller 1979). Males of *esculenta* and *lessonae* maintain mixed leks, in which *esculenta* males do most of the calling but seldom try to mate with approaching females (Blankenhorn 1977). *Rana lessonae* males mate with *esculenta* females. Because the *lessonae* genome will be lost there should be selection on *lessonae* males to avoid matings with *esculenta* females, but they do not discriminate between types of females. Since *esculenta* females may be common this can result in considerable wasted effort. For *lessonae* the absence of a choice mechanism in males is maladaptive. Similar mating patterns are observed in *Poecilopsis* fish (e.g., Schultz 1967) and *Ambystoma* amphibians (Uzzell 1964). This may be a genetic constraint, in that males are unable to evolve discrimination despite selection pressure for them to do so, but it should be noted that the hybrid females are under strong selection to be mated by the males (Uzzell 1964).

Heritability of total fitness

There is a constraint peculiar to the good genes models of sexual selection which could be assessed to test their applicability. A particular feature of these models is the requirement that total fitness is heritable. Absence of heritability in total fitness would be an important constraint in these models only. From the review of Charlesworth (this volume) it seems clear that mutational pressure can give rise to some genetic variation in total fitness. This may be amplified by gene flow and also temporally varying selection. Whether it is sufficient to place a significant selective pressure on mating preferences is unclear.

In principle, field measurements of genetic variance in fitness would test a basic assumption of the good genes models. In practice, total fitness is likely to be greatly affected by environmental variance (see papers in Clutton-Brock 1986), making heritabilities low and their detection a substantial genetic problem. Furthermore, causes of variation in fitness will almost certainly vary across generations due to temporal fluctuations in the environment. Measurements of genetic variance in total fitness may be extremely difficult to obtain. It is clear that measurements of a restricted set of the components of fitness are insufficient. Because of negative genetic correlations among components, incomplete measures provide little predictive information about total fitness.

SUGGESTIONS FOR FUTURE RESEARCH

In this final section we summarize some of the suggestions for future research that have arisen in previous sections and briefly note some specific topics that have not previously been discussed. Other papers in this volume have emphasized the importance of obtaining measurements of the heritability of female preferences. However, it is currently unclear how consistent an individual's choice is; repeated mating trails should be used to estimate repeatabilities, which place an upper bound on heritabilities (Falconer 1981) before any attempt is made to directly estimate them. Within-individual behavioral variation may be an important constraint.

An understudied aspect with respect to sexual selection is that of signal/receptor properties and also their relationship with the environment (Endler 1986, and in preparation). A related question is the general shape of preference functions which can have drastic effects on model results (Lande 1981; Seger 1985). Measurements of preference functions, and observations on exactly how females choose, are badly needed.

There are few quantitative measurements of costs of choice and cost functions for sexually selected characters. A particular cost we have not discussed at all is selection associated with avoidance of inbreeding, which appears to be very common.

Finally we mention a useful peripheral study which would strengthen the assumptions of the parasite model, as well as other good genes models. Surveys of genetic variation at the molecular level, as by electrophoresis, might be made to reveal fluctuating selection. Fluctuating selection may be revealed even if the allelles surveyed are merely linked to those on which selection is acting. Changes in both linkage disequilibrium and gene frequencies are of interest from the point of view of the models.

CONCLUDING REMARKS

In this paper we have attempted to discuss generally the potential importance of constraints and also specific types of constraints that require further investigation. Virtually all our conclusions are tentative for lack of empirical research and indicate that much is to be done. Our group experienced difficulty in devising discriminating and realistic tests for most of the problems we have noted in this report. Partly this is a reflection of the likelihood that all the constraints we have discussed will be of some significance in some systems, and partly a reflection of what we see as the purpose of the models. Theoretical models have been extremely useful in demonstrating the validity (or otherwise) of verbal models. We feel that more theoretical research is needed (particularly on good genes models, fertility effects associated with variation in preference (including search costs), and monogamy). We encourage workers to develop tests of the more specific and complex models. However, questions derived from empirical in-depth studies, like that of Ryan (1985), are likely to provide increasing insights into the biology of sexual selection. The models are also useful as an aid in answering such questions.

We conclude with a note on the difficulties of assessing past intensities of sexual selection in different species by using the comparative method, on which many tests rely. First, sexually selected characters may have different visibility to researchers than to the animals involved and may consequently be given incorrect weightings; pheromones, for example, strike humans less forcefully than bright plumages. Second, intense sexual selection may not necessarily lead to exaggerated dimorphism. In primates and some birds, large male advantages in intrasexual competition tend to lead to males being larger than females, and the degree of size dimorphism is correlated with polygamy (but see Leutenegger and Cheverud 1982). In anurans, fish, and insects, however, males are often smaller than females despite there being a large male advantage in sexual competition. This is explicable in terms of the strong positive relationship between female body size and fecundity (Ralls 1976), and, in some anurans at least, reduced growth rate in males as a consequence of remaining on the breeding site (see Halliday, this volume). Also selection for characteristics such as earlier sexual maturity in males may be more powerful than sexual selection for large size. In fishes and insects, small males are favored in mating competition because of their greater agility. All this indicates that the degree of sexual dimorphism should be used with extreme caution as a measure of the intensity of sexual selection in such organisms. This is likely to be generally applicable to other sexually selected traits.

Acknowledgements

We thank J. Bradbury and B. Kraus for comments on the manuscript, and the staff of the Dalem Konferenzen for their cheerful logistic support.

REFERENCES

Andersson, M. 1982. Female choice selects for extreme tail length in a windowbird. Nature 299: 818–820.

Andersson, M. 1986. Evolution of condition-dependent sex ornaments and mating preferences: sexual selection based on viability differences. Evolution 40: 804–816.

Barton, N.H., and Hewitt, G.M. 1981. Hybrid zones and speciation. In: Evolution and Speciation: Essays in Honour of M.J.D. White, eds. W.R. Atchley and D. Woodruff, pp. 109– 145. Cambridge: Cambridge Univ. Press.

Berger, L. 1971. Viability, sex and morphology of the F2 generation within forms of *Rana esculenta* complex. Zoolog. Polon. 21: 345–393.

Berger, L. 1977. Systematics and hybridization in the *Rana esculenta* complex. In: The Reproductive Biology of Amphibians, eds. D.H. Taylor and S.I. Guttman, pp. 367–388. New York: Plenum Press.

Blankenhorn, H.J. 1977. Reproduction and mating behavior in *Rana lessonae – Rana esculenta* mixed populations. In: The Reproductive Biology of Amphibians, eds. D.H. Taylor and S.I. Guttman, pp. 389–410. New York: Plenum Press.

Burley, N. 1981. Mate choice by multiple criteria in a monogamous species. Am. Natural. 117: 515–528.

Cade, W.H. 1981. Alternative mating strategies — genetic differences in crickets. Science 212: 563–564.

Carvello, H., and Cameron, G. 1987. The effects of sexual selection on the foraging behavior of the Gulf Coast fiddler crab (*Uca Panacea*). Anim. Behav., in press.

Carson, H.L. 1978. Speciation and sexual selection in Hawaiian *Drosophila*. In: Ecological Genetics: The Interface, ed. P.F. Brussard, pp. 93–107. New York: Springer-Verlag.

Charlesworth, B.; Coyne, J.A.; and Barton, N.H. 1987. The relative rates of evolution of sex chromosomes and autosomes. Am. Natural., in press.

Clarke, C.A., and Sheppard, P.M. 1955. A preliminary report on the genetics of the machaon groups of swallowtail butterflies. Evolution 9: 182–201.

Clutton-Brock, T.H., ed. 1987. Lifetime Reproductive Success. Chicago: Univ. of Chicago Press.

Clutton-Brock, T.H.; Albon, S.D.; and Guinness, F.E. 1985. Parental investment and sex differences in juvenile mortality in birds and mammals. Nature 313: 131–133.

Crook, J.H. 1964. The evolution of social organization and visual communication in the weaver birds (Ploceinae). Behaviour Suppl. 10: 1–178.

Darwin, C. 1871. The Descent of Man and Selection in Relation to Sex, 2nd ed. London: John Murrary.

Dobzhansky, T. 1970. Genetics of the Evolutionary Process. New York: Columbia Univ. Press.

Dominey, W.J. 1984. Effects of sexual selection and life history on speciation: species flocks in African Cichlids and Hawaiian *Drosophila*. In: Evolution of Fish Species Flocks, eds. A.A. Echelle and I. Kornfield, pp. 231–249. Orono, Maine: Univ. of Maine Press.

Endler, J.A. 1983. Natural selection on color patterns in poeciliid fishes. Env. Biol. Fish. 9: 173–190.

Endler, J.A. 1986. Defense against predators. In: Predator-prey Relationships, Perspectives and Approaches from the Study of Lower Vertebrates, eds. M.E. Feder and G.V. Lauder, pp. 109–134. Chicago: Univ. of Chicago Press.

Falconer, D.S. 1981. Introduction to Quantitative Genetics, 2nd ed. London, New York: Longman.

Fisher, R.A. 1931. The evolution of dominance. Biol. Rev. Camb. 6: 345–368.

Fisher, R.A. 1958. The Genetical Theory of Natural Selection, 2nd ed. New York: Dover.

Graf, J.–D., and Müller, W.P. 1979. Experimental gynogenesis provides evidence of hybridogenetic reproduction in the *Rana esculenta* complex. Experientia 35: 1574–1576.

Grant, P.R. 1975. The classical case of character displacement. Evol. Biol. 8: 237–337.

Hamilton, W.D., and Zuk, M. 1982. Heritable true fitness and bright birds: a role for parasites? Science 218: 384–387.

Harvey, P.H., and Mace, G.M. 1982. Comparisons between taxa and adaptive trends: problems of methodology. In: Current Problems in Sociobiology, eds. Kings' College Sociobiology Group, pp. 343–361. Cambridge: Cambridge Univ. Press.

Haskins, C.P.; Young, P; Hewitt, R.E.; and Haskins, E.F. 1970. Stabilised heterozygosis of supergenes mediating certain Y-linked colour patterns in populations of *Lebistes reticulatus*. Heredity 25: 575–589.

Kallman, K. 1975. The platyfish, *Xiphophorus maculatsu*. In: Handbook of Genetics, vol. 4., Vertebrates of Genetic Interest, ed. R.C. King, pp. 81–132. New York: Plenum Press.

Kirkpatrick, M. 1985. Evolution of female choice and male parental investment in polygamous species: the demise of the sexy son. Am. Natural. 125: 788–810.

Kirkpatrick, M. 1986. Sexual selection and cycling parasites: a simulation study of Hamilton's hypothesis. J. Theor. Bio. 119: 263–271.

Kirpichnikov, V.S. 1981. Genetic Basis of Fish Selection. New York: Springer-Verlag.

Krebs, J.R., and Dawkins, R. 1984. Animal signals: mind-reading and manipulation. In: Behavioural Ecology, 2nd ed., eds. J.R. Krebs and N.B. Davies, pp. 380–402. London: Blackwell.

Lande, R. 1981. Models of speciation by sexual selection on polygenic traits. Proc. Natl. Acad. Sci. USA 78: 3721– 3725.

Leutenegger, W., and Cheverud, J. 1982. Correlates of sexual dimorphism in primates: ecological and size variables. Int. J. Prim. 3: 387–402.

Levin, D.A. 1985. Reproductive character displacement in *Phlox*. Evolution 39: 1275–1281.

Mayr, E. 1963. Animal Species and Evolution. Cambridge, MA: Belknap Press.

McKinney, F.; Craig, K.M.; and Bruggers, D.J. 1984. Sperm competition in apparently monogamous birds. In: Sperm Competition and the Evolution of Animal Mating Systems, ed. R.L. Smith, pp. 423–454. New York: Academic Press.

Paterson, H.E.H. 1982. Perspective on speciation by reinforcement. S. Afr. J. Sci. 78: 53–57.

Payne, R.B. 1984. Sexual selection, lek and arena behavior, and sexual size dimorphism in birds. Ornithological Monographs 33. Washington, D.C.: American Ornitholo gists Union.

Price, T.D. 1984. The evolution of sexual size dimorphism in Darwin's finches. Am. Natural. 123: 500–518.

Ralls, K. 1976. Mammals in which females are larger than males. Q. Rev. Biol. 51: 245–276.

Ratcliffe, L.M., and Grant, P.R. 1983. Species recognition in Darwin's finches (*Geospiza*, Gould). II. Geographic variation in mate preference. Anim. Behav. 31: 1154–1165.

Rice, W.R. 1984. Sex chromosomes and the evolution of sexual dimorphism. Evolution 38: 735–742.

Ryan, M.J. 1985. The Tungara Frog, a Study in Sexual Selection and Communication. Chicago: Univ. of Chicago Press.

Schultz, R.J. 1967. Gynogenesis of triploidy in the viviparous fish *Poecilopsis*. Science 157: 1564–1567.

Seger, J. 1985. Unifying genetic models for the evolution of female choice. Evolution 39: 1185–1193.

Sheppard, P.M.; Turner, J.R.G.; Brown, K.S.; Benson, W.W.; and Singer, M.C. 1985. Genetics and the evolution of mullerian mimicry in *Heliconius* butterflies. Phil Trans. Roy. Soc. Lon. B 308: 433–613.

Slatkin, M. 1984. Ecological causes of sexual dimorphism. Evolution 38: 622–630.

Spencer, H.G.; McArdle, B.H.; and Lambert, D.M. 1986. A theoretical investigation of speciation by reinforcement. Am. Natural. 128: 241–262.

Sved, J.A. 1981. A two-sex polygenic model for evolution of premating isolation. 1. Deterministic theory for natural populations. Genetics 97: 197–215.

Turner, J.R.G. 1981. Adaptation and evolution in *Heliconius*: a defense of Neo-Darwinism. Ann. Rev. Ecol. Syst. 12: 99–121.

Uzzell, T.M. 1964. Relations of the diploid and triploid species of the *Ambystoma jeffersonianum* complex (Amphibia, Caudata). Copeia 1964: 257–300.

West-Eberhard, M.J. 1983. Sexual selection, social competition, and speciation. Q. Rev. Biol. 58: 155–183.

Wiley, R.H. 1983. The evolution of communication: information and manipulation. In: Communication (Animal Behaviour, vol. 2), eds. T.R. Halliday and P.J.B. Slater, pp. 156–189. London: Blackwell.

Winge, O., and Ditlevson, O. 1947. Colour inheritance and sex-determination in *Lebistes*. Heredity 1: 65–83.

Yamamoto, T.-O. 1975. The Medaka, *Oryzias latipes*, and the Guppy, *Lebistes reticulatus*. In: Handbook of Genetics, vol. 4, Vertebrates of Genetic Interest, ed. R.C. King, pp. 133–149. New York: Plenum Press.

List of Participants with Fields of Research

R.V. ALATALO
Dept. of Biology
University of Jyväskylä
Yliopistonkatu 9
40100 Jyväskylä
Finland

Female mate choice in birds

M.B. ANDERSSON
Dept. of Zoology
University of Göteborg
P.O. Box 25059
40031 Göteborg
Sweden

Evolution of behavior, sexual selection

S.J. ARNOLD
Dept. of Biology
University of Chicago
940 E. 57th Street
Chicago, IL 60637
USA

Behavioral genetics and evolution

C.R. BOAKE
Hawaiian Evolutionary Biology Program
University of Hawaii
3050 Maile Way, Gilmore Hall 310
Honolulu, HI 96822
USA

Evolutionary genetics of sexually selected traits

G. BORGIA
Dept. of Zoology
University of Maryland
College Park, MD 20742
USA

Field studies and models of sexual selection; evolution of social behavior

J.W. BRADBURY
Dept. of Biology, C–016
University of California, San Diego
La Jolla, CA 92093
USA

Behavioral ecology/sociobiology

B. CHARLESWORTH
Dept. of Biology
University of Chicago
1103 E. 57th Street
Chicago, Il 60637
USA

Population genetics and evolutionary theory

T.H. CLUTTON-BROCK
Large Animal Research Group
34A Storeys Way
Cambridge CB3 0DT
England

Behavior and ecology

N.B. DAVIES
Dept. of Zoology
University of Cambridge
Downing Street
Cambridge CB2 3EJ
England

Behavioral ecology: mating systems, cuckoos and their hosts

J.A. ENDLER
Dept. of Biological Sciences
University of California
Santa Barbara, CA 93106
USA

Ecological genetics, evolution of animal color patterns, evolution of complex morphological traits

P.-H. GOUYON
CNRS-B2P
B.P. 5051
34033 Montpellier Cedex
France

Evolution of sex in plants, resource allocation into male/female growth functions, nuclear vs. cytoplasmic genes, ecological succession's effect on genetic evolution

A. GRAFEN
Dept. of Zoology
ABRG
1 South Parks Road
Oxford OX1 3PS
England

Evolution theory and sexual selection

T.R. HALLIDAY
Animal Behaviour Research Group
The Open University
Milton Keynes MK7 6AA
England

Animal behavior: sexual behavior of amphibians, reproductive physiology of amphibians

W.D. HAMILTON
Dept. of Zoology
South Parks Road
Oxford OX1 3PS
England

Evolution of sex and sexual selection

P. HAMMERSTEIN
Max-Planck-Institut für Verhaltens-physiologie
Abt. Wickler
8131 Seewiesen
F.R. Germany

Models in behavioral ecology and sociobiology; evolutionary game theory

P.H. HARVEY
Dept. of Zoology
University of Oxford
South Parks Road
Oxford OX1 3PS
England

Comparative studies of life histories

G. HAUSFATER
Division of Biological Sciences
University of Missouri
110 Tucker Hall
Columbia, MO 65211, USA

Sexual selection in relation to host-parasite dynamics; reproductive strategies

L. HEISLER
Dept. of Ecology and Behavioral Biology
109 Zoology Building
University of Minnesota
318 Church St. SE
Minneapolis, MN 55455, USA

Population and quantitative genetics (theoretical and empirical studies of sexual selection)

K.-G. HELLER
Universität Erlangen-Nürnberg
Institut für Zoologie, Lehrstuhl II
Staudtstr. 5
8520 Erlangen
F.R. Germany

Communication and parental investment in bushcrickets

R.D. HOWARD
Dept. of Biological Sciences
Purdue University
West Lafayette, IN 47907, USA

Sexual selection; measurement of reproductive success

M. KIRKPATRICK
Dept. of Zoology
University of Texas
Austin, TX 78712, USA

Theoretical population genetics; quantitative genetics; sexual selection; evolution of shape and growth

B. KÖNIG
Zoologisches Institut der Universität Basel
Rheinsprung 9
4051 Basel
Switzerland

Behavioral ecology, mutualism

R. LANDE
Dept. of Biology
University of Chicago
920 E. 58th Street
Chicago, IL 60637
USA
Population genetics and evolution

K.E. LINSENMAIR
Zoologisches Institut III
Universität Würzburg
Röntgenring 10
8700 Würzburg
F.R. Germany
Behavioral and physiological ecology

J. MAYNARD SMITH
School of Biological Sciences
University of Sussex
Brighton BN1 9Q6
England
Evolution theory

M. MILINSKI
Arbeitsgruppe für Verhaltensforschung
Fakultät Biologie
Ruhr-Universität
Postfach 102 148
4630 Bochum
F.R. Germany
Behavioral ecology of host-parasite interactions

P. O'DONALD
Emmanuel College
University of Cambridge
Cambridge CB2 3AP
England
Theoretical and experimental studies of sexual selection

G.A. PARKER
Dept. of Zoology
University of Liverpool
P.O. Box 147
Liverpool L69 3BX
England
Evolutionary biology

L. PARTRIDGE
Dept. of Zoology
Edinburgh University
West Mains Road
Edinburgh EH9 3JT
Scotland
Sexual selection and life history evolution in fruitflies

J. PARZEFALL
Zoologisches Institut und Museum
Universität Hamburg
Martin-Luther-King-Platz 3
2000 Hamburg 13
F.R. Germany
Alarm reaction, schooling and aggressive behavior in cave fish populations and their hybrids with epigean relatives

K. PESCHKE
Zoologisches Institut III
(Lehrstuhl Tierökologie)
Universität Würzburg
Röntgenring 10
8700 Würzburg
F.R. Germany
Mate choice and competition for mates in insects

H.J. POETHKE
Lehrstuhl für Biologie V
(Ökologie) der RWTH Aachen
Kopernikusstr. 16
5100 Aachen
F.R. Germany
Evolution of mating strategies

T.D. PRICE
Dept. of Biology, C-016
University of California, San Diego
La Jolla, CA 92093, USA
The role of sexual selection in the evolution of plumage patterns in birds; quantitative genetics and speciation

D.C. QUELLER
Dept. of Biology
Rice University
P.O. Box 1892
Houston, TX 77251, USA
Evolution of social interactions

H.-U. REYER
Max-Planck-Institut für Verhaltens-physiologie
8131 Seewiesen
F.R. Germany
Cooperative breeding in birds: ecological causes and fitness consequences

N. SACHSER
Universität Bayreuth
Lehrstuhl für Tierphysiologie
Postfach 3008
8580 Bayreuth
F.R. Germany
Social organization, reproductive strategies, behavioral endocrinology

T. SLAGSVOLD
Zoological Museum
University of Oslo
Sarsgate 1
0562 Oslo 5
Norway
Sexial selection in birds

W.J. SUTHERLAND
School of Biological Sciences
University of East Anglia
Norwich NR4 7TJ
England
Growth patterns of plants; genetic variation within individual plants

A.R. THORNHILL
Dept. of Biology
University of New Mexico
Albuquerque, NM 87131
USA
The nature of the traits favored by sexual selection and the evolution of mating systems

F. TRILLMICH
Max-Planck-Institut für Verhaltens-physiologie
8131 Seewiesen
F.R. Germany
Parental investment, sex allocation and conflict between the sexes

A.J. VAN NOORDWIJK
Zoologisches Institut
Rheinsprung 9
4051 Basel
Switzerland
Genetic ecology, quantitative genetics, evolutionary ecology

C. VOGEL
Institut für Anthropologie
Universität Göttingen
Bürgerstrasse 50
3400 Göttingen
F.R. Germany
Primate behavior, sociobiology

M.J. WADE
Dept. of Biology
University of Chicago
915 E. 57th Street
Chicago, IL 60637, USA
The role of population structure in evolution

R. WARNER
Dept. of Biological Sciences
University of California
Santa Barbara, CA 93106
USA
Evolution of mating systems and life histories, especially in fishes

F.J. WEISSING
Universität Bielefeld
FSP Mathematisierung
Postfach 8640
4800 Bielefeld 1
F.R. Germany
Population genetics, behavioral ecology, evolutionary game theory

M.J. WEST-EBERHARD
Escuela de Biologia
Universidad de Costa Rica
Ciudad Universitaria
Costa Rica
Social behavior (including sexual behavior); evolutionary consequences of social competition and ontogenetic flexibility; field studies of tropical social wasps

G.S. WILKINSON
Institute for Behavioral Genetics
Campus Box 447
University of Colorado
Boulder, CO 80309
USA

Evolution of behavior

P. WIRTZ
Zoologisches Institut
Universität Freiburg
Albertstrasse 21a
7800 Freiburg
F.R. Germany

Evolution of social behavior

Subject Index

Author Index